MISSING DANIEL

THE TRUE STORY OF A YOUNG SAILOR'S DEATH IN VIETNAM

By

Daniel James Roberts

Lieutenant Commander, USN (Ret.)

ISBN

Hardcover: 979-8-90190-443-5

Paperback: 979-8-90190-442-8

BOOK PUBLISHING GROUP LLC

Book Publishing Group LLC

42 Broadway 12th floor, New York, NY 10004

www.bookpublishinggroupllc.com

Dedication

There are countless people I could name and dedicate this book to. Numerous people played a role in bringing this book to life, far more than I could ever fully acknowledge here. So many friends and family who stood beside me and helped make this book possible. Yet as I think about all who deserve recognition, my heart always returns to one name.

Mary Katherine Davidenko

Therefore, I would like to dedicate this book to my Grandmother Mary. Grandma Mary, thank you for the love, support, and wisdom you have given to our entire family. I dedicate this book to you and for all the pain you endured.

About the Author

Lieutenant Commander Daniel James Roberts, USN (Ret.), the author of this book, is the proud nephew of CMCN Daniel Joseph Davidenko (USN), and he feels deeply honored that his mother, Jean Frances Roberts, named him after her youngest brother, Daniel. His Uncle Daniel, or Danny as the family liked to call him, was 11 years old when the author was born in August 1962. Sadly, in November 1969, the author's uncle, Danny, was killed in Vietnam when the author was just seven years old, growing up in Oak Forest, Illinois. So, the author never had the opportunity to truly get to know the man he was named after. Ironically enough, about 11 years after his Uncle Danny's death in Vietnam, he also joined the U.S. Navy when he turned 18 years old. When the author first enlisted in the U.S. Navy in November 1981, he didn't think much about how his mother Jean, felt about his decision to join the military, especially after she lost her baby brother to the Vietnam War. Looking back, he realized that he had followed in his Uncle Danny's footsteps, something that must have deeply unsettled his mother. The author's grandmother, Mary (Danny's mother), sadly passed away in January 1981, just 1 month before he decided to join the

U.S. Navy. In hindsight, this may have been a blessing, because as he thought back on his decision, he didn't stop to think about his grandmother Mary, and if she would have approved of his decision to join the U.S. Navy, especially after she had lost her youngest son, Danny, to the Vietnam War. In retrospect, he understands now how his decision to join the U.S. Navy would have placed unnecessary stress on her already broken heart. The thought of her grandson Danny, her son's namesake, possibly facing the same fate in some godforsaken war would have been unbearable for her. Even more troubling is the fact that he also never got the chance to ask his own mother Jean, how she felt about his decision to join the U.S. Navy. Reflecting back on his decision to join the U.S. Navy, he knew deep down inside that his mother had very serious reservations and probably spent most nights worrying about his safety while serving on active duty.

However, in the end, joining the U.S. Navy proved to be one of the best decisions of his life. What started as a single enlistment to fund his college tuition evolved into a successful 23-year military career. During his second enlistment in the U.S. Navy, he was selected for a college commissioning program called the Enlisted Commissioning Program (ECP). Following his selection, he received orders to report to the Naval Science Institute (NSI) in Newport, Rhode Island. After successfully graduating from NSI, he was assigned to the Naval Reserve Officer Training Corps (NROTC) at Auburn University in Auburn, Alabama. While attending Auburn University, he was a full member of the NROTC unit, remaining on active duty as a Midshipman and actively participating in all NROTC events. During his senior year, he was selected as Auburn University's NROTC Battalion Commander, leading the unit until his graduation on March 20th, 1992. He graduated from Auburn University with a degree in Criminology, specializing in crime scene analysis, forensic analysis, evidence collection, and photography. After graduating from Auburn University, he continued serving his country as an active-duty Naval Officer, attending Naval Flight Officer (NFO) training at Naval Air Station Pensacola, learning to fly. However, he later transitioned from Naval Aviation to Naval Intelligence, ultimately finishing his career as a Naval Intelligence Officer. After 23 years of military service, he retired from active duty in November 2004. Throughout his multiple tours of duty, he gained extensive experience in multi-source intelligence collection, intelligence photography, and conducting detailed intelligence analysis and assessments. After retiring from the U.S. Navy, he continued applying his intelligence expertise as a civilian intelligence analyst, supporting the Joint Special Operations Command Joint Capabilities (SOCJC) and assisting theater Special Operations Commands (SOC) in counter-terrorism operations worldwide. After accumulating over 30+ years of military and civilian intelligence experience, he returned to his roots and started his own construction business in Virginia Beach, Virginia. Becoming an independent business owner afforded him the opportunity to write this book about his Uncle Danny. For more than two decades, the writing of this book remained on his mind, and now it is finally finished. He hopes that through this book, you will gain insight into Danny's short life and, based on the limited evidence available, form your own determination about his manner of death (MOD). Let your thoughts and reasoning guide you in understanding what truly happened to CMCN Daniel Joseph Davidenko (Danny) on November 10th, 1969, in Cam Ranh Bay, Vietnam.

Table of Contents

Introduction

Missing Daniel
(The True Story of a Young Sailor's Death in Vietnam)

By Lieutenant Commander Daniel James Roberts, USN (Ret.)

"Missing Daniel" is a true story about the brief, yet impactful life of a young man named Daniel, who was killed in Cam Ranh Bay, Vietnam, on November 10th, 1969. To fully understand this story, it is essential to first explore the relationship between its main character, Daniel, and the author. Daniel, the central figure, was both the author's namesake and Uncle. To share Daniel's story, I will begin in the summer of 1962, when Daniel was just eleven years old.

On August 30th, 1962, expecting parents Jean and Herbert Roberts welcomed the newest addition to the Roberts family. Long before this special day, Jean and Herbert had already chosen a name for their newborn son. On that joyful summer day, Jean proudly named her son Daniel after her youngest brother, Daniel Joseph Davidenko. That newborn baby boy was me, the author of this book. This is a true story about my uncle Daniel, fondly known as 'Danny' by his family. Danny's life was cut short by a tragic event that took place in Cam Ranh Bay (CRB), Vietnam, while serving his first tour of duty with Construction Battalion Maintenance Unit THREE ZERO TWO (CMBU-302). Danny was a beloved son, brother, uncle, and friend, as well as the youngest of eight children. My Uncle Danny was approximately 1 month away from his 19th birthday before he died in Vietnam. I was only seven years old when Danny passed away. His tragic early death robbed our family of the chance to truly know him or see the person he would have become. I know firsthand how deeply my uncle's death affected my mother. At just seven years old, I could feel the grief and pain that consumed my mother and our entire family during his funeral. Many years later, in my early thirties, I came to fully understand the depth of their sorrow, especially my mother Jean's, on that fateful day, November 11th, 1969, when our family first learned of Danny's death. Looking back, I can't help but wonder how my mother Jean, coped with the immense sadness and stress of losing her baby brother Danny. I remember growing up with my brothers, sisters, and cousins, all of us calling him Uncle Danny. Since I was named after him, my entire family also called me Danny. To help keep things clear in the reader's mind, from here on out, I will refer to my Uncle Daniel Joseph Davidenko as Danny. I will use his full name as needed when referring to official military documents and records.

In my distant childhood memories, I can still hear my mother Jean calling my name, whether it was to wake me for school or to call me to dinner. Now, some forty years later, I often wonder, each time she said my name, did it bring back all the painful memories of losing her youngest brother Danny? I find it troubling to think about how my mother Jean, must have felt when her 18-year-old brother Danny, joined the U.S. Navy back in 1968. She must have believed he would be safe on a U.S. Navy ship, far from the dangers on land in the Vietnam War. Then my thoughts turned to Danny's death, and I wondered how she endured the heartbreaking reality that her baby brother never made it home from his first tour of duty in the Vietnam War. From the stories told about my Uncle Danny after his death, I knew that my entire family was overcome with grief. Sadly, I never asked my mother Jean, how she coped with the sadness, stress, and pain of hearing the news of Danny's death.

As a young kid, I did not take into consideration my mother's feelings before I joined the U.S. Navy in 1981. Initially, it never occurred to me that I was following in my Uncle Danny's footsteps by joining the U.S. Navy. As painful as my enlistment news must have been, my mother Jean never let her emotions show, at least not to me directly. But deep down, I believe she suppressed those feelings, not wanting to stand in the way of my plans for the future. Why didn't she try to stop me from joining the U.S. Navy? I never had the chance to ask my mother Jean, how she felt about my decision to join the U.S. Navy. My mother Jean, passed away long before I began writing this book about her brother Danny. Looking back, I wish I had asked her more about Danny and his personality, his likes and dislikes, and how she and the family coped with his loss. Her insight would have profoundly shaped the way this story is told. I wish I had asked her more about how she truly felt about me joining the U.S. Navy after experiencing the loss of her brother. In hindsight, I regret not realizing the emotional toll it must have taken on my mother. For her, the connection between my uncle's naval service and my own decision to enlist in the U.S. Navy must have been an immense emotional burden. I never once considered my mother's feelings about me (her son) joining the U.S. Navy; what a missed opportunity. I can only imagine the intense emotions she must have felt, never knowing for certain if I was safe or even alive. How I wish I had written more letters to my mother. How I wish I had taken more time to call her, so she could hear my voice and be reassured that her son was safe and out of harm's way. I wish I could tell her how sorry I am for not considering her feelings. There's a country song titled "Til You Can't" by Cody Johnson, and one of the verses (see below) breaks my heart every time I hear it:

"So take that phone call from your mama…
And just talk away
'Cause you never know how bad you wanna…
'Til you can't someday
Don't wait on tomorrow, 'cause tomorrow may not show
Say your sorry's, your I love you's…
'Cause man, you never know
If you got a chance, take it…
Take it while you got a chance
If you got a dream chase it…
'Cause a dream won't chase you back
If you're gonna love somebody…
Hold 'em as long and as strong and as close as you can
'Til you can't."

The Authors Mother Jean Francis Roberts and the Author Daniel James Roberts

I am deeply thankful and incredibly proud that my mother gave me Danny's name, and that, like him, I served honorably in the United States Navy (USN). To this day, it is not exactly clear how or why Danny died in Vietnam. The family was told that his manner of death (MOD) was listed as an accidental non-combat death. Danny died through non-hostile action while on base in Cam Ranh Bay, Vietnam. Initial reporting revealed that Danny experienced a traumatic event, described as a gunshot wound (GSW) to the chest, which resulted in the loss of his life on November 10th, 1969. Over the years, I have made several requests for more information about this event and his military service from the Department of the Navy (DON) through the Freedom of Information Act (FOIA), the National Archives, and through numerous attempts to contact and locate members of Construction Battalion Maintenance Unit THREE ZERO TWO (CBMU-302) that served with Danny in Vietnam. It is my presumption, based on my preliminary investigation and research, that Danny's manner of death may not have been properly investigated. I know for a fact that physical evidence at the scene was not collected, processed, or recorded by his command or other legal entities. In my research, I have discovered numerous facts that raise questions about how Danny's manner of death (MOD) was determined. I was alarmed when I learned that there was no concern for evidence collection, processing, and preservation surrounding Danny's death. In my estimation, countless mistakes were made during this tragedy, making it difficult for any investigator to determine the actual manner of death in this case. It is important to make the following clear: The intent of this book is NOT to discredit or diminish the legacy of the outstanding men who served in the elite CBMU-302 unit in Vietnam. My goal is to present the facts as they were discovered and to share the life story of Daniel Joseph Davidenko (Danny), while highlighting the events surrounding his untimely death.

In this book about Daniel Joseph Davidenko (Danny), I will attempt to chronicle his brief but meaningful 18 years of life. This true story takes you on a journey through the life and times of Danny and the family that still loves and misses him to this day. Danny died too young, and his family continues to grieve his loss. He is gone but never forgotten. We will begin with a brief look into Danny's life journey, his childhood, his friends, his enlistment in the U.S. Navy, and his first and only tour of duty in Cam Ranh Bay, Vietnam. With only eighteen years of memories to draw from, I have gathered stories, photographs, and documents from family members, friends, and the men who served alongside Danny in CBMU-302. Through their recollections, I hope to reconstruct his life and bring him to life in the heart and mind of the reader. The intent behind writing this book is not to become the author of a bestseller or to seek accolades and fame. Initially, I had no desire for widespread publication, other than to have it professionally produced in book form for my family. My original plan for this book was simply to uncover the truth surrounding Danny's death and to share it with my family. I was not writing this book for bookstore shelves or for profit. The purpose of this book is to honor the life and memory of Daniel Joseph Davidenko (Danny). If, by chance, this book were to make it to the bookstore shelves, I would be humbled and grateful, for that would mean Danny's memory would live on beyond our immediate family. The main purpose of this book is to help our family and friends begin the healing process from the pain endured after Danny's death. The idea of writing this book has been on my mind since 2003, shortly after my mother Jean passed away from cancer. Over the past 20 years, I have gathered information and struggled to find enough data to create and complete the book. In 2013, I made the decision to leave my civilian government service job and start my own business. Owning my own company gave me the opportunity and flexibility to begin this book. However, my business kept me busy, and finding the time to write became increasingly difficult. Fast forward to 2020, fifty-one years after Danny's death in Vietnam, I dedicated my time and energy to my family and the writing of this book about Danny.

My research plan included hundreds of information requests, reaching out to friends and family, conversations with CBMU-302 personnel, as well as phone calls with forensic pathologists and crime scene investigators. My goal was to gather and process all available information in order to piece together the facts and finally understand what truly happened to Danny on November

10th, 1969. Ultimately, this book is being written for my family, so they can complete the healing process, offer forgiveness where it's needed, and find closure, mentally, spiritually, and physically, on the truth behind Danny's tragic death. I'm not a deeply religious man, but I believe in a higher spiritual power. That's why I can't help but feel Danny's presence when I write about his life. I feel my grandmother Mary's presence (Danny's mother) every time I uncover new information surrounding the events that took Danny's life. And, most notably, I experience an overwhelming sense of my mother Jean's presence whenever I type Danny's name, as I attempt to tell his story. Sensing their spiritual presence made me feel like they were guiding and directing me toward the truth and compelling me to write this book to ensure that all information surrounding his death is told.

The information used to write this book was gathered from official records, letters from Danny, letters from family, and conversations with family and friends, all of which helped me to recreate his life story. Danny's military journey was reconstructed from his official military records and military documents, as well as recollections and memories from former CBMU-302-unit members. Knowing that we truly do not know the actual circumstances surrounding Danny's death, I will use all available information to provide the reader with plausible scenarios of what may have happened on a secure military base in Cam Ranh Bay, Vietnam, on November 10th, 1969. I want the reader to understand who Danny was and how this eighteen-year-old young man from Brookfield, Illinois, died in a non-combat setting on the relative safety of a U.S. Naval base. I believe the totality of all the collected information will help me develop several fictitious, yet very plausible scenarios of the events that unfolded that fateful day, ultimately allowing the reader to come to their own conclusion. Based on my research, it is my belief, that the Department of Defense (DOD), along with all the other branches of the military stationed in Cam Ranh

Bay, Vietnam, failed to provide Lieutenant (LT) Paul F. Jacobs, the lead Judge Advocate General (JAG) investigator from CBMU-302, with the proper legal support and investigative tools he needed to uncover the facts surrounding the death of one of their own shipmates.

From my research, I discovered that Danny was the only CBMU-302 casualty in Vietnam during the command's extensive tour from 1967 through 1972. A majority of my initial research into Danny's death focused on a series of one-on-one interviews (phone conversations), emails, and letters with Danny's Executive Officer (XO), LT Paul Jacobs. Through these conversations, I learned that LT Jacobs was appointed as the sole JAG investigator into Danny's death. After discovering multiple documents that contradicted LT Jacobs' faded memory of his JAG investigation findings, my research focus shifted to finding answers to a multitude of questions. The main document that drew my attention immediately was the death notification letter from the Commanding Officer (CO) of CBMU-302. In this letter the CO prematurely determines Danny's manner of death (MOD) within 24 hours of discovering Danny's body. The following questions kept running through my mind: Why did the CO prematurely send a letter to Danny's parents determining Danny's MOD before all the facts and evidence were collected, processed, and before the investigation was completed? Why did the CO rush to judgment regarding the manner of death? Why didn't the CO wait for the completed Judge Advocate General JAG investigation, the Naval Investigative Service (NIS) investigation, the Military Police Report, the official autopsy, and the coroner's report before sending that letter. All of these investigations and reports are used to determine manner of death (MOD) and time of death (TOD). From all the available information, official records, and my personal interviews with CBMU-302 members, I will attempt to find answers to these questions. Based on these findings, it will be up to the reader to conclude, beyond a reasonable doubt, the manner of death (MOD) in Danny's case. It's important to note that in subsequent chapters, I will outline the basic investigative steps required to determine the MOD in all death investigations. What I am asking the reader to do is this: Examine all the information presented regarding the events leading up to the shot(s) fired on that fateful day that took the life of my Uncle Danny. I will present all information gathered from interviews, letters, and the discoverable records and documentation obtained during my research and allow the reader to determine the most logical conclusion. By the end of this book, the reader will make their own determination about the manner of death (MOD) of Construction Mechanic (CMCN) Daniel Joseph Davidenko on November 10th, 1969, on the relative safety of the U.S. Naval base in Cam Ranh Bay, Vietnam.

Chapter 1

Growing Up in Brookfield

On December 14th, 1950, just eleven days before Christmas, the town of Brookfield welcomed its newest resident. Mary Davidenko gave birth to her youngest son, Daniel (Danny) Joseph Davidenko, at Saint Anthony Hospital in Chicago, Illinois. Elmer (Danny's father) and Mary named their son after Elmer's father, Daniel P. Davidenko. Danny spent most of his life in his parents' home at 9527 Congress Park Avenue in Brookfield, Illinois. From the moment he was born until his first day of school, Danny's parents ensured his every need was met, providing him with a joyful and secure life. Danny was blessed to be born into a large family with 7 other siblings, three sisters, and four brothers. As a baby, Danny was content and didn't demand much attention. With his older sisters and brothers, Danny always had someone looking out for him. There was always someone home to care for him, or whenever their parents needed a break. During their early years, Danny and his brothers mostly played around the house, allowing their mother, Mary, to keep an eye on them. As they grew older, Mary became more relaxed, letting them venture off and explore the local community on their own. Danny and his family lived in Brookfield, a village in Cook County, Illinois, that is located about 13 miles west of downtown Chicago and is home to the renowned Brookfield Zoo. The zoo was just 7 miles from the Davidenko home, and in fact, Danny and his brothers could walk there in about 45 minutes. The Burlington Northern Santa Fe Railway (BNSF), formerly the Chicago Burlington & Quincy Railway, had tracks just a few blocks from their house. Several family members have told me that Danny and his twin brothers, Albert and Charles, would often hop onto a railcar for free rides around town. The zoo, along with many other great locations and attractions along the BNSF tracks, might have been the perfect spots for them to hop off and explore.

While researching content for this book, I found it fascinating to learn about the history of Brookfield. I certainly hadn't known, and I doubt Danny and his brothers did either, that the area was once inhabited by Native Americans, whose way of life revolved around agriculture, corn cultivation, and the building of small villages. The local Brookfield train station, once known as the Grossdale Train Station, was a 20-minute walk from the Davidenko home. In 1981, the original station was relocated across the tracks to the corner of Brookfield Avenue and Forest Avenue, where it was transformed into the Brookfield Historical Society. The train station, built in the 1880s before the city was founded, is the oldest building in Brookfield. Most of the area is flat, with small hills and rises along Salt Creek, a tributary of the Des Plaines River, the Illinois River, and eventually the Mississippi River. Salt Creek is wide and long, offering excellent fishing and abundant wildlife. Just a 30-minute walk from Danny's home, I can imagine that if the Davidenko boys were anything like their father, Elmer, they likely enjoyed fishing along the riverbanks during the summer.

The Davidenko house sat on a corner lot, with small front and side yards and a modest (fenced) backyard. In the back and side yards, Mary had a few mulberry trees that provided delicious fruit during the late summer months. I'm sure that when Danny and his siblings were kids, they feasted on those berries every summer, just as I did when I was a young boy. Many times, over the years, my mother Jean (Danny's sister), took my other siblings and me to my grandmother's house to visit. I remember a few summers when my brothers, Bob and Mark, and I were climbing those trees and eating every berry we could reach. Behind the house stood an old two-car garage, adjacent to an alleyway. The paved alley was used by homeowners to access their garages and also served as the route for the city's weekly trash collection. The front of the Davidenko house faced north on Congress Park Avenue, situated between Blanchan Avenue to the east and Eberly Avenue to the west. On this small section of Congress Park Avenue, four grassy islands with trees and landscaped areas divide the road. The avenue itself consisted of two one-way streets: one heading east and the other west. As Danny grew up in Brookfield, these grassy strips of land became great places to play catch, tag, football, and a variety of other activities with friends and family. I remember playing in those same grassy fields with my brothers and cousins when our families visited our grandparents, Elmer and Mary (Danny's parents).

At age five, Danny began elementary school in Brookfield, attending Saint Barbara Elementary School with his twin brothers. Saint Barbara's was a private school, separate from the Brookfield public school system. The school is co-located with Saint Barbara's Church and the Holy Guardian Angels Parish, located at 8900 Windemere Avenue in Brookfield. About 1.4 miles from the Davidenko home, Albert, Charles, and Danny either took the bus or were driven to school by their mother, Mary. During my research into Saint Barbara Elementary School, I discovered that the parish's pastor had announced the school's closure, which officially occurred in 2012. The school had been active for nearly a century but struggled with declining enrollment and mounting financial challenges. Unfortunately, the parish found itself in a difficult position, with too few students and insufficient financial support to remain operational. As a result, the parish listed the school building for sale in January 2022.

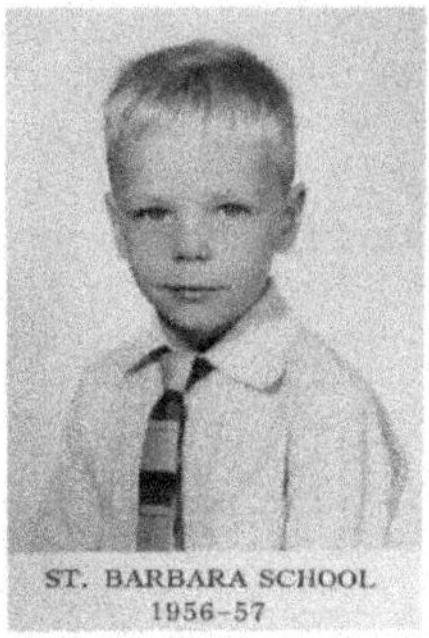

Daniel's Class Picture 1956-57 & Saint Barbara Elementary School

During one of my searches, combined with a bit of luck, I was able to locate a classmate of Danny's named Pamela (Mrs. Pamela Malysiak Kipfer), who attended Saint Barbara's with Danny. Pamela told me that Saint Barbara's was a typical Catholic school during the 1950s and 1960s. Pamela recalls that at Saint Barbara's, boys and girls were kept separate on the playground during recess and lunch. She explained that due to the Catholic Church's strict separation of males and females in school, the girls didn't have much of an opportunity to interact with the boys in their class or during recess. Pamela also mentioned that she and Danny were assigned to the same classroom, which had about 50 students, all seated alphabetically. Because their last names were on opposite ends of the alphabet, they were never seated close enough to form a close friendship. Pamela does, however, remember that Danny was a quiet boy with dark blonde hair. She also recalled that Danny lived on Congress Park Avenue, not far from where she lived. Pamela mentioned that they both attended high school together, but due to the large size of their high school class, she never had the chance to become close friends with Danny. When I attempted to obtain Danny's school records, I was denied access since I was not an immediate next of kin. As a result, all of Danny's official school records remain closed. However, based on saved memorabilia, I believe Danny attended Saint Barbara's Elementary School from 1955 to 1964.

Lyons Township High School (LTHS)

In 1964, Danny started his high school education at Lyons Township High School, District 204, located in La Grange, Illinois. The history of the LTHS dates back to September 4th, 1888, when it first opened its doors with the enrollment of 39 students. LTHS is a public high school and is often referred to as LT. The LTHS vision for all their students was to have them graduate, go on to college, and from there, enjoy a good life and career. LTSH was divided into two campuses, one North and one South. The South Campus was located at 4900 South Willow Springs Road and was designed for first-year freshmen and second-year sophomores. The North Campus, located at 100 South Brainard Avenue, was designed for third-year junior and fourth-year senior students. The South Campus was about 2.4 miles from Danny's home, requiring a 50-minute walk each way. In contrast, the North Campus was approximately 1.2 miles away, making for a much shorter 25-minute walk. Like most young adults attending LTHS, Danny and his brothers, Albert and Charles, either walked or took the bus to the South Campus during their first two years of high school. In their junior and senior years, they had more transportation options like walking, taking the school bus, or driving themselves once they obtained their driver's license.

The only high school picture I could find of Danny was in the LTHS yearbook titled 'Tabulae '67'. This was the yearbook for the class of 1967, which would have been Danny's junior year of high school. I first learned about the 1967 yearbook and Danny's photo from Craig Langele, a childhood friend of Danny and his brothers, Albert and Charles. After receiving this information, I searched the LTHS website and found a link to all the school's yearbooks. This led me to additional details about the school, its sports teams, and the school's mascot, named Noil. I didn't learn much about Danny's school experience from the yearbooks, but they did give me insight into what high school was like for him in 1967. While the LTHS yearbooks are available online, their readability is limited due to their small size. To gain a deeper understanding, I reached out to the school and alumni staff in hopes of connecting with people who knew, loved, and cared about Danny. However, I quickly learned from several sources that Danny wasn't particularly fond of high school and did not complete his senior year. Instead, he dropped out with plans to join the U.S. Navy, looking ahead to a different path for his future. I reached out to LTHS administration and alumni, hoping to find someone who remembered Danny. Someone he might have left a lasting impression on, or simply someone who interacted with him during his freshman, sophomore, and junior years at LTHS. However, my research was limited due to Danny's shy nature and tendency to keep to himself, as well as the fact that he had attended LTHS over 50 years ago. Only a few people still remembered him.

Danny received his driver's license on December 7th, 1967, just seven days before his 17th birthday. According to his 1967 driver's license, Danny was five feet nine inches tall, weighed 136 pounds, had brown hair, and green eyes. This is strange, as every picture I've seen of Danny shows him as a very good-looking young man with blonde hair. This wouldn't be the first time I've encountered discrepancies in descriptions of Danny's hair color during my research. I believe that when Danny was a child, his hair was light blond, and as he got older, his hair color began to turn more to a darker shade of blond.

Sometime at the beginning of Danny's sophomore year of high school, he began working at Art Knight's Auto and Body Shop in Lyons, Illinois. During his time there, he learned a great deal about car repairs and auto body work. Danny worked at the shop until he enlisted in the U.S. Navy on December 16th, 1968. The

owner, Art Knight, had kind words to say about Danny after learning of his death in Vietnam. In a newspaper statement, Art said, "*He was a real good kid and very close here. It's a real tragedy,*" referring to Danny's passing. Another employee at the auto shop was Joe Wydra, a co-worker and friend of the Davidenko brothers. Joe signed a sympathy letter that was sent to Danny's parents after the auto shop learned of his death. Danny's work and training at Art Knight's Auto and Body Shop provided him with the automotive skills that secured him a guaranteed seat in the U.S. Navy's Construction Mechanic school, part of the U.S. Navy SEABEE community.

Danny's Yearbook Photo, Yearbook Cover, LTHS North Campus

He worked at the auto shop until the day arrived for him to begin his enlistment with the United States Navy. In an effort to find more information about Danny and his time working at Art Knight's Auto Body Shop, I located the shop at 8604 Plainfield Road in Lyons, Illinois. I called and spoke with a woman named Ann, who told me she was the wife of Art's son. Sadly, she informed me that both Art and her husband (Art's son) had passed away and would not be contributing to the book. I left my contact information with Ann, hoping that someone might still remember Danny. Danny learned a great deal about mechanics and auto body repair while working at Art Knight's Body Shop.

Based on Danny's automotive experience and his desire to work on cars, the U.S. Navy recruiter most likely directed him toward the Construction Mechanic (CM) career path, as the job focused on all aspects of automotive repair, perfectly aligning with the skills Danny had developed. I'm sure it was a sad day for Danny when he had to say goodbye to his co-workers at Knight's Auto Shop. It must have been difficult for him, a young man leaving the job he loved, his home, his family, and his girlfriend, to embark on a similar occupation in the military, one that would take him far away from everything and everyone he knew. However, Danny didn't join the U.S. Navy blindly. He had numerous conversations with his brother-in-law, Peter (RMC Peter Stanko, USN – who at that time was currently on active duty), his father, Elmer (a former U.S. Navy sailor), and his brother, Jimmy (also a former U.S. Navy sailor), about serving in the U.S. Navy. Peter Stanko, Danny's brother in-law and my uncle, never really went by his first name Peter, and preferred to be called Fred. From these conversations, Danny learned both the pros and cons of enlisting in the U.S. Navy. The pros outweighed the cons, and Danny began to dream about the excitement of foreign travel, choosing to focus on the adventure rather than the risks that come with military service. Danny was focused on the benefits the military would offer: a steady income, a place to live, foreign ports of call, the chance to meet new friends, and the opportunity to do what he loved, working on cars and trucks. In Danny's mind, this military job must have felt like a dream come true. From my own experience joining the U.S. Navy, I know that Danny's mind had to be racing and that he could feel his heart pounding harder and harder as his enlistment date approached. The excitement, combined with nervous anticipation, must have been building up inside him as he waited for his report date on December 16th, 1968.

As a young man with his whole life ahead of him, I believe Danny was very excited to join the U.S. Navy. However, I don't think he fully considered that his new U.S. Navy job also came with the possibility of working in dangerous environments. Since Danny was too young to join the U.S. Navy on his own, he needed the consent of his parents. Through my numerous requests for Danny's military records, I was able to uncover the following information. On the morning of September 19th, 1968, Danny's parents, Elmer and Mary, took him to the recruiting station in La Grange, Illinois, to sign his DD-373 Form. This form was for Consent and Declaration of the Parent or Guardian for the Enlistment of a Minor in the U.S. Armed Forces. On the form, the recruiting personnel urged Danny's parents to keep him in school and have him graduate from high school before his enlistment date arrives. This form also stated the following:

"I/we further understand that if he (Danny) *has reached his 18th birthday during the ensuing 120-day delay* (delayed entry into the U.S. Navy), *he would be enlisted in the Regular U.S. Navy in the program for which he is qualified for a minimum period of four years".*

Both Elmer and Mary signed the agreement. Sadly, Danny decided not to finish high school, and just two days after his 18th birthday, on December 16th, 1968, Danny departed for active duty without his high school diploma.

From my numerous conversations with family members, I believe this is how the morning of December 16th, 1968, began in Danny's Brookfield home that cold Monday morning. At 6 AM (0600) Danny got out of bed and shuffled to the bathroom to make himself presentable. Looking in the mirror, he splashed water on his face and combed his blonde hair. He then threw on a pair of pants and a t-shirt before heading to the kitchen. As Danny entered the kitchen, he saw his mother, Mary, pouring a second cup of coffee. Mary had been up since 4:00 AM (0400) because she couldn't sleep, knowing Danny would be leaving home early that morning.

Before Danny could ask, she poured him a hot cup of coffee. Mary didn't need to ask; she knew exactly how Danny liked his coffee, with a generous helping of sugar and some whole milk. Mary sat down next to Danny and sipped her morning coffee. Before any conversation could start between them, Elmer walked into the kitchen, grabbed a cup of coffee, and sat across from Danny. The three of them sat there, drinking their coffee, engaging in light small talk. Not one of them was ready to address the elephant in the room; Danny's enlistment day had arrived. Mary had already prepared breakfast, and the smell of biscuits filled the kitchen. As the small talk continued, she began cooking bacon and eggs to go with the biscuits. While Mary cooked, the conversation shifted to Elmer's to-do list and the grocery list she was preparing. After breakfast, the dreaded conversation about Danny having to leave finally began. At 7:30 AM (0730), Danny took his last sip of coffee and said words to the effect of, *"Well, it's time, I'd better get dressed and ready because you know the U.S. Navy recruiters will be here soon."*

Danny went to his room to get dressed and to finish packing. Mary was cleaning up the kitchen while Elmer sat at the kitchen table reading the morning paper. Elmer wasn't particularly concerned about the news; he was just trying to keep himself from thinking about his son Danny's departure. Danny could hear Mary humming a tune in the kitchen, trying her best not to break down and cry. Mary's soft hum seemed to fill the silence, a quiet attempt to hold herself together in the midst of the emotional moment. As Danny appeared in the kitchen with his luggage in hand, the conversation turned emotional as each of them said "I love you" multiple times. At approximately 8:00 AM (0800) on that cold December morning, the doorbell rang, and two sharply dressed U.S. Navy men were standing at the door. This would be the first time Mary and Elmer had two Naval personnel standing on their doorstep, but unbeknownst to them at the time, it would not be the last.

As for my Uncle Danny's experience on December 16th, 1968, I can only speculate that it had to be an exciting, sad, and possibly quite scary day for him as well. Danny was leaving home for the first time and starting this new phase of his life. However, this particular morning had to be extremely terrifying for Mary; her baby boy was leaving home to join the military during an active war in Vietnam. Elmer, a former U.S. Navy man from WWII, was just as worried as Mary. However, according to family members, Elmer never really showed his emotions, especially to his boys, because he was raised with the belief that a father must remain strong and not show such emotions. With tears in her eyes, Mary gave Danny a big hug and kiss before saying goodbye. Elmer remained calm, giving Danny a hug and a firm handshake as they each said their goodbyes. From their front porch, they watched Danny as he walked down the steps towards the U.S. Navy vehicle. As Danny opened the car's back door, he stopped and paused for a moment, then he slowly turned around and looked up at the porch to see his parents' waving goodbye for the last time. While holding back his tears, Danny got into the vehicle with the two U.S. Navy personnel. Due to the cold, Mary and Elmer went back inside the house and pulled back the curtain to look through the front window. They both peered out at Danny, hoping to get one last look at him before heading off to boot camp. Then Mary's attention turned to the 'Blue Star Flag' with three stars hanging in the living room window. Mary gazed at the flag as she watched her Danny being driven away. Mary took down the flag with the three blue stars and looked at Elmer. Immediately, they both knew that they would soon be hanging a new flag in their window, one with four blue stars.

Each Blue Star Signifies that a Family Member is Actively Serving in the Military

Each Gold Star Signifies that an Active Duty Military Member Died Serving their Country

As the car drove out of sight, reality set in for both Mary and Elmer. They knew that the Vietnam War was raging out of control, but they believed that Danny would be safe at sea on a U.S. Navy ship. They also knew that Danny had wanted to join the U.S. Navy and that this was the beginning of a new chapter in his life. This day also marked the moment Mary and Elmer realized that four of their sons, Ray, Alby, Chuck, and Danny, were all currently serving on active duty, each doing their part for America by serving their country.

Author's Note: On November 1st, 1981, my father (Herbert) and mother (Jean) had a very similar experience to Elmer and Mary when my naval enlistment day finally arrived. On that November morning in 1981, around 7:00 AM (0700), my mother made me coffee and breakfast while we all sat and talked in the kitchen until the U.S. Navy recruiter arrived. At approximately 9:00 AM (0900) that morning, the doorbell rang, and one U.S. Navy recruiter stood at the door in his dress uniform with orders to take me to boot camp. I remember my mother's tears falling and the deep hugs she gave me as we said our goodbyes. I remember my father standing there, trying to be strong and stoic. I got the typical firm handshake and a great big hug from my father before we said our goodbyes. There were other family and friends there, but their presence was all a blur to me due to my nerves and seeing the worry and stress on my mother's face. All I remember was my mom and dad saying goodbye. I'm sure my whole family gave me hugs and said goodbye, but I blanked that memory out and could only think of my mother's face and her tears. The last thing I remember was my family standing on the driveway as the car drove away.

Chapter 2

Danny's Parents and the Vietnam Conflict

Danny's mother, Mary Katherine Davidenko, didn't live a pampered life. For most of her years, she resided in poor, lower-middle-class neighborhoods. However, that didn't mean she lacked a happy or fulfilling life. On the contrary, families in Chicago's lower-middle class were skilled at stretching a dollar and living within their means. If I could go back in time and ask Mary about her life, I believe she would say that it was good in many ways, though at times, it felt like a living hell. In the 1940s, times were tough, choices were limited, and Mary did the best she could to raise her children. Mary's first marriage was to Frank Nichols (Gondolla), which took place on January 2nd, 1936. Mary was only 16 years old when she married Frank. Just seven months later, at the age of 17, she became a mother for the first time, giving birth to her daughter, Joan, on August 8th, 1936. Over the next five years, by the time she was 23, Mary gave birth to four more children. Mary and Frank had a total of five children, three girls and two boys. They named their children Joan, Jean, Jan, Jim, and Ray. However, Mary and Frank's marriage ended after approximately 11 years. While I wasn't able to locate their divorce papers to determine the exact date, I believe their relationship ended sometime in 1947.

Like all marriages, Mary's first marriage to Frank had its share of good times and bad. According to family members, she was happy and living a fulfilling life until the marriage ended due to irreconcilable differences. Over the next year or so, Mary met and began dating a man named Elmer Arthur Davidenko. On July 24th, 1948, Mary married Elmer, and together, as husband and wife, they began raising Mary's five children. Raising five children surely took its toll on both Mary and Elmer, both physically and mentally. Mary must have had a heart of gold, boundless energy, and an abundance of love to raise five children on a shoestring budget. Mary was the typical Irish mother, always worrying about her children. Raising five kids in Chicago during a time of high crime and civil unrest must have been incredibly difficult. In the 1930s, Chicago was a major hub for mobsters and gangsters, also known as the Mafia. By the 1940s, formal and informal policies had pushed Chicago's Black residents into segregated neighborhoods. This created widespread uncertainty and unrest within the Black communities, ultimately sparking the civil rights movement. As time went on, the unrest in the city remained constant, and Mary knew she had to raise her children to be kind, to show love, and to respect authority, even in the midst of such an unsafe and violent time in Chicago. When her children were young, crime and violence in Chicago posed the greatest threat. For Mary, not a moment was spent thinking about the wars and turmoil around the world. These wars didn't directly affect her or her family. However, when her boys came of age, all of that would change.

On the day that Mary and Elmer married, Mary's children from her first marriage were all school-aged, ranging from six to twelve years old. Their wedding wasn't the typical romantic affair with a grand ceremony or a glamorous honeymoon. Instead, on that day, the two became one family, with seven members embarking on their new life journey together. Through this marriage, Elmer became an instant father with a household full of children, along with all the responsibilities that come with being a parent. One year later, on July 26th, 1949, the family grew by two when Mary gave birth to twin boys, Albert and Charles, bringing the total number of children to seven.

At this time, the economy was struggling. Although World War II ended in 1945, the United States entered into a Cold War with the Soviet Union (USSR/Russia). The Cold War was largely a political display of force and power; however, it had a significant impact on the U.S. economy, placing a heavy burden on most American families. The Cold War also instilled fear in American households, with many believing the USSR was infiltrating U.S. politics, policies, culture, and daily life with communist ideology. Like most Americans at the time, Mary and Elmer likely paid little attention to the Cold War or world news. However, they felt the economic strain it caused, struggling to meet their monthly bills and facing rising food costs that limited the groceries they could afford to feed their seven children.

As the economy worsened, Mary and Elmer's budget was stretched to its limit, making it increasingly difficult to make ends meet and put food on the table. By December 1950, Mary was eight and a half months pregnant, living in a small home with Elmer and their seven children. A few weeks later, on December 14th, 1950, Mary gave birth to their last child, a baby boy named Daniel, a welcomed new addition and an early Christmas present for the entire family. However, with Christmas right around the corner, Elmer had to work overtime to make ends meet financially and to save money for Christmas presents. While Elmer was busy at work, Mary stayed at home caring for the new baby, cleaning the house, doing the laundry, and preparing meals for her family.

Mary, Elmer, and the kids lived in a small house with limited space. Somehow, they managed to make it work, with two adults and eight children sharing a single bathroom and three small bedrooms. These living conditions would place a stressful burden on any family, but somehow, Mary and Elmer made it work; they had to, as the livelihood of the entire family was at stake. Both Mary and Elmer needed a great deal of patience, along with love, to endure the sweat and tears of keeping a family this size together. Mary was the backbone of the family, and she knew exactly how to hold it all together. Mary's desire to provide for her children brought her happiness and fulfillment, but this monumental task also created a great deal of stress and anxiety.

Despite the challenges, both Mary and Elmer did their best to care for their family.

From 1948 to 1955, two of Mary's daughters left home by falling in love and getting married. On May 22nd, 1953, Joan, Mary's oldest daughter, became the first to marry and move out, joining her new husband, Peter Stanko (aka Fred Stanko), as he began his career in the U.S. Navy. On November 22nd, 1955, Mary's second-oldest daughter, Jean, married and moved out of the house with her new husband, Herbert Roberts, who was in the painting and decorating business. In that same year, 1955, Mary and Elmer moved into their new home located in Brookfield, Illinois. A few years after the move to Brookfield, Mary's youngest daughter, Janice, married Bill Stockdale on October 11th, 1958. Bill served one tour of duty with the U.S. Army before starting a career in business sales as he and Janice made plans for their future. By November 1958, Mary and Elmer must have felt that the house seemed a little bigger, and their monthly bills a bit smaller. Raising five young boys was no easy task, but with all three of their daughters living on their own, Mary and Elmer's home life became a little easier, and Elmer's paycheck seemed to stretch a bit farther.

I received mixed reviews from various family members about how Elmer accepted his role as a father and how he treated his children. Several relatives shared that he was kind and supportive toward his daughters, but a bit rough and tough on his sons. From my perspective, there should be no difference in how a father raises his boys and girls when it comes to love and support. But I understand what Elmer was trying to do with his boys; he wanted them to grow up to be strong and successful men. That said, it's important to remember that Elmer was still a very young man, likely with little knowledge of parenting or experience in raising eight children. Imagine being 25 years old and a father of eight children, struggling to make ends meet financially. When you consider Elmer's role as a father from this perspective, it offers a different understanding of his relationships with his children. Remember, Elmer married Mary when he was 22 years old, and she was 29 with five children from a previous marriage.

Elmer's love for Mary must have been very strong, as he was willing to marry her, knowing he would become an instant father and raise her five children from her previous marriage. At just 22, Elmer took on the role of head of the household, bearing the responsibility of providing for his new, sizable family. By the time Elmer turned 25, his family had grown with the addition of his three sons, Albert, Charles, and Daniel. I'm sure Elmer experienced both happy and difficult times, but through it all, he and Mary made it work, keeping their family safe, healthy, and together. Elmer may not have won any Father of the Year awards or shown much physical affection to his sons, but from my research, I can confidently say that Elmer was a good man who provided for his family, ensuring they had food to eat, clothes to wear, and a roof over their heads. Like Mary, Elmer endured the pain and grief of losing his youngest son, Danny, the son he had named after his own father. Sadly, after 33 years of marriage, Mary and Elmer's marriage came to an end on January 29th, 1981, after Mary passed away from cancer.

Vietnam War

To truly understand Mary's loss and the loss that every mother of a son in the Vietnam War experienced, I feel it's important to provide a brief history of how the United States became involved in a war 8,500 miles away from Brookfield, Illinois. The Vietnam War began on November 1st, 1955, and ended with the fall of Saigon on April 30th, 1975. The conflict engulfed the neighboring countries of Laos and Cambodia, which were fighting their own Civil Wars at the time. Casualty estimates for the Vietnam War vary depending on the source. Most estimates combine both civilian and military deaths for

North and South Vietnam, as well as the deaths for Laos and Cambodia. According to research by Guenter Lewy, they estimate that the Vietnam War resulted in approximately 1,353,000 military and civilian deaths [20].

So how did the United States get involved in the Vietnam War? In 1950, the U.S. sent economic aid to French forces in Vietnam; however, by 1954, the French forces were defeated, and Vietnam was divided into North Vietnam (Communist) and South Vietnam (Non-Communist). Three years later, in 1957, the North Vietnamese Army (NVA) was fighting for control of South Vietnam. The U.S. and our allies were fearful that communism would spread throughout the world. Therefore, according to historical data, the fear of communist expansion around the world was one of the reasons behind the U.S. entering the Vietnam War.

Fueled by the fear of communist expansion and the reported incident involving the USS Maddox (DD-731) in the Gulf of Tonkin on August 2nd, 1964, the United States increased its military involvement in the Vietnam conflict. The story behind the USS Maddox incident is that the ship was conducting a surveillance and intelligence mission in the Gulf of Tonkin when it observed North Vietnamese patrol boats attacking Hon Me Island. Reports state that the USS Maddox fired a salvo of weapons at the patrol boats in an effort to stop their attack on the island. Moving forward to the year 1965, the U.S. deployed combat troops into South Vietnam, and the U.S. increased bombing missions against North Vietnam targets [22]. By 1975, the U.S. suffered more than 58,000 U.S. casualties, and South Vietnam surrendered to the communist North. All U.S. troops are pulled out, and the U.S. embassy is evacuated. Twelve years after losing the Vietnam War, in 1982, the U.S. built the Vietnam Veterans Memorial in Washington, D.C. and dedicated the memorial in honor of all the men and women who lost their lives in this war. CMCN Daniel J Davidenko was among those 58,000 casualties; the photo and the pencil etching below show his name as it appears on panel W16, Line 52 on the Vietnam Memorial Wall.

Daniel's (Danny) Name on The Vietnam Veterans Memorial

Tragically, after a 20-year conflict and a decade of U.S. troops engaged in a brutal ground war, South Vietnam ultimately surrendered to the North. If the U.S. government's goal was to stop the spread of communism, then it failed when it pulled out of South Vietnam and allowed the communists to take control. It makes you wonder what this war was really about. What did the U.S. have to gain? What did the U.S. accomplish? And was it worth the sacrifice of so many lives? To find the answer, you need only look into the eyes of a mother who lost her son or daughter in the Vietnam War, for in her eyes, the answer is clear. The Vietnam War was not worth the sacrifice of a single American life. To all the mothers who lost a son or daughter in Vietnam, I cannot begin to imagine the pain of your loss. I feel pain and sadness at the loss of my Uncle Danny, but my pain and that of my entire family could never come close to the pain, grief, and heartache endured by my grandmother Mary, Danny's mother.

Mary's Boys

I don't think Mary and Elmer ever knew exactly why Danny chose to join the U.S. Navy, but I'm sure they had their reservations before signing the papers for his enlistment. So why did Danny decide to join the U.S. Navy? Did he enlist out of patriotism, for the opportunity to travel overseas, for financial stability, or simply to leave home and get out of Brookfield? Danny knew that his father Elmer, had been drafted into the U.S. Navy during WWII and that his brother Jim, had volunteered, serving a four-year enlistment in the U.S. Navy. I can only speculate that Danny asked his father and brother about their time in the U.S. Navy and, hopefully, he took away some valuable advice. After speaking with my cousin Kim (Stanko) Koran, I learned that Danny had planned to visit his sister Joan and her family in Connecticut. Joan (Kim's mother) was married to Fred Stanko, who was serving as an active-duty U.S. Navy Chief at the time. My cousin Kim remembers that our Uncle Danny visited them in Connecticut for a week with a friend from Brookfield, sometime before he decided to join the U.S. Navy. We believe Danny planned this visit to have extensive talks with his brother-in-law, Fred Stanko, so that Danny would gain current firsthand knowledge about the U.S. Navy, Navy boot camp, and what he could expect during his tour of duty. Fred served on board submarines and was stationed in Groton, Connecticut. Fred was a Chief Radioman (RMC) assigned to the USS Francis Scott Key (SSBN-657), a nuclear-powered, ballistic missile submarine, that was homeported in Groton, Connecticut. It was crucial for Danny to visit Fred because Fred was nearing his 20-year retirement date from the U.S. Navy. Having served on active duty for over 18 years, Fred had a wealth of information to share about his tours of duty with Danny. Fred was a senior enlisted man with multiple tours of duty under his belt and would be a great source for "Ground Truth" information about boot camp and U.S. Navy life in general. I'm sure Fred gave Danny 'The Gouge,' as we say in the U.S. Navy, which means 'The Important Stuff' you really need to know.

RMC (SS) Peter Stanko, USN, (Ret.) / aka Fred Stanko

Author's Note: I know firsthand that RMC Peter Stanko (my Uncle Fred) had valuable information that would have helped Danny in his decision to join the U.S. Navy. I know this for a fact because, in 1980, long after Fred retired from the U.S. Navy, he gave me 'The Gouge.' That information proved essential to my successful 23-year naval career. Fred shared with me all of his stories about foreign travel and seeing the world. I remember how his eyes would light up when he talked about the foreign ports of call that he visited. I'm sure Danny's eyes lit up, too, just as mine did, as Fred described in great detail his U.S. Navy experiences and adventures.

On March 8th, 1965, the conflict in Vietnam turned to war when President Johnson placed 3,500 U.S. Marines on the ground near Da Nang, South Vietnam. The Vietnam conflict had been in the news now for over ten years, and by December 16th, 1968, Mary had four of her boys (Raymond, Albert, Charles, and Danny) who either volunteered to serve or were drafted into this war. Mary followed the nightly TV news and read the local papers, and by now, she must have been fully aligned with the anti-Vietnam protesters. Each night, the top news networks reported on the war in Vietnam and the growing anti-war protests across the United States. On the morning of December

17th, 1968, the day after Danny left for boot camp, Mary sat at her kitchen table with a cup of coffee in her hand, struggling to hold back her tears. As she sipped her coffee, she tried to calm her fears, her mind racing with thoughts of her boys' whereabouts and whether they were safe from harm.

For the most part, Mary was a quiet woman who rarely discussed world politics or the Vietnam War. However, whenever she gathered with her daughters, the conversation inevitably turned to the war and the safety of her boys. During these moments, she would vent her frustrations about the government, the conflict, and her deep fears for her sons being in harm's way. Mary's fear and pain were compounded by the relentless coverage of U.S. troop casualties in Vietnam, which were reported almost daily in the local papers and on the evening news. Both the front page and the obituary section of the local newspaper often listed the names of young men from Brookfield and across Illinois who had lost their lives in the war. Each day, Mary found it increasingly difficult to pick up the local paper, dreading the possibility that one day, one of her sons might appear on this casualty list. Every morning, Mary could hear her heart pounding as she forced herself to unroll the newspaper and scan the front-page headlines for the casualty report. She would breathe a huge sigh of relief when there was no new casualty report, especially when there were no Brookfield or Illinois sons on the list. Mary hated to read these reported casualty totals. Her heart would go out to the mothers as she read the names of the lost. Each time she read the paper, one question weighed heavily on her mind: Will my sons be next? According to the Vietnam Veterans Home Page, approximately 2,930 service members from the state of Illinois lost their lives during this protracted war in Vietnam.

Amidst all the worry, Mary was also a proud mother, deeply proud of her sons who were serving their country overseas. I keep referring to Mary's sons as 'boys' to emphasize the mother-son relationship, but in reality, they were all young men of military age, either drafted by Uncle Sam or they chose to volunteer. Regardless of whether her sons were drafted or volunteered, Mary undoubtedly felt what most mothers did about the war: disgust, sadness, and anger. During my research, I found stories from various mothers who shared their grief and thoughts on how these terrible wars claimed their sons' lives. Most of these mothers described war as senseless and unnecessary, emphasizing that it only brought death and destruction. The majority of these mothers described their loss as devastating, with the death of their sons leaving them feeling empty inside and heartbroken. I was too young to remember my grandmother Mary sharing her feelings about Danny's death. However, I learned that Mary did confide in my mother Jean, as well as my Aunts Joan and Jan, sharing both her emotions and the limited information the U.S. Navy provided about Danny's passing. Through family interviews, I learned that Mary had confided in her daughters and shared all of her feelings, the stories, and the rumors she endured surrounding Danny's death. After delving deeply into all the available information surrounding Danny's death, I've come to the conclusion that neither Elmer nor Mary, nor the entire family, for that matter, ever received a clear answer about how Danny was killed. As the Vietnam War dragged on, Mary's disapproval of the conflict grew stronger. Despite her worries for her sons, she remained relatively quiet, stoic, and deeply proud of them for serving their country overseas. Mary grew up in the Irish-Catholic faith, and I'm sure her nightly prayers ended with this simple request, *"Lord, please keep all my boys safe and bring them home to me soon."*

As the days passed after the news of Danny's death, Mary's heart remained broken. Simple items around the house became daily reminders of her loss, and the heartache and pain consumed her every day. Danny's room, his clothes, his collectibles, each a stark reminder of her devastating loss. Mary saved every piece of memorabilia that reminded her of Danny. As a mother, letting go of his personal effects was unbearably hard. She did everything she could to help cope with the pain, even holding on to a few condolence letters sent to her and Elmer after Danny's death. As I read each of these letters that Mary saved, I wondered if they offered her any peace, or if they only fueled her sadness and contempt for the Vietnam War. One week after Danny's death, the U.S. Navy sent Mary two new military flags for her windows. The first flag featured three blue stars on a white field, edged on all four sides with a red banner, gold fringe, and tassels, representing her three boys on active duty (Ray, Alby, and Chuck). The second flag displayed one large gold star on a white field, trimmed with a red border, gold fringe, and tassels, symbolizing her son Danny who was killed in Vietnam. I wondered if she displayed the flags proudly in her window, or if she kept them stored in a box, unable to bear the sight of them, as it reminded her of the tragic day that took Danny's life. In an effort to understand the emotional pain Mary was enduring, I talked with several Gold Star mothers who had lost their sons to war. I wanted to know what these mothers truly felt about the sympathy letters and the informal sympathy statements made by family, friends, and even strangers. In recent wars like Afghanistan and Iraq, as well as in older conflicts like the Vietnam war, I learned from these mothers that most were deeply offended by comments from strangers about the loss of their sons in these wars. These Gold Star mothers expressed that they hated when complete strangers would say things like, *"Thank you for your son's service," "Your son's a hero," "Your son's a true patriot," or "God bless you."* While these people were simply trying to be sympathetic and polite, the majority of mothers found that such comments only made the pain and heartache resurface. The majority of these Gold Star mothers didn't care to hear such words about their sons from complete strangers. One mother expressed that, deep down, she wanted to scream, *"What service? Don't you mean SACRIFICE?" "What hero? Don't you mean VICTIM?" "What patriot? He was DRAFTED; he had no choice. God bless me. No, GOD BLESS MY SON, because he's never coming home."* Needless to say, these mothers, who had lost their sons in war, found it difficult to talk about their loss. But I felt it was necessary to listen to their stories to truly understand their pain and emotions, so I could do my best to capture what I believed my grandmother Mary was feeling. After listening to countless stories from these mothers and learning of their pain and grief from the loss of their sons, I had to pause for a moment to collect my thoughts and to suppress my feelings and emotions. After regrouping my thoughts, my focus returned to Mary and the painful reality that she would never stop grieving the loss of her son, Danny. To compound her grief and stress even further, she would lie awake at night, haunted by the fear of losing her other sons who were serving overseas, possibly heading to Vietnam. Mary's fear of losing another son to war was unfathomable. The loss of another son to the Vietnam War would have devastated both Mary, Elmer, and our entire family.

Author's Note: As I reflect back on my 23 years of active duty in the U.S. Navy, I understand the dangers that came with the job. Though I didn't like to think about it, I accepted the possibility that I might be killed while serving my country. Like almost every person who ever served their country, there comes a time when you contemplate the possibility of your death while serving on active duty. I am confident that my Uncles Jimmy, Raymond, Albert, Charles, and Danny all contemplated the possibility of their deaths. It's not something service members dwell on, but we do think about it. We are trained to compartmentalize these feelings, carry on, and to follow orders because, in our minds, death is not an option—only victory.

Between 1959 and 1972, Mary had five of her sons serving on active duty in the service of their country. While collecting information to write this book, I couldn't help but think about all five of my Uncles serving this great nation on active duty.

With my five Uncles' military service fresh in my mind, my attention turned briefly to the story of the Sullivan brothers (Joseph, Francis, Albert, Madison, and George). All five Sullivan brothers were U.S. Navy sailors assigned to the USS Juneau (CL-52), a U.S. Navy light cruiser that fought in the Naval Battle of Guadalcanal on November 13th, 1942. During this battle, the USS Juneau was struck on the port side by a torpedo launched by the Japanese destroyer Amatsukaze. Later that same day, the USS Juneau was struck again by two Japanese submarine torpedoes. Immediately after the second attack, a large explosion occurred, and the USS Juneau broke into two large pieces, disappearing beneath the surface of the ocean within 20 seconds [51]. Ten survivors were rescued from the water eight days later. In total, 687 men, including the five Sullivan brothers, were killed in action as a result of the ship sinking. All five of the Sullivan brothers were from Waterloo, Iowa, and their deaths were the biggest blow to any one family in U.S. wartime history. Since that tragic event, the U.S. Navy has built two ships named "USS The Sullivans" to honor the five fallen Sullivan brothers. Due to this tragic event, the U.S. Navy changed its policy and no longer allows multiple family members to serve on the same ship, in an effort to prevent such an incident from happening again. Sadly, as you can imagine, the Sullivan boy's mother, Alleta Sullivan, was immediately informed about the loss of her five sons. In one article I read about this event, it describes Alleta reading the formal U.S. Navy letter detailing the tragedy that took the lives of her five sons. Alleta Sullivan also received two letters from Franklin D. Roosevelt (FDR) in February of 1943. The first was a condolence letter regarding the death of her five sons in the line of duty. The second, sent later that year, invited her to attend the christening of the USS The Sullivans, a destroyer named in honor of her five sons.

All service members eventually contemplate the possibility of their death once they enter active duty. Accepting the fact that they may die in the line of duty is never easy. Unfortunately, during times of war, too many service members made this ultimate sacrifice. After Danny's death in Vietnam, and in an effort to prevent another tragic event like the loss of the five Sullivan brothers, my Uncle William R. Stockdale, a U.S. Army Veteran and loyal patriotic American, took matters into his own hands to protect Mary and Elmer's other three sons still on active duty in the U.S. Army.

CPL William R. Stockdale (Served in U.S. Army during the Korean War)

My Uncle Bill Stockdale helped Mary and Elmer get in contact with their congressman, Harold R. Collier, who represented Brookfield, Illinois, in the 4th Congressional District.

Approximately one week after Danny's interment at Saint Mary's Cemetery, my Uncle Bill drafted a letter on behalf of Mary and Elmer Davidenko, requesting that their three other sons, Raymond, Albert, and Charles, all of whom were serving overseas, be exempt from assignment to Vietnam. Below is a copy of the original letter.

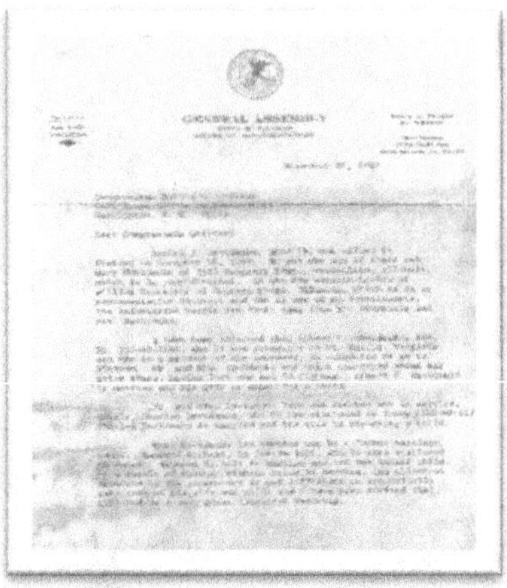

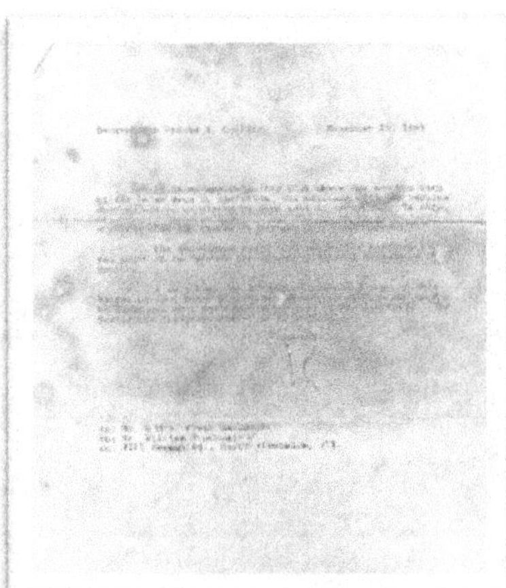

Letter to Congressman Harold R. Collier (Illinois 4th Congressional District)

Due to the fact that this document is old, in poor condition, and difficult to read, below is an easier-to-read, transcribed copy of the letter.

Transcribed Letter (Page 1)

"November 26, 1969
Congressman Harold R. Collier
2438 House Office Building
Washington D.C. 20515
Dear Congressman Collier:

Daniel J. Davidenko, aged 18, was killed in Vietnam on November 10, 1969. He was the son of Elmer and Mary Davidenko of 9527 Congress Park, Brookfield, Illinois, which is in your district. He was the brother-in-law of William Stockdale of Chicago Ridge, Illinois, which is in my Representative District and who is one of my constituents. The information herein set forth came from Mr. Stockdale and Mrs. Davidenko.

I have been informed that Albert E. Davidenko, Ser. No. (###-##-####), who is now presently in Ft. Eustis, Virginia and who is a brother of the deceased, is scheduled to go to Vietnam. Mr. and Mrs. Davidenko are quite concerned about his going there, having lost one son in Vietnam. Albert E. Davidenko is married, and his wife is expecting a child. Mr. and Mrs. Davidenko have one further son in service, namely, Charles Davidenko, who is now stationed in Korea. (###-##-####) Charles Davidenko is married, and his wife is expecting a child.

Mrs. Davidenko has another son by a former marriage, namely, Raymond Nichols, SS (###-##-####), who is also stationed in Korea. Raymond Nichols is married and has one infant child. As a result of Raymond Nichols being in service, the allotment received by the government is not sufficient to economically take care of his wife and child, and I have been advised that they endure a very great financial hardship.

Transcribed Letter (Page 2)

Congressman Harold R. Collier, November 26, 1969

It is my understanding that where one son has been killed in an area of operation, the military will not require another son to be placed in that area of operation. *If this is correct, I as well as Davidenko's would appreciate anything that you can do to prevent this from happening.*

The Davidenko's would also appreciate anything that you might do to further the return of Raymond Nichols to this country.

I am taking the liberty of sending a copy of this letter to your North Riverside address, as well as one copy to Elmer and Mary Davidenko and a copy to MR. Stockdale. The Davidenko's telephone number is HU5-5413. Cordially, (signature)

RJP: jb
cc: Mr. & Mrs. Elmer Davidenko
cc: Mr. William Stockdale
cc: 8711 Cermak Rd., North Riverside, Ill".

This letter, drafted by my Uncle Bill Stockdale with help from my grandfather Elmer Davidenko, was instrumental in keeping Mary's other three sons from receiving follow-on orders to Vietnam. All three of Mary's sons who were drafted into the U.S. Army (Raymond, Albert, and Charles) were initially subject to follow-on orders to Vietnam. However, those orders were rescinded, and each was allowed to finish their tours of duty in non-combat zones abroad or stateside in the United States. If it had not been for the efforts of my Uncle Bill Stockdale, Mary's other sons might have been sent to Vietnam and subjected to the possibility of becoming yet another sad statistic in this senseless war.

For all those who are not familiar with the military code of conduct, I feel it's important to point out Article 1 of the U.S.

Code of Conduct, which all men and women must understand before they sign up and join the military. This article must not be taken lightly, and serious consideration must be given before making such an important decision. Article 1 of the U.S. Code of Conduct reads as follows:

> *"I am an American, fighting in the armed forces which guard my country and our way of life. I am prepared to give my life in their defense."*

Author's Note: Every time I read Article 1 of the U.S. Code of Conduct, the hair on the back of my neck stands up. I still remember the feeling in my gut when I first read it, and how Article 1, in particular, gave me pause. Before signing the document, I looked around the room at the other recruits as they signed their names. Then I looked back down, closed my eyes, and took a deep breath, not out of fear, but to fully consider what I was about to agree to. By signing that document, I was making a solemn pledge to my country, acknowledging that I was prepared, if necessary, to give my life in its defense. Without hesitation, I signed my name on the dotted line.

I think about the men and women who served before me and those who have served since, and I hope and pray they never have to invoke this article. There is a saying, often found on memorials, headstones, and even in some song lyrics: *"All gave some, and some gave all."* Giving one's life for one's country is considered the ultimate sacrifice. I don't know of any Soldier, Sailor, Marine, or Airman who would want any part of this ultimate sacrifice, but when put into a combat situation, brave men and women will do what they have to do, no matter the consequences.

Mary was proud and happy for her son Danny, after he completed Navy boot camp and construction mechanic school. She was thrilled when she learned that his first duty assignment would be with CBMU-302 in California. Her joy, however, came to a sudden and painful halt when Danny learned that his new command was deploying to Vietnam. When Mary heard this news, she was in shock and disbelief. When the day arrived for Danny to leave for CBMU-302, she knew that once he arrived in California, he would be put on a plane to Vietnam. The day my Uncle Danny shipped out on orders to Vietnam, I often wondered what his parting words were to his mother, Mary, and his father, Elmer. Did Danny say something like, *Don't worry about me, I'll be fine, or I'll be okay, I'll be on a Naval Base where it's safe?* Or perhaps something like, *I love you, Don't worry about me, I'm not on the front lines, or Please write me as often as you can?* I also wondered what parting words Danny heard from his mother and father as he left for Vietnam. Did Mary say something like, *I wish you didn't have to go… I wish I never signed your enlistment papers… I love you very much… I will miss you… or Please write me to let me know you're safe?* What words did Elmer have for his son, knowing that he was heading off to a war zone? Did Elmer say, *Keep your head down and don't volunteer for anything… Don't forget to write your mother…* or perhaps something like, *Well, until I see you again, I love you, write soon?* We will never know what was truly said between Mary, Elmer, and Danny as he departed for Vietnam. I just hope that whatever they said to each other, it was meaningful, genuine, and full of love, something that would provide a lasting memory implanted in their hearts and minds, because, unbeknownst to them, this was the last time they would ever see each other again. Vietnam took its toll on over 58,000 U.S. families who had a loved one die while fighting in this senseless war. To my family and to the other 57,999 families who lost their son or daughter, all I have to say is this: **THEY WILL NEVER BE FORGOTTEN!**

Chapter 3

Wars and the Tremendous Loss of Life

Since the founding of this great nation on July 4th, 1776, America has either been directly engaged in or supported wars by providing troops, training, funding, and weapons to protect its interests. There has been no significant period of peace in which the U.S. was not involved in some form of war or conflict. Regardless of our personal views on war, as Americans, we have little control over the decisions the U.S. government makes regarding entering or supporting conflicts. In this chapter, I will provide a brief overview of the major wars the United States has fought in or supported throughout its history. As you read these historical snapshots, consider the tremendous loss of life and the devastation each war causes. Think of the mothers around the world who sit quietly beside a gravesite, mourning a son lost to a foreign war, or who approach each holiday with heaviness in their hearts because someone will always be missing. Pause for a moment and consider their pain. Now imagine if that war had come to American soil, and you not only lost your son, but everything you had spent your life building. In the blink of an eye, you could lose your family, your land, your home, and your assets. Americans often fail to realize how fortunate we truly are. As a nation, we have not experienced a major foreign war on U.S. soil since the American Revolution began on April 19th, 1775. Because of this fact, Americans have been spared from the death and devastation that comes with such wars. After the Revolution, the United States would face another tragic war on its own soil during the American Civil War, which began on April 12th, 1861. The conflict claimed more than 625,000 lives, more American soldiers than all other wars in the nation's history combined. As you continue to read, you'll learn about the staggering loss of the most precious gift, life itself. This brief historical summary of wars and their death tolls should serve as a warning to all Americans that it's not a matter of if war will come to the U.S., but rather when.

From the beginning of our nation's history, Danny's family, both past and present, has answered the call to serve in these wars, defending our way of life, our freedoms, and our country's national interests. On his father's side, there is a long line of Davidenko men who fought for this country, and the same can be said for his mother's side of the family. When Danny's mother, Mary, was a young girl, she had no way of knowing, nor was it a concern, that numerous men from both sides of her family would honorably serve this great nation in times of war. From her perspective, she never gave much thought to these wars until she got married and had children. As Mary became a wife and mother, her concerns about the United States' involvement in wars grew, as these conflicts now affected her entire family. Her husband, Elmer, served honorably in World War II, and her five sons (Jimmy, Raymond, Albert, Charles, and Danny) all served with honor during the Vietnam War. Mary did not pay much attention to World War II, the Korean War, or other conflicts around the world. However, in 1968, everything changed because she would have four sons all serving on active duty during the highly contested Vietnam War. The possibility of losing her sons during this war must have weighed heavily on Mary's heart and mind. She placed her faith in God, praying each night for their safety and a swift return home. But her faith was put to the test on November 10th, 1969, when she learned that her youngest son, Danny, had been killed in Vietnam. The following list (though not a complete list) of wars, paints a grim picture of the death toll suffered by both American military personnel and civilians, as well as the countless casualties endured by other countries. As you read this brief list of historical wars in which the U.S. participated, please keep in mind the human cost and the total number of deaths attributed to each war. The death toll is incomprehensible.

1861 – 1865 The U.S. Civil War

"The Civil War erupted from a variety of long-standing tensions and disagreements about American life and politics. For nearly a century, the people and politicians of the Northern and Southern states had been clashing over the issues that finally led to war: economic interests, cultural values, the power of the federal government to control the states, and, most importantly, slavery in American society" [53]. The Civil War in the United States began on April 12th, 1861, and ended when the Confederate states surrendered on November 6th, 1865. The Civil War was the deadliest of all American wars. According to the History.com website, the total number of deaths is estimated at 618,222, with 360,222 Union deaths and 258,000 Confederate deaths [52].

1914 – 1918 / World War One (WW I)

World War I started on July 28th, 1914, and ended on November 11th, 1918. The fighting during WWI occurred throughout Europe, the Middle East, Africa, the Pacific, and parts of Asia. What started World War I in 1914 was the growing tensions throughout Europe and the assassination of Archduke Franz Ferdinand of Austria and his wife, Sophie. The assassination of the Archduke and his wife set off a rapidly escalating chain of events that led many countries into the battle. In WWI, men from both Elmer's and Mary's families served in the military and fought in the war. The British ocean liner Lusitania was sailing from New York to Liverpool with hundreds of American passengers onboard when it was attacked by a German U-boat on the 7th of May 1915. The sinking of the Lusitania, coupled with the Germans' sinking of four U.S. Merchant Ships the following month, led President Woodrow

Wilson to go before Congress to call for a declaration of war against Germany. The U.S. officially entered the war on April 6th, 1917, two and a half years after the war began in Europe. It is estimated that during WWI, over 28,000,000 military personnel and civilians were killed during the war, and this number included over 100,000 American troops.

1939 – 1945 / World War Two (WW II)

Prior to WWII, the world knew that Hitler was planning to invade Poland, so Great Britain and France, in an effort to prevent this invasion, entered into a support agreement with Poland, thinking this would prevent Hitler from invading. Then, in late August 1939, Hitler (The German Leader) and Joseph Stalin (The Soviet Leader) signed the German/Soviet Non-Aggression Pact. Hitler knew this Pact would prevent the Soviet Union from interfering with his plans on invading Poland. This Pact worried Great Britain and France because both countries agreed to support Poland if it were attacked by Germany. World War II began on September 1st, 1939, when Hitler invaded Poland from the West. Two days later, France and Britain declared war on Germany, beginning World War II. The U.S. had been providing support to our allies before officially entering the war. The U.S. maintained neutrality until Pearl Harbor was attacked on December 7th, 1941. The day after the attack on Pearl Harbor, the United States entered World War II. Danny's father, Elmer Davidenko, was drafted into the U.S. Navy and served in WWII. The U.S. Navy trained Elmer to serve on board large ships that offloaded tanks, equipment, and personnel. These ships were called LST's (Landing Ship, Tanks) that pulled right up to the shore so the tanks and personnel could go ashore quickly. World War II came to an end with the unconditional surrender of Germany on May 7th, 1945, and when Japan surrendered to the United States on board the USS Missouri on August 14th, 1945. The Japanese surrender occurred only after the U.S. dropped atomic bombs on Hiroshima and

Nagasaki on August 6th, 1945, and again on August 9th, 1945. The bombs' devastating results forced Japan to an unconditional surrender. World War II involved more than 50 nations, and the war was fought on land, sea, and air in nearly every part of the world. There are numerous historical reports as to the death toll of World War II. It is estimated that approximately 80,000,000 deaths occurred, mostly among civilians. Tens of millions died due to genocides (including the Holocaust), starvation, massacres, and disease. It is estimated that the U.S. suffered approximately 407,000 military deaths and approximately 12,000 civilian deaths [53].

1947 – 1991 / The Cold War

The term *"Cold War"* was first used by the English writer George Orwell in an article published in 1945 to refer to what he predicted would be a nuclear stalemate between *"two or three monstrous super-states, each in possession of weapons (nuclear weapons) that could kill millions of people in a few seconds"* [49]. The United States of America (USA) and the Union of Soviet Socialist Republics (USSR) were once Allies during World War II; however, after WWII, the U.S. and the USSR became competitors on the world stage and engaged in what was called the "Cold War". It was called the Cold War because it never resulted in an overt, physical war; the two nations used espionage, political subversion, and proxy wars. The Cold War began shortly after WWII on March 12th, 1947, and ended on December 26th, 1991 [1]. The Cold War had several effects on America. It shaped American foreign policy and political ideology and impacted our domestic economy. The Cold War led to destructive conflicts like the Vietnam War, the Korean War, and the Soviet-Afghan War. But what is most disturbing is the fact that it led to both the USA and the USSR building huge arsenals of atomic (nuclear) weapons and ballistic missiles. Both countries feared an attack from each other, adhering to mutual mass destruction [50]. The exact number of people who died in the Cold War is difficult to determine. However, estimates suggest that the total number of casualties, including both military personnel and civilians, ranges from around 20,000 to 30,000 [54].

1950 – 1953 / The Korean War

The Korean War began on June 25th, 1950, when North Korea invaded South Korea following clashes along the border in South Korea. North Korea was supported by China and the Soviet Union, while South Korea was supported by the United Nations, principally the United States. The fighting ended with an armistice on July 27th, 1953. Japan ruled over Korea until its surrender at the end of World War II. The United States and the Soviet Union divided Korea along the 38th parallel into two zones of occupation. The Soviets supported the northern zone, and the Americans supported the southern zone. In 1948, as a result of Cold War tensions, the occupation zones became the sovereign states of North Korea and South Korea. Both governments of the two new Korean states claimed to be the sole legitimate government of all of Korea, and neither accepted the border as permanent. It is estimated that approximately 150,000 troops from South Korea, the U.S., and participating U.N. nations were killed in the Korean War. Within that number, the Pentagon estimates that 36,914 of those deaths were American troops. It is also estimated that as many as 1,000,000 South Korean civilians, 800,000 communist soldiers, and more than 200,000 North Korean civilians were killed during the Korean War [1].

1955 – 1975 / The Vietnam War

The U.S. entered the Vietnam War based on the U.S. government's fear of the "Domino Theory," which suggested that a communist government in one nation would quickly lead to communist takeovers in neighboring countries, each falling like a row of dominoes. In Southeast Asia, the U.S. government used this domino theory to justify its involvement in the

Vietnam War and its support for a non-communist dictator in South Vietnam. The conflict in Vietnam, Laos, and Cambodia began on November 1st, 1955, and ended with the fall of Saigon on April 30th, 1975. The Vietnam War lasted almost 20 years, with direct U.S. involvement until 1973. All of Mary's sons served on active duty during the Vietnam War, but Danny was the only one stationed in-country, in Cam Ranh Bay, Vietnam. During my research, I learned that if President John F. Kennedy (JFK) was NOT assassinated on November 22nd, 1963, he would have begun a troop withdrawal from Vietnam, with a goal to have all troops out of Vietnam by the end of 1965.

National Security Action Memorandum (NSAM) 263

President John F. Kennedy (JFK) approved the recommendations by Secretary of Defense Robert McNamara and Chairman of the Joint Chiefs of Staff General Maxwell Taylor on the withdrawal from Vietnam. The main recommendations by the McNamara-Taylor report were that a phased withdrawal would be completed by the end of 1965. This recommendation became known as JFK's NSAM-263, which approved a plan to withdraw U.S. military forces from Vietnam. JFK's plan was to begin by withdrawing 1,000 U.S. soldiers from Vietnam, followed by a complete withdrawal from South Vietnam after his re-election in 1964. President John F. Kennedy's national security directive was approved on October 11th, 1963 [56]. JFK had made the formal decision to withdraw from Vietnam, regardless of whether we were winning or not. Robert McNamara believed the U.S. was not winning the war and supported JFK's decision. The first stage of withdrawal had been ordered, with a final withdrawal date set for two years later. However, after JFK's assassination, President Johnson's administration authorized Operation Plan Thirty-Four Alpha (OPLAN 34A), which permitted commando raids against targets in the North. This decision ultimately committed the U.S. to the Vietnam War, a conflict that was nearly impossible to win [57]. President Kennedy intended to withdraw all U.S. troops from Vietnam by 1965. It is impossible to know for certain, but I often wonder whether my Uncle Danny might still be alive today had JFK not been assassinated. Below is a copy of President John F. Kennedy's National Security Action Memorandum 263 approving the withdrawal from Vietnam. This document was originally classified "TOP SECRET" but was declassified many years later.

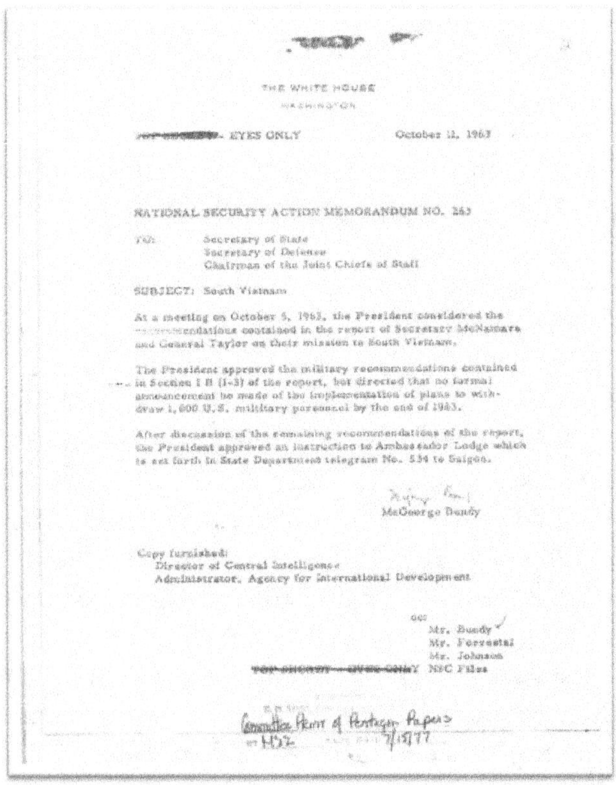

THE WHITE HOUSE

WASHINGTON

~~TOP SECRET~~ - EYES ONLY October 11, 1963

NATIONAL SECURITY ACTION MEMORANDUM NO. 263

TO: Secretary of State
Secretary of Defense
Chairman of the Joint Chiefs of Staff

SUBJECT: South Vietnam

At a meeting on October 5, 1963, the President considered the recommendations contained in the report of Secretary McNamara and General Taylor on their mission to South Vietnam.

The President approved the military recommendations contained in Section I B (1-3) of the report, but directed that no formal announcement be made of the implementation of plans to withdraw 1,000 U.S. military personnel by the end of 1963.

After discussion of the remaining recommendations of the report, the President approved an instruction to Ambassador Lodge which is set forth in State Department telegram No. 534 to Saigon.

McGeorge Bundy

Copy furnished:
Director of Central Intelligence
Administrator, Agency for International Development

cc:
Mr. Bundy
Mr. Forrestal
Mr. Johnson
NSC Files

~~TOP SECRET - EYES ONLY~~

DECLASSIFIED - TOP SECRET National Security Action Memorandum (NSAM) 263

The main purpose of this book is to focus on the short life of my Uncle Danny, his tour of duty in Vietnam, and his tragic death in Cam Ranh Bay. However, I believe that by adding these staggering statistics, as well as the many myths and facts about the 20-year conflict listed below, I can enhance the reader's understanding of this emotional story. Vietnam War death tolls are estimated to be approximately 1,353,000 during the duration of the war [21]. According to the website called "*Vietnam War Facts, Stats and Myths,*" they report some staggering numbers and information most Americans probably never heard before. According to this website on Vietnam statistics, 9,087,000 military personnel served on active duty during the Vietnam War. Of the 9 million service personnel, only 2,709,918 Americans served in uniform in Vietnam. Of those 2.7 million service members, only 240 men were awarded the Medal of Honor during the Vietnam War. The first man to die in the Vietnam War was James Davis in 1961. The youngest service members killed in Vietnam were just 16 years old, with a total of five 16-year-old boys giving their lives in support of this war. The oldest service member killed in Vietnam was 62 years old. A total of 58,000 U.S. service members were killed in Vietnam. Approximately 75,000 service members left this war severely disabled. Over 5,283 service members lost limbs, and 1,081 of those service members sustained multiple amputations. Of those killed, 61% were younger than 21 years old, and 11,465 of those killed were younger than 20 years old. The average age of the men killed in Vietnam was 23.1 years. According to the POW/MIA Affairs Committee update from June 2024, there are 1,237 Americans still unaccounted for (Missing in Action) from the Vietnam War.

There are numerous myths about the United States involvement in the Vietnam War. The myth that most Vietnam veterans were drafted into the war is false. The fact is that only 1/3 of the men who served in Vietnam were drafted; the other

2/3 were all volunteers. Another myth was the belief that a disproportionate number of black men were killed in the Vietnam War. However, the facts reveal that 86% of the men who died in Vietnam were Caucasians, 12.5% were black, 1.2% were other races. Another Vietnam fact was that U.S. military personnel serving in Vietnam coined the word "Fragging" during the War. The term "fragging" means the deliberate or attempted killing by a soldier of a fellow soldier, usually a superior. The term came about when such killings were most often attempted with a fragmentation grenade, making it appear that the killing was accidental or during combat with the enemy. Now, the term fragging encompasses any deliberate killing of a military member. Documented and suspected fragging incidents totaled nearly 900 from 1969 to 1972. Fragging should not be confused with the unintentional killing and/or wounding of personnel; such incidents are referred to as friendly fire.

As we begin to examine the recorded documents and conversations about what happened to Danny, we will closely examine and consider four of the five possible manners of death (MOD): accidental, suicide, homicide, and murder. Since we know that Danny's cause of death (COD) was a gunshot wound to the chest, the fifth manner of death, 'Natural Causes,' will not be considered, as it does not pertain to this situation. Additionally, death by fragging will not be considered, as Danny was not a senior enlisted member or officer, and the collected data does not support this scenario.

U.S. Wars from the 1980s to the Present

After Vietnam, the U.S. participated in multiple conflicts around the world, including but not limited to Lebanon, Grenada, Libya, the Persian Gulf, Panama, Iraq (Operation Desert Storm), Somalia, Bosnia, Croatia, Haiti, Kosovo, Syria, Afghanistan (Operation Enduring Freedom), and Yemen. As a retired Military Intelligence Officer, I served in or supported, in various capacities, all of the wars and conflicts listed above, both during my active-duty service and later as a civilian contractor and government employee. As a young man, when I joined the U.S. Navy, I didn't fully understand, or even consider, how dangerous military service could be.

I knew that death was a possibility, no matter what the job or mission was, but I didn't want to think about the reality that I might die or that I might witness the death of one of my shipmates. The reality of death in the military hit me hard on January 26th, 1983, when I witnessed the deaths of three of my shipmates during flight operations aboard the USS Nimitz aircraft carrier. I was only 20 years old at the time, and it was difficult for me to cope with the loss of these good men. I didn't know it then, but that tragic event would haunt me for the rest of my life. Unbeknownst to me at the time, these deaths wouldn't be the last I'd witness of fellow Sailors, Soldiers, Airmen, or Marines.

The wars of the 1980s to present were supposed to be the last wars mentioned in this chapter; however, as I write this book, a war between Ukraine and Russia began on February 24th, 2022, and continues with no end in sight and with death tolls rising each day. Additionally, on October 7th, 2023, a series of coordinated armed attacks led by the Palestinian Islamist militant group Hamas from the Gaza Strip conducted an unprovoked attack against the country of Israel. In a single day, Hamas terrorists massacred approximately 860 Israeli civilians and over 350 Israeli soldiers and police. These unprovoked attacks on Israel led to an all-out war against the Hamas terrorist group and its leaders. This war against Hamas in Gaza eventually expanded into other countries. On October 1st, 2024, Israel invaded Southern Lebanon because Hezbollah joined the fight with Hamas and began to attack Israel from the North. These hostilities between Hezbollah and Israel began just after the Hamas attack on October 7th, 2023. This war expanded yet again when Israel began retaliatory strikes against Yemen on July 20th, 2024, and then again on September 29th, 2024, when Israel targeted the Iran-backed Houthis in response to missile attacks by the militant group. The Houthis have long called for the destruction of Israel, but did not take direct action against the Jewish state until Hamas's October 7th attacks, which initiated the Israel–Hamas war [64].

Based on my military intelligence experience, my research, and reporting on this war, I share the concerns of many around the world that the Iranian government plays a central role in fueling the conflict with Israel and in financing terrorist operations globally. As the war rages on between Israel and Hamas in Gaza, the role of Iran will remain a central factor. Tehran is not only Israel's top regional foe but also the leading provider of military aid and training for Hamas [65]. Currently, the United States of America is not actively engaged in these wars in Ukraine and Israel, but the U.S. Government is supplying weapons and financial support to both. Sadly, with this ever-expanding war raging overseas, the fears of World War III are on the horizon and in the hearts and minds of all Americans. What will history record as the death toll from these two wars? Only time will tell.

Chapter 4

Mary's Magnificent Six

My family and I were deeply saddened by the senseless loss of my Uncle Danny in the Vietnam War. However, we were immensely grateful that his brothers were diverted from the same fate and returned home safely. In this chapter, I have gathered as much information as possible, including stories and pictures, to honor the service of Mary's husband and sons. Each story is based on details provided by family members, letters, military records, and verbal accounts from friends and relatives of the six men listed below. Every member of Mary's "Magnificent Six" answered the call to serve this great nation, and all served their country with honor. Every one of them was honorably discharged from their respective branches of service.

Elmer, Danny, Jimmy, Alby, Ray, & Chuck

Only two were alive to give me their firsthand accounts of their military tour(s) of duty. Sadly, my Uncle Chuck (Danny's brother) passed away on November 18th, 2024, before I could complete my interviews and questions concerning Danny's tragic death. Chuck provided me with a wealth of information about Danny's time in the U.S. Navy and his efforts to bring Danny's body home to rest. Retelling the story of his brother's death and his role in escorting Danny's body home was incredibly difficult for my Uncle Chuck. During one of our many conversations, Chuck finally said he was done; he couldn't talk about it anymore; it was just too painful. I chose to give him time, hoping he might eventually change his mind and resume our conversations. Unfortunately, he became ill, and his focus understandably shifted to his health. A few months later, he passed away. I was devastated when I learned of my Uncle Chuck's passing, especially knowing I still had so many unanswered questions. Rest in peace, Uncle Chuck. I hope you're reunited with Danny and the rest of your family in heaven.

With all the information I gathered, I decided to title this chapter Mary's Magnificent Six. While details about my grandfather, Elmer, and my Uncles Jimmy, Raymond, Albert, Charles, and Danny's military service were limited, I believe they still paint a clear picture of our family's deep patriotism and unwavering love for America. Before outlining the military service of my grandfather and my Uncles, I'd like to share a brief story about the going-away party held for me before I left for U.S. Navy boot camp. A week before my enlistment date on November 2nd, 1981, my parents threw me a going-away party at the American Legion in Chicago Ridge, Glenn Maker Post 1160. They invited all of my friends and family to attend. This was the same location where my grandfather, Elmer, held a going-away party for his son, Danny. Later, I learned that it was actually my grandfather, Elmer A. Davidenko (aka JaJa); my uncle, William R. Stockdale (aka Bill); and my father, Herbert V. Roberts (aka Rocky) who had planned the entire party at the American Legion.

Herbert V. Roberts, Jr., Elmer A. Davidenko, & William R. Stockdale

At my party, I remember my grandfather Elmer telling me about his tour of duty in the U.S. Navy. He shared that he was trained in boat landing operations and learned to drive and operate small U.S. Navy boats. At the time, I had no idea what a U.S. landing boat (ship) was or how dangerous these boat operations had been during WWII. By the end of my first year of military service, I had learned more about my grandfather's military experience and the role these landing boats played during the war. Through my military history training, I learned that these WWII ships called "Landing Ship Tanks (LST)" were used to deploy tanks, troops, equipment, vehicles, and supplies from ship to shore during offensive operations. I don't recall which specific LST name or number my grandfather, Elmer, told me he served on; however, the most common and recognizable type during WWII was the LST, which had a bow that opened to offload tanks, servicemen, and equipment. I don't fully remember all the stories Elmer shared with me about his time in the U.S. Navy during World War II as we shared a few beers at my going-away party. However, my FOIA request for his military records provided his assignments and identified the ships he served aboard during his tours of duty. Long after that party, while on active duty, I learned how the U.S. Navy used LSTs extensively during World War II. During the D-Day invasion, also known as "Operation Overlord," the largest seaborne invasion in history took place when Allied forces stormed the beaches of Normandy on June 6th, 1944. Elmer would have been 18 years old at the time of this invasion, just out of boot camp and in the middle of his shipboard training at Camp Bradford U.S. Navy Base in Norfolk, Virginia. As a result, Elmer did not participate in the D-Day operation. It wasn't until I obtained Elmer's military records that I learned of his contribution to some other significant military operations.

Mary's U.S. Navy Men

Mary's husband, Elmer, and sons Jimmy and Danny all served in the United States Navy. Though at different times, each of them went through the U.S. Navy boot camp at Great Lakes, Illinois. Each of them was issued a U.S. Navy uniform, received dental and medical exams, and had their heads shaved, leaving barely a hair visible. During the first two weeks of training, they were taught naval history, how to swim, how to march, how to fight fires, how to use various weapons, how to wear their uniforms, and how to conduct themselves as sailors. Everything they had learned in civilian life was discarded, and they had to relearn how to do things the U.S. Navy way. U.S. Navy boot camp was no easy feat. The U.S. Navy's goal was to instill confidence, provide hands-on training with shipboard systems and weapons, and teach sailors how to think, act, and respond to emergency situations. Upon graduation from boot camp, each of them attended advanced training in a specific field, either by choice or based on the needs of the U.S. Navy. Elmer was drafted and trained in shipboard operations. Jimmy volunteered for the U.S. Navy and chose to become a radioman, training in naval communication systems. Danny also volunteered and selected the job of Construction Mechanic, training with the U.S. Navy Construction Battalions, which the U.S. Navy fondly calls the SEABEES.

Seaman (SN) Elmer Arthur Davidenko, USN (Danny's Father)

In February 1944, Elmer turned eighteen. His service record describes him as six feet tall, 168 pounds, with blue eyes and brown hair. Before entering the military, he worked as a truck driver for the Chicago Sewerage Company, earning $30 a week. On March 20th, 1944, he was drafted into the U.S. Navy. At the time, an E-1 sailor earned $50 per month, a salary that also covered uniforms, meals, and lodging. That same day, his records show a change in status from inductee to voluntary enlistment, suggesting the U.S. Navy offered him an incentive to sign on. His title shifted from U.S. Navy Apprentice Seaman, USN-I (USN-Inducted) to U.S. Navy Apprentice Seaman, USNR (USN Reserve). Elmer's Boot Camp began on March 20th, 1944, at Naval Station Great Lakes in Illinois. Just over a month later, on April 25th, 1944, Elmer graduated and received orders to the Amphibious Training Base (ATB) Camp Bradford in Norfolk, Virginia (now known as Joint Expeditionary Base Little Creek). There, he trained and qualified on the .40 caliber anti-aircraft machine gun, the M1903 Springfield rifle, and the standard .45 caliber Colt-1911 pistol. After completing his training at Camp Bradford, he was assigned temporary duty in Boston, Massachusetts, before receiving follow-on orders to the USS LST-1028, a tank landing ship commissioned on July 7th, 1944. His duties included standing security watch, serving as a lookout, and working as a small boat coxswain. While serving aboard the USS LST-1028 in January 1945, Elmer took part in the invasion of Lingayen Gulf in the Luzon, Philippine Islands. On January 10th, 1945, the ship bow was driven into the shoreline at Orange Beach to offload troops and equipment when it came under attack by a Japanese torpedo boat. The port side was hit and severely damaged, and although the crew carried out their mission, the damage proved beyond repair. Elmer's medical records were not available, so it is unknown whether he sustained any injuries during the attack. After this attack, the USS LST-1028 was damaged beyond repair and was later decommissioned on November 19th, 1945. This ship was later sold for scrap on August 29th, 1947. For the ship's service in World War II, USS LST-1028 earned one battle star. During Elmer's deployment aboard the LST-1028, he received the Amphibious Forces Insignia, the Combat Action Medal (and Ribbon), the American Campaign Medal (and Ribbon), the Asiatic-Pacific Campaign Medal (and Ribbon), and the World War II Victory Medal (and Ribbon).

USS LST-1028 Awards Received

USS LST-1048 (Same Class as 1028) & USS LST-1028 During the Invasion of Luzon PI

On October 4th, 1945, Elmer transferred to the U.S. Naval Receiving Station in Seattle, Washington, where he awaited further assignment. A month later, on November 12th, 1945, he received orders to the U.S. Naval Receiving Station at the Naval Reserve Armory in Chicago, Illinois. After completing his duties there, he was assigned to the USS LSM-356, reporting aboard on January 19th, 1946. Just a few months later, on April 17th, 1946, he transferred to another ship, the USS LSM-303. This move to USS LSM-303 was likely tied to the USS LSM-356's deployment schedule and Elmer's approaching separation from active duty, scheduled for May 31st, 1946. According to his records, Elmer was honorably discharged on June 2nd, 1946, but reenlisted the very next day for an additional four years. Unfortunately, my FOIA request did not include details of this enlistment. The only document provided shows that he received an honorable discharge from the U.S. Navy on June 2nd, 1950.

Elmer Aurthur Davidenko & Elmer Davidenko Headstone

Radioman Second Class (RM2) James William Nichols, USN (Danny's Brother)

By late 1959, Mary's eldest son, Jimmy, decided to join the U.S. Navy, fully aware that he could be assigned to a ship or duty station supporting the Vietnam conflict. By the time I began writing this book, my Uncle Jimmy had passed away, leaving me unable to interview him about his first and only tour of duty in the U.S. Navy. I deeply wish I could have heard his thoughts and memories about his brother Danny, especially how he first learned of his youngest brother's death in Vietnam. Danny's brother, James William Nichols (Jimmy), joined the U.S. Navy on April 23rd, 1959, approximately five years after the United States became involved in the Vietnam conflict.

It is unclear how Jimmy traveled from his home in the Chicagoland area to the Great Lakes Training Center for boot camp. However, from some limited military records, I discovered that he completed all required U.S. Navy medical screenings and administrative paperwork during his bootcamp training. Jimmy successfully graduated from boot camp and went on to Radioman School, where he qualified as a radio operator. Jimmy's service record reveals that he served aboard the USS Dash (MSO-428), a wooden-hulled ocean minesweeper that conducted operations in the Mediterranean and the South China Sea. Currently, these old minesweepers are no longer part of the U.S. Navy's inventory. According to Wikipedia, there were once more than 50 minesweepers in the fleet, including the USS Dash. It was one of five wooden-hulled minesweepers known as "The Dirty D's," alongside the USS Detector, USS Direct, USS Dominant, and USS Dynamic. These MSO-class ships operated in the Mediterranean and the South China Sea, tasked with protecting U.S. and Allied shipping by locating and neutralizing mines laid by enemy forces. Although Jimmy's ship, the USS Dash, was not operating off the coast of Vietnam during his assignment, many MSO-class ships took part in OPERATION MARKET TIME, a mission aimed at intercepting the maritime flow of supplies to North Vietnam.

Jimmy served in the U.S. Navy for three years, ten months, and twenty days before being honorably discharged from his active-duty contract on March 12th, 1963. After completing his primary enlistment contract, Jimmy was transferred to the US Naval Reserve to fulfill the second part of his service obligation, which ended on April 22nd, 1967. Jimmy attained the rank of Petty Officer 2nd Class (E-5) during his time in the U.S. Navy. He was awarded the National Defense Service Medal and the Good Conduct Medal for his service.

RM2 James Williams Nichols

USS Dash MSO-428

Construction Mechanic Constructionman (CMCN) Daniel Joseph Davidenko, USN

Mary's youngest son, Danny, joined the U.S. Navy at the height of the Vietnam War. After completing his U.S. Navy boot camp and Construction Mechanic School, he was assigned to his first active-duty command, CBMU-302, home ported in California. CBMU-302 fell under the command of the Commander U.S Pacific Fleet (USPACFLT). The USPACFLT commander ordered CBMU-302 to deploy to Cam Ranh Bay, Vietnam, long before Danny was assigned to the unit. When Danny arrived at CBMU-302, the commanding officer assigned him to the mechanics shop to perform body repair and mechanical maintenance on all vehicles and heavy equipment within the unit's inventory. His detailed military service will be covered in greater depth in Chapter 7.

CMCA Daniel Joseph Davidenko (Bootcamp Photo) & SEABEE Logo

Mary's U.S. ARMY Men

Mary's sons, Raymond (Ray), Albert (Alby), and Charles (Chuck), all served in the United States Army. Each of them completed U.S. Army boot camp, fully aware that it would be tough, consisting of approximately six to ten weeks of rigorous and demanding training. U.S. Army boot camp pushed Ray, Alby, and Chuck to their limits, both mentally and physically. During the Vietnam War era, the U.S. Army established multiple training centers, such as Fort Campbell, to prepare newly drafted soldiers for combat. When Mary's boys arrived at their respective U.S. Army boot camps, they were issued olive drab fatigues and a pair of boots, marking the beginning of their transition from civilians to soldiers. They all knew that basic training (boot camp) would be no walk in the park. They learned everything the U.S. Army required them to know, including how to march, shoot, and maintain various weapons, engage in hand-to-hand combat, uphold U.S. Army core values and traditions, and, ultimately, become combat-ready soldiers.

Specialist Four (SP4) Raymond Garold Nichols, USA (Danny's Brother)

Mary's second-oldest son, Raymond (Ray), was drafted into the U.S. Army. When I set out to write this book about Danny, I had the opportunity to hear Ray's account of his time in the U.S. Army and his memories of the day he learned about Danny's death. Ray's story became crucial in the development of this book. Getting Ray's story was a bit of a challenge because my Uncle Ray lives in Illinois, while I live in Virginia. To bridge the distance, I asked for help interviewing Ray, gathering his thoughts, memories, and the memorabilia he had saved from the day his younger brother, Danny, was killed in Vietnam. The first person who came to mind was my cousin, Mary (Stockdale) McEldowney, who lives close to Ray. I knew she would be the perfect person to interview him and record his story. With a quick phone call, I asked Mary if she would be willing to help me gather Uncle Ray's story. Without hesitation, Mary said yes, she was ready and willing to help. Mary and I reviewed a list of topics I wanted her to discuss with Ray, including his time in the U.S. Army and, more specifically, the death of his brother Danny. She worked tirelessly to gather Ray's story, generously volunteering countless hours of her time and typing skills to document his account for this book. The following information was collected by Mary during her interviews with Ray and served as the foundation for drafting his story.

Danny's brother, Raymond Garold Nichols (Ray), was drafted into the U.S. Army on April 4th, 1968. Early that spring morning, around 5:00 AM (0500), Elmer (Ray's stepfather) drove him downtown to the U.S. Army induction center. There, Ray spent the entire day in training classes, undergoing a full medical screening, receiving all necessary shots, and completing the required paperwork. After a full day of induction, Ray received his orders and a plane ticket to Louisiana. That evening, he boarded a commercial flight in Chicago, heading for Fort Polk, Louisiana, where he would undergo basic training. In 1962, Fort Polk began conducting basic training and advanced individual training for U.S. Army soldiers. A small portion of the base is covered with dense, jungle-like vegetation, and the region's heat, humidity, and precipitation helped commanders acclimatize new infantry soldiers for combat in Vietnam. The climate and terrain of Louisiana served as a simulation of the conditions soldiers would face in Southeast Asia.

After Ray's training at Fort Polk, he was sent to Fort Irwin in California. Fort Irwin is located in the Mojave Desert of California, and at that time, the base was the home to the U.S. Army's main military training center. At Fort Irwin, Raymond performed various Kitchen Patrol (KP) duties before being trained and assigned to the U.S. Army's 226th Military Police (MP) Detachment. From Fort Irwin, Ray was then transferred to Fort Lewis, Washington, which, during the Vietnam War, became a personnel transfer and training center. Ray was sent there to wait for his transfer to Camp Greaves in South Korea. Camp Greaves was located north of the Imjin River, just across Freedom Bridge in South Korea. It was the U.S. Army's most forwardly deployed battalion, and the only U.S. battalion stationed north of the Imjin River, approximately 2,500 meters from North Korea. The Imjin River flows from north to south, crossing the Demilitarized Zone (DMZ) before joining the Han River downstream of Seoul, near the Yellow Sea.

SP4 Raymond Garold Nichols & Ray's Unit Crest

Raymond Garold Nichols served just over two years in the United States Army, with three months and four days of that time spent overseas in Korea. He likely would have completed his tour in Korea, or possibly been sent to Vietnam, if not for the efforts of his brother-in-law, Bill Stockdale, along with Elmer and Mary. Together, they drafted a family letter to the government, requesting that the rest of Mary's sons be kept out of Vietnam. The letter worked, and Ray was sent back stateside to complete his tour of duty. Although he was not particularly fond of the U.S. Army, he achieved the rank of Specialist Four (SP4/E-4) and was honorably discharged on April 14th, 1970. During his time in the U.S. Army, Ray earned the National Defense Service Medal, the Armed Forces Expeditionary Medal, the Good Conduct Medal, the Sharpshooter Badge for the M-14 rifle, and multiple letters of commendation.

Private (PV1) Albert Elmer Davidenko, USA (Danny's Brother)

Albert was one of Mary's twin boys who served in the U.S. Army. Sadly, my Uncle Albert (Alby) passed away before I could hear his firsthand account of his memories of his brother Danny or how he first learned about his younger brother's death in

Vietnam. Albert Elmer Davidenko was drafted on July 22nd, 1969, approximately three and a half months before Danny died in Vietnam. It's unclear whether Elmer drove Alby to the U.S. Army induction center or if the U.S. Army picked him up from his home. However, we do know that Alby completed all the induction medical screenings and administrative paperwork. Alby successfully completed U.S. Army boot camp and graduated from the U.S. Army's Cargo Handling School before being assigned to USARPAC's 412th Transportation Company stationed in Okinawa, Japan. Alby served a total of 1 year, 11 months, and 23 days on active duty, and a majority of that time was spent in Japan, in support of the Vietnam War.

PV1 Albert Elmer Davidenko, Albert's Unit Patch, & Albert's Head Stone

According to family members, Alby was deeply troubled by the loss of his baby brother, and coupled with his dislike of the U.S. Army, he experienced a minor mental breakdown. He refused to return to his U.S. Army unit in Japan. In fact, Alby went Absent Without Official Leave (AWOL) for several weeks after his brother's funeral. However, after speaking with other family members, he ultimately decided to return to his unit in Japan. Upon Alby's return to Japan, his commanding officer (CO) charged him under Article 86 of the Uniform Code of Military Justice (UCMJ), which covers absence without leave.

Article 86 states that AWOL occurs when a service member fails to report back to their unit or place of duty at the time prescribed by a higher authority. It's a serious offense that can lead to severe penalties, including dishonorable discharge, forfeiture of pay, and even confinement. In some cases, a defense strategy may be used against AWOL charges under Article 86. One defense strategy often involves proving that the absence was not intentional or was caused by factors beyond the service member's control, such as a family emergency or medical issues. Presenting evidence of such circumstances can significantly influence the outcome of the case. However, Alby didn't need a defense. The CO understood that the underlying reason for Alby's decision to go AWOL was the tragic loss of his brother in Vietnam.

As a result, the CO took this into account when determining Alby's punishment. Alby knew he had to plead guilty to going AWOL and hoped the CO would show him some mercy, given the circumstances. As it turned out, the CO understood Alby's distress over the loss of his brother and rendered the following judgment: Albert Elmer Davidenko, according to records, was hereby charged with being AWOL. Alby pleaded guilty to the charge. His commanding officer read aloud the following punishment: *"Albert Elmer Davidenko, your punishment for the charge of AWOL will be one reduction in rank, with no forfeiture of pay, and two weeks of extra duty."* Alby accepted the one-rank reduction without hesitation and committed to staying out of trouble until his honorable discharge on August 5th, 1971. As a result of the rank reduction, Alby was discharged at the rank of Private (PV1/E-1). During his time in the U.S. Army, he was awarded the National Defense Medal.

Sergeant (SGT) Charles Fredrick Davidenko, USA (Danny's Brother)

Danny's brother, Charles Frederick Davidenko (Chuck), was drafted into the U.S. Army on October 21st, 1968. On the morning of the 21st, Elmer (Chuck's father) drove him downtown to the U.S. Army induction center, where he, like his brothers Ray and Albert, spent the entire day in training classes, undergoing a full medical screening, and receiving all the necessary shots for overseas duty. The following day, Chuck boarded a plane bound for Fort Campbell, Kentucky, to begin his U.S. Army basic training. About two months later, Chuck completed his basic training and received orders to report to Fort Leonard Wood, located in the South-Central Missouri Ozarks.

After reporting to and settling in at Fort Leonard Wood, Chuck began his training as a U.S. Army Field Lineman. He learned the basic maintenance of communications systems. Upon completing his lineman training, Chuck received orders to report to Fort Gordon, located in Richmond County, Georgia. Fort Gordon is known for providing U.S. Army soldiers with advanced training in Signal Corps specialties, particularly in Signals Intelligence. While stationed there, Chuck received training as a Pole Lineman and a Telecommunications Center Specialist. Pole Lineman training taught Chuck how to install, operate, and maintain wire and cable communications systems. His Telecommunications Center Specialist training focused on maintaining U.S. Army communications systems. Upon completing his advanced training, Chuck received orders to report to the 51st Signal Battalion, stationed in Uijongbu, South Korea.

PV1 Charles F. Davidenko

SGT Charles F. Davidenko

Shortly after arriving in South Korea, Chuck discovered that his military duty station was just north of the capital city of Seoul. He was assigned to Camp Pililaau (from my research, I believe this Camp was also known as Camp Red Cloud), situated in the city of Uijongbu, between Seoul and the Korean Demilitarized

Zone (DMZ). This base was a key choke point along the main traditional invasion route from North Korea into Seoul and housed both U.S. and Korean military bases, crucial for the defense of the Korean capital. Sergeant Charles Frederick Davidenko served for 8 years, 2 months, and 25 days in the United States Army. Within that time period, 3 years, 4 months, and 22 days were spent on overseas tours in Korea and Hawaii. Unlike his brothers, Chuck found military life tolerable and even somewhat enjoyable, deciding to reenlist for multiple tours of duty. He excelled in his career field and went on to become a U.S. Army Drill Instructor. Chuck rose to the rank of Sergeant (SGT/E-5) and was honorably discharged from the U.S. Army on November 22nd, 1976. During his time in service, Chuck received the National Defense Service Medal, two Good Conduct Medals, the Armed Forces Expeditionary Medal, and the Marksman (Rifle) Badge.

Throughout my research and interviews with numerous friends and family members, everyone mentioned how Mary frequently expressed her pride in her boys and their service to this great country, the United States of America. From everything I read and heard, it became clear to me that Mary was a true American patriot in her own right. Looking back at everything Mary went through, it became clear to me that while she never gave up on America or the American Dream, she did lose hope in America's future during the Vietnam War, the war that took her youngest child's life. Mary loved her children deeply and would do whatever she could to support their hopes and dreams. She remained steadfast in her support for her military men. But sadly, that support ended when Mary passed away in 1981. She left this world with a broken heart, devastated by the loss of her baby boy, Danny, to a senseless war. I can only hope and pray that Mary is now reunited with all her children in heaven, and that her broken heart has been healed and renewed.

Chapter 5

Memories from Family & Friends

For this chapter, I reached out to friends and family who remembered Danny, asking them to share their stories and memories so I could better understand who he was and how he is remembered. Unfortunately, I did not receive as many responses as I had hoped. While many initially expressed a willingness to share their stories for the book, most found that their memories were brief and had faded with time. This is understandable, given that Danny passed away just shy of his nineteenth birthday. Many of his nieces and nephews were too young to remember him, and the majority of his siblings and other family members have passed away. As a result, the stories that follow are drawn from those who spent cherished time with Danny, as well as from others who lovingly passed down memories they heard about his short life. As the author of this book, I took the liberty of sharing my own memories first; though they are limited, they remain fond and meaningful to me.

Memories from Daniel James Roberts (Danny's Nephew / Namesake)

As I mentioned in the introduction, my mother named me after her youngest brother Danny. He was 11 years old when I was born. I was very young when he passed away, and though I had seen him many times, I can only remember three events from family gatherings. Two of the three memories I have of Danny are warm and loving, reminding me of how fun he was to be around. The last memory, however, is sad and filled with pain.

My first vivid memory is from when I was about five years old. My siblings, cousins, and I were playing with our Uncle

Danny and his brothers, Chuck and Alby, on the front lawn of my Aunt Jan and Uncle Bill Stockdale's house. I believe it was the summer of 1967 (possibly the 4th of July), during one of our usual family gatherings. Danny, Chuck, and Alby would play with their young nieces and nephews on the front lawn and in the backyard of my aunt and uncle's house in Chicago Ridge. Danny, Chuck, and Alby would play tackle football with us, and the game often turned into a full-blown wrestling match between my uncles and a crowd of excited nephews. I remember my uncles taking off their collared shirts and joining us for games of tag, wrestling, football, or wiffle ball, dressed in their white tank top T-shirts and blue jeans.

In the backyard of the Stockdale residence, we often played wiffle ball on a makeshift field that my cousin Billy Stockdale created, where all the bases, including home plate, were muddy holes filled with water. For a five-year-old, it was a blast, having fun with all my cousins and uncles, sliding into each muddy base and laughing uncontrollably. When it was my turn at bat, I could hear my uncles yelling, *"Swing, Batter, Swing!"* Of course, I would swing and miss the ball completely, but it didn't matter. My family would cheer me on, telling me to run and slide headfirst into the muddy hole at first base. What five-year-old wouldn't love playing in the mud, especially when our parents didn't really seem to mind? Or did they? Our parents were so busy socializing, eating, and drinking that they probably didn't notice the mess we were making until it was time to leave. I guess our parents were just happy to have some adult conversation and time away from their kids. To put this into perspective, between Joan, Jean, and Jan's families, there were at least 14 children running around Jan's average-sized, middle-class suburban home in Chicago Ridge, IL. We had a blast, and our parents seemed to ignore the fact that we all were getting our clothes dirty and muddy. It wasn't until it was time to head home that our parents finally noticed us kids getting into the family's new

Ford station wagon. I can still remember my dad not being too happy about all of us kids tracking mud all over the new family car.

The second fond memory I have of my Uncle Danny is when he took us for a ride in his new car. I remember it being a black convertible; I believe it was a Ford Mustang. My Uncle Danny packed four of us kids into the back seat. Back then, seat belts were optional. I can still picture all of us crammed into that back seat. I sat right behind my Uncle Danny as he drove, and if my memory is correct, I remember my cousin Jerry next to me, followed by my brothers Bob and Mark. I remember my cousin Billy riding shotgun in the front seat as we drove around Chicago Ridge with my Uncle Danny. Danny would spin the back tires, accelerating rapidly as soon as the streetlight turned green. Looking back on that car ride, I can still picture the convertible top down and the rush of the warm evening breeze blowing through my hair as Danny sped down Ridgeland Avenue. Each of us kids in the back seat hoped the streetlights would turn red so Danny would have to stop, and then we could hear the tires screech as the light turned green.

I remember the ride very well because when it was over, we returned to my aunt's house, and Uncle Danny started closing the top and rolling up the windows. Somehow, as the windows were going up, my left thumb got stuck between the glass and the metal frame. I started screaming because my thumb was stuck in the window. Danny immediately rolled it down to free my thumb, but the damage was done. My thumb had a huge cut on it that bled all over his car. I'd like to say I took the pain like a champion, but if I remember correctly, I cried like a typical little five-year-old kid. I remember Danny being very kind and gentle with me. He took out a hanky and wrapped it around my thumb. Danny quickly took me into the house to get a band-aid for my thumb and to explain to his sister Jean (my mother) what had happened. I remember how compassionate and concerned

Danny was, making me feel safe and okay. After a good cry and the application of a band-aid, I sat at my aunt's kitchen table with Uncle Danny, and we shared a cold bottle of Pepsi.

The last memory I have of my Uncle Danny is of his funeral on November 20th, 1969. I can still recall stepping into the funeral parlor, the scent of roses filling the air. I remember my mom with tears streaming down her face, while my dad quietly told my brothers and me to stay silent and to behave ourselves. I remember walking up to Danny's casket with my mother, kneeling beside it, and gazing at his lifeless body dressed in his U.S. Navy uniform. As we approached the casket, my mother motioned for me to kneel beside her. I felt nervous and confused, as this was the first time I had ever seen a lifeless body, especially the body of my Uncle Danny, someone I loved and admired. Everything felt like it was in slow motion. I watched my mom as she folded her hands and bowed her head in prayer, tears continuing to fall from her eyes. I have to admit, I was scared. I don't remember closing my eyes or even saying a prayer. My mom and I were on our knees for only a minute or two, but it felt like hours. I remember my mom staring down at me after her prayer. Why was she looking at me? I just stared back into her tear-filled eyes, unsure of what to say or do. Then, she turned back to the casket for one last look at her baby brother Danny. My mom touched her forehead with her right hand, saying, *"In the name of the Father."* She moved her hand down to her chest, saying, *"and of the Son,"* then touched her right and left shoulders, saying, *"and of the Holy Spirit,"* and then whispered, *"Amen."*

After my mom's prayer, she took my hand, and we walked back to our seats while other family members stood in line to pay their respects and pray by the casket. Looking back on the day of Danny's funeral, I can't help but feel a deep sadness knowing his life was cut short. I hate that I never really had the chance to get to know him or see the kind of man he would have become. I often wondered what our relationship would have been like. After all, I was named after him, so I believe there was a special place in his heart for me. Over the years, I've wondered what would have happened if that tragic event had never occurred, and Danny had been alive when I went off to U.S. Navy boot camp. What kind of 'U.S. Navy Gouge' would he have shared with me about his time in the U.S. Navy? I know Danny's death was incredibly hard on my mother, and now, years later, I often wonder what was going through her mind every time she called my name. I never got the chance to ask my mother these questions, so I will never know for sure what was going through her mind. However, I am extremely thankful for my first name, and I am honored to share this name with my Uncle Danny.

Memories from Richard Allen Roberts (Danny's Nephew)

When I talked to my oldest brother Richard (Rich), about my decision to write this book about our Uncle Danny, he agreed to share his own memories and stories about our Uncle Danny. At the time, I didn't know it, but I later learned that Rich was extremely close to our Uncle Danny. Rich was a few years younger than Danny, but he remembers spending a lot of time with him in Brookfield and at family gatherings. One of Rich's fondest memories of our Uncle Danny took place in Brookfield, Illinois. They liked to play around the house, but Danny preferred to explore the town of Brookfield. Rich remembers one particular day when he and Uncle Danny walked to the train tracks, and Danny taught him how to hop onto a moving train. They both jumped onto an open boxcar and rode it into town, where they eventually jumped off together. Luckily for them, neither of them was hurt while hopping on and jumping off a moving train. Rich didn't have any money, but Danny had a couple of dollars in his pocket. Danny took Rich to the local

Woolworths store, where they enjoyed lunch at the luncheonette (diner).

Woolworth Store with Soda Shop Luncheonette (Diner)

Back then, Woolworth's "soda shop" was immensely popular with middle-and high-school kids. Danny and Rich sat at the counter, each ordering a hot dog and a Pepsi-Cola. Shortly after finishing their meal, Danny grabbed Rich by the hand, and they walked out of the soda shop. Once outside, still holding Rich's hand, Danny took off running. Rich was a bit confused. Why are we running? Rich doesn't remember seeing Danny pay for their lunch because he was busy eating his hot dog and drinking his soda. All Rich remembered was that Danny seemed to be in a hurry. Years later, Rich would look back on that day and wonder if they had just committed their first dine-n-dash, or if they were running because they were late for the movie they planned on seeing. Back in the late 60's, the average hot dog and soda cost around 25 cents. So, the cost of their lunch, or dine-n-dash meal, was less than one dollar. After lunch, Danny and Rich went to the local movie theater, where they watched a film and shared a box of popcorn. Rich remembered almost every detail of that day as if it were yesterday. He cherished the time spent hanging out with his cool Uncle Danny. After the movie, with bellies full of popcorn, Danny and Rich headed to the train tracks, hoping for a train to roll by so they could hop on a boxcar and ride the train back to Danny's house.

Memories from Sheri (Davidenko) Ramey (Danny's Niece)

Sheri (Davidenko) Ramey is the daughter of Albert Elmer Davidenko (Alby), who is one of Danny Davidenko's brothers. At my request, Sheri took the time to go through her mom and dad's photos and saved memorabilia. She provided a wealth of useful information that helped me in writing this book. I cannot thank Sheri enough for putting in the time and energy to help me gather family letters and memorabilia in support of this book. Sheri's father Alby, passed away before we could gather any firsthand details from his memory of the day he first learned of his brother Danny's death. However, Sheri found an old scrapbook her father had kept, along with some letters he wrote while on active duty in the U.S. Army. One letter revealed that Alby wasn't fond of the U.S. Army and couldn't wait for his tour of duty to end. In another letter, Alby mentioned he had just 362 days left until he was out of the U.S. Army for good. Alby was stationed in Japan, loading ships with supplies for the war effort in Vietnam, when he first learned of his brother Danny's death. Like his brothers Ray and Chuck, his commanding officer sent him home to attend Danny's funeral. The story below was shared by Sheri, based on the information she found and the stories her father Alby, told her about her Uncle Danny.

Sheri and I spoke numerous times over email and the phone, sharing bits of information about Alby and Danny. She then sent me a package containing everything she could find from the boxes her parents left after they both passed away. The photos and documents will be incredibly helpful in telling Danny's life story. Sheri found her father's photo album with several pictures of Danny, ones I had never seen before. In one of our conversations, Sheri mentioned that she didn't find any letters directly from Danny to her father Alby. She speculated that this was likely because Danny had only been in the U.S. Navy for 11 months before his death. Sadly, despite my research and reaching out to all our family members, no letters were found from Danny to Alby, or vice versa. I find it hard to believe that they didn't write to each other. Both being overseas and away from home, letters would have been a welcome comfort to help ease the pain of their separation. I can't help but feel that these letters exist, either tucked away in some old attic or lost in the move from Chicago to Wisconsin, where Alby retired. One story Sheri recalls vividly is the one her dad told her about the day he learned that his younger brother had been killed in Vietnam. Sheri's father told her that he was on active duty in the U.S. Army, stationed in Okinawa, Japan, in support of the war effort when he received the news. Alby said he was torn up inside upon hearing about the death of his younger brother. Alby was so emotionally distraught over his brother's death that when he returned home on military leave for the funeral, he mentally and physically couldn't bear the thought of returning to his active-duty U.S. Army post in Japan. Alby went AWOL (Absent Without Official Leave) for approximately three weeks before reluctantly returning to his U.S. Army post overseas. Sadly, the U.S. Army held Alby accountable for his AWOL actions. In a letter my cousin Sheri wrote to me, she made the following direct and heartfelt statement: *"At the time of Danny's death, my father (Alby) was only 20 years old, drafted into a war he didn't support and never wished to be involved in."*

As I write Alby's story, I thought about how difficult it would be for any distraught Soldier, Sailor, Airman, or Marine, regardless of rank, to return to their military post after the funeral of their younger brother. To truly understand what Alby was going through mentally, you have to put yourself in his shoes. He was a young 20-year-old man, drafted into a ruthless and bloody war with no end in sight. The thought of being shipped off to Vietnam from Japan was a very real possibility, and I'm sure it weighed heavily on his mind. Alby was fully aware of the death and carnage the Vietnam War was inflicting on American troops in-country. Imagining myself in Alby's position, I could feel his resentment toward the U.S. Army and the pain he felt over the loss of Danny. With that level of stress, it would be easy to justify the thought of going AWOL and not returning to the U.S. Army. Alby was depressed and emotionally distraught, and therefore, the decision to go AWOL seemed like the right decision at the time. My heart goes out to my Uncle Alby, and I don't fault him for going AWOL. At twenty years old, I can't say for certain what I would have done in his place. The mental stress of military life is very demanding on every Soldier, Sailor, Airman, and Marine. I believe this stress is only compounded when you've been drafted into a war you didn't support, and from reading the daily newspapers reporting the large numbers of American servicemen and women losing their lives. Alby reluctantly returned to Okinawa, Japan, and completed his tour of duty and was honorably discharged a year later.

Memories from Pamela Malysiak Kipfer (Danny's Grammar School Classmate)

I located Pamela via my research into Danny's education and the schools he attended. I found Pamela on the Lions Township High School (LTHS) alumni website. Here is what Pamela had to say about what she remembered about Danny. Pamela stated the following:

"I knew that Danny had died in Vietnam but did not realize how tragically soon. I actually went to grammar school with Danny at St. Barbara's in Brookfield. I am sorry, I cannot pass on any more information. In the 50's and 60's, Saint Barbara was like all Catholic schools, the boys and girls were on separate playgrounds during recess and the lunch hour, so you really couldn't get to know your male schoolmates. We had 50+ students in our classroom, so unless you sat next to the person, you did not really get to know that person. As I recall, we sat in alphabetical order, so I was half a classroom away from Danny. I remember him as being quiet, not tall, and with dark blond hair. I wish I could tell you who he hung out with, but again, those Catholic rules and regulations did a lot to keep boys and girls separated! I think he lived not far from me; I want to say on Congress Park Avenue. I did not know Danny at all in high school. We had a class of about 1,000 as well as all the other classes being that large. It is amazing we made any friends at all! I am so happy that you are going to write a book about him. I would definitely like to purchase a copy of it. Please keep my contact information to let me know when the book gets published. What a lucky person Danny was to have such a great nephew and family. If I can help in any way, please let me know. Best of luck to you, and as I said, please keep in touch with me. Sincerely, Pamela Malysiak Kipfer."

Memories from Michael Walen (Danny's Neighborhood Friend)

Michael is another friend that I was lucky enough to find during my research. His name initially came to me from a sympathy card signed by him and other co-workers from

Knights Auto Shop, the same auto shop where Danny worked. We communicated a few times via email and phone, and he agreed to tell a few of his memorable stories about his friendship with Danny. Michael remembers traveling to Wisconsin with Danny and another friend, Todd Thompson, along with Danny's parents, Elmer and Mary. According to Michael, the three of them (Danny, Todd, and Michael) took Elmer's boat out in the early afternoon to spy on the girls' camp across the lake. The next morning, they heard Elmer screaming and yelling from his boat about a hundred yards from the lake house because the boat had run out of gas. Michael recalls this happening two days in a row, forcing Elmer to row the boat back to the dock each time to refill the tank.

Michael recalls another story about him and Danny spending the night in his parents' pop-up camper. Once it got dark, they would sneak out and walk down the alley behind the house to visit a man they knew as Popeye. Popeye was well known by most of the high school kids in the neighborhood. Popeye was a homeless older Black man who spent most of his time behind the small local gas station in town. Being under the legal drinking age, Danny and Michael would ask Popeye to buy them booze. They gave him all the money they had, including a little extra, so he could buy a bottle for himself. Once Popeye handed over their whiskey, they ran back to the camper to drink it. That night, Michael drank far too much whiskey and ended up sick like a dog, vomiting his guts out. Michael stated, *"From that experience, I still don't like the taste of whiskey."*

On another adventure, Michael and Danny met up with their friend Greg Delveckeyo, who drove an old 1953 Volkswagen Karmann Ghia. The three of them drove the car to the Lemont Quarry, where they spent the day hill climbing and exploring the backroads. However, at some point, Greg lost control of the car and wrecked the Karmann Ghia at the quarry, leaving the trio stranded until local authorities came to their rescue. Michael stated, *"That turned out to be one hell of an evening."* Because of the wreck, the three of them didn't make it home until extremely late. Michael recalls getting into big trouble with his parents, and he was sure Danny and Greg faced the same fate.

Memories From Carol Lynn Myers (Danny's Neighborhood Friend)

On the 51st anniversary of Danny's death (November 10th, 2023), my cousin Sheri posted a heartfelt memory on Facebook, honoring his life and acknowledging the loss of our Uncle Danny. She expressed that he will never be forgotten. While reading this post, I noticed that a woman named Carol Lynn Myers responded, saying that Danny was a good guy and that she had many fond memories of him and the friends they all spent time with. I reached out to Carol on Facebook and asked if she would be willing to share her memories of Danny, and she agreed to tell me her stories. To my surprise, Carol shared many memorable stories about Danny, their friends, and their youthful antics. Each story captivated me and revealed details that brought my Uncle Danny to life as a teenager. Carol has held these memories close to her heart for the past 50 years. She shared how she and her friends enjoyed hanging out with Danny, along with his brothers Chuck and Alby. Carol was the same age as Danny, though they didn't attend the same school; Danny went to Lyons Township High School, while Carol attended Riverside High School. Carol lived on the Northside of Brookfield, and in 1965, she met Danny's brother Chuck, who was dating her friend Joanne. Chuck set Carol up on a blind date with his twin brother, Alby. However, according to Carol, she and Alby didn't hit it off as a couple. Despite that, they remained good friends, and soon, everyone began hanging out together. Not long after Carol met the Davidenko brothers, Chuck broke up with Joanne, and Carol began dating him. This relationship with Chuck led to the development of friendships among all their friends, including both Alby and Danny.

As the friendships grew within this group of young Brookfield teenagers, they started hanging out together regularly, meeting up at various places around town. One of their favorite hangouts was the Congress Park Bowling Alley on Ogden Avenue. This old six-lane bowling alley featured oiled blonde maple lanes and dark knotty pine paneling on the walls. Once inside, Danny and his friends would leave the hustle and bustle of Ogden Avenue behind and enjoy bowling on the simpler, older lanes, less flashy and more laid-back than the modern alternatives. In fact, this bowling alley had an employee who manually reset all the bowling pins and returned the bowling balls. Although automated pin-setting equipment was common in most bowling alleys by the 1950s, this one was not automated. Carol describes the bowling alley as an old, dingy, and crummy little place, in very rough shape, with only SIX bowling lanes.

Former Location of the Congress Park Bowling Alley (Currently a Window Company)

This six-lane bowling alley was owned by a guy named Mike. Danny and Carol, as well as other Brookfield friends, would laugh and joke around, and they often joked with the owner Mike, by asking him to turn on lane SEVEN. On one occasion, Danny and his group of friends turned off all the lights in the bowling alley and started bowling in the dark on all six lanes. It was a wild, carefree moment that probably added to the rebellious fun of their youth. On a few occasions, Danny and his friends got a little out of control, and Mike, the owner, would threaten to call the cops if they didn't settle down. After a few threats like this, someone in the group would distract Mike while another person removed the mouthpiece from his phone. In the 1960s, telephones were designed with removable mouth and earpiece speakers, which could be easily unscrewed by twisting them counterclockwise.

When Mike tried to make a call, no one on the other end could hear him, making it impossible for him to follow through on his threats. Danny and his friends would laugh uncontrollably whenever they heard Mike getting louder and louder as he yelled into the mouthpiece of the phone. I have to admit; this was a clever and funny joke to play on the owner of the bowling alley. It's clear that these jokes were all just good teenage fun, and hopefully, Mike, the owner, saw it that way. Even though Carol called it a *"crummy old bowling alley,"* the owner, Mike, knew that Danny and his friends were frequent customers. In fact, the bowling alley was one of their favorite hangouts.

Carol described several other hangouts that Danny, Chuck, Alby, and their friends frequented. One of their other favorite hangouts was the famous Route 66 Drive-In Theater in Countryside, Illinois. The single-screen Drive-In Theater opened on May 28th, 1948. The drive-in was located off LaGrange Road, near Route 66 (Joliet Road), in the city of LaGrange, Illinois. The Drive-In Theater closed after the 1976 season and was afterward torn down to make way for the Countryside Plaza shopping center. Carol recalls how they would all pile into the trunk of a friend's car to sneak into the drive-in theater to watch a movie.

Route 66 Drive-In Theater, Pete and Tessie's Hamburgers, & Jimmy's Hot Dogs

After a movie or a few games of bowling, Danny, Carol, and their friends would head to a friend's house to party in the basement. When they were done, they'd head to one of their regular food hangouts. One they regularly visited was called Pete and Tessie's. It was a small restaurant that served great hamburgers and milkshakes, which everyone seemed to enjoy. Carol remembers that Pete was an older, heavy-set Greek man who owned the hamburger restaurant. Both Pete and his wife, Tessie, lived in a small apartment above the restaurant. Danny and his friends would visit, order hamburgers and milkshakes, and play records on the old jukebox Pete had in the corner of the restaurant. After they finished eating and listening to music, they'd walk up and down Ogden Avenue, chatting, joking around, and just killing time.

Another food spot Danny and his friends liked to visit was Jimmy's Hot Dog Stand in Brookfield. It was a popular hangout where the gang could grab a bite to eat and just chill. Carol remembers that you could get a hot dog, fries, and a Pepsi for thirty cents. If they weren't at the bowling alley or their favorite eating spots, they'd often hang out at John Enko's gas station, just down the road from Danny's house in Brookfield. Danny and his friends would often spend time together at the gas station or roam up and down Ogden Avenue. Carol recalls how Danny would hang out with her and her boyfriend, Frank Docstader, who went by the name 'Doc.' In 1967, Carol married Doc and started working at a place called Mister Donut. While she worked there, all of Carol's friends, including Danny, would hang out at the donut shop.

Another favorite pastime was going to the Arie Crown Forest Preserve to drink beer and tell scary ghost stories, each trying to outdo the other. No one in the group really admitted to being scared by the stories, but after a particularly creepy tale, if someone needed to relieve themselves in the woods, they wouldn't go alone and would have a friend tag along. Another hot spot for Danny and his friends was the Lemont quarries, where they would hang out and go swimming. Carol, Doc, and Danny became good friends and would often hang out at the local billiard room (pool hall) in Hodgkins, Illinois. Both Carol and Danny were fairly good pool players and would often team up to hustle unsuspecting people who challenged them to a game. Carol stated that she never dated Danny, but from the first day they met, they became very good friends and liked hanging out together. She remembers that Danny was fun to be around and loved to joke around. Carol also said that Danny was talkative with people he knew, but quiet around those he didn't. While playing billiards at the pool hall, Danny would put his arm around Carol's shoulder and tell people she was his sister. They'd challenge people to play pool games and would often win. Carol stayed in touch with Danny after he joined the U.S. Navy and wrote him a few letters. She also met his parents and spoke highly of his mother, Mary. Carol recalled that Mary told her she was very proud of her sons serving in the military.

Carol was saddened by the news of Danny's death in Vietnam. Carol remembers when she first heard the news about Danny's death; no one knew what exactly happened to Danny. Carol does remember being called by her friend Craig's mother (Mrs. Langele), and she was told that Danny had been killed while guarding a supply area. Days later, Carol heard rumors going around suggesting that Danny may have committed suicide; however, she did not believe it to be true. She believed the suicide rumor to be false because that was not the Danny she knew. The Danny she remembers was fun, loving, kind, happy, and full of life. But sadly, even to this very day, Carol still does not know the truth behind his death. I asked her if she had attended Danny's funeral and if she ever visited his gravesite at Saint Mary Catholic Cemetery, and after a short pause, I heard her say in a soft and sad voice, *"No, I have not."* Lastly, I asked Carol if she knew Danny's girlfriend, Helen. Carol stated that she did remember Helen and that she was nice, bubbly, and fun to be around. I am thankful for Carol taking the time to share her story about Danny. From her story, I personally feel like I went back in time to see Danny as a young man spending time together with his brothers and friends at the pool hall, burger restaurant, hot dog stand, and the gas station. Her story gave me greater insight into Danny's short life.

A Posting from CMCN Howard Decker (Danny's Roommate in CRB Vietnam)

Another message that I found on Danny's memorial wall had me searching for a man named Howard Decker. Howard posted the following statement about Danny on his Vietnam memorial page:

"I STILL REMEMBER YOU. Dan was my roommate. Over the years, I have often thought of him. I have been to the wall four times over the years and each time, I have touched his name and remembered him. Just wanted everyone to know, that to me, Dan, is not forgotten. Howard / Sep 4, 2007".

Initially, I had trouble locating Howard Decker. I searched the internet to find him so I could hear his story and memories of Danny. Sadly, on December 14th, 2023, (this day would have been Danny's 73rd birthday) while searching for Howard, I came across an obituary for a Howard Decker. I had finally found him, but I was seven months too late. Howard Decker's Obituary stated that he passed away on Friday, May 5th, 2023. I learned that when Howard mentioned that he was Danny's roommate, that he was talking about being roommates in Cam Ranh Bay Vietnam. Howard retired from the U.S. Navy at the rank of CM1 (1st Class Petty Officer / E-6) and he was honorably discharged from the U.S. Navy after serving his country for 20 years. Howard's funeral service was held on Friday, July 28th, 2023, at the National Cemetery of the Alleghenies.

CMCN Howard Anderson Decker (Danny's Roommate in CRB Vietnam)

In an effort to find more information about Howard and Danny's relationship while roommates in Cam Ranh Bay, Vietnam, in January of 2024, I located Howard's wife, Lee

Decker, and she agreed to share some of Howard's memorabilia and the story Howard told her about Danny. Lee stated that Howard talked about the loss of Danny and how he made it a point to touch Danny's name on the Vietnam Memorial Wall every time he went to visit the memorial. According to the story that Howard told his wife, Howard believed that Danny received a 'Dear John' letter from his girlfriend (Helen). According to his wife, Howard never really knew how Danny was killed, but rumors circulating around the command suggested that he was shot in the chest. From that rumor, Howard thought that Danny might have committed suicide, which only fueled the rumors about the mysterious 'Dear John' letter. However, according to Lee, Howard's story about the suicide was just a rumor.

During my conversation with Howard's wife, she stated that Howard never mentioned actually seeing or reading the rumored 'Dear John' letter. How I wish I could have talked to Howard to ask him if he saw or read this rumored 'Dear John' letter that Danny supposedly received from Helen. Howard's theory of suicide is plausible, but it's based solely on the belief that a 'Dear John' letter existed. The possibility of Danny committing suicide was feared by other family members and friends; it was all based on a rumor about the unsubstantiated story of a 'Dear John' letter. No one that I have interviewed believes that Danny was even capable of such a thing, and that no one has ever seen proof of this mysterious 'Dear John' letter. It is unfortunate that if this letter exists, that this letter was not among Danny's personal effects that were returned home to his family. Not one family member has ever mentioned the existence of this letter among saved family memorabilia. Lee Decker went through all of Howard's Vietnam photos and documents. She provided me with several photos of Howard in his uniform in Vietnam. Pictured above is Howard Decker's photos from CBMU-302 and a photo Howard from Facebook.

Finding out about Howard's untimely death felt like a painful punch to my stomach. What if I had started this book 25 years earlier? I would have had more of Danny's siblings, friends, and shipmates still alive to share their stories about him. I would have been able to ask Howard directly: "Did you see a 'Dear John' letter, or was this just a rumor?" All of their memories would have been fresher and clearer in their minds. Maybe some of their photos and memorabilia would still be in a box somewhere in the attic. Many friends, family members, and shipmates mentioned that some of their boxes were misplaced during moves or tucked away in an attic, buried under numerous other boxes filled with years of personal items. Regardless, I am grateful for their time and efforts in sharing what they remembered and the memorabilia they kept safe over the years. I want to give a big thank you to all the contributors of stories and memorabilia, as your memories and keepsakes helped the readers and me to gain a better understanding of who Daniel Joseph Davidenko was during his short time on this planet.

Chapter 6

Danny's Girlfriend Helen

Over the years, I heard my family mention that Danny had a high school girlfriend, but I never knew her name, until one day, my cousin Suzie (Stockdale) Eppley recalled that Danny's girlfriend's first name was Helen. During my search for a girl named Helen, I discovered numerous Helens who had attended grammar school and high school with Danny. However, even after compiling a list of Helens, with different last names, finding the right Helen, the one my Uncle Danny fell in love with, felt like searching for a needle in a haystack. To complicate matters, each Helen I identified as a potential match likely had a different last name due to marriage. This search was becoming increasingly difficult, and despite my efforts, I eventually resigned myself to the conclusion that I might never find the Helen I was looking for.

During my search for the right Helen, I encountered one dead end after another, crossing off name after name as I realized each Helen was not Danny's high school girlfriend. My research efforts regarding Danny's girlfriend were focused primarily on his high school years at Lyons Township High School (LTHS). The list of Helen names was long until I came across a young woman named Helen Jablonsky from the class of 1970. Immediately, this name triggered something in my head, where have I seen this name before? Then, in a flash, it hit me. I remembered seeing this name on Danny's Vietnam Memorial page. This woman, Helen Jablonsky, left the most heartfelt and moving message for Danny, expressing how much she loved and missed him. The following is the message left on Danny's memorial wall dated March 30th, 2009:

"Love Forever - Forget Never. It broke my heart to lose you, but you did not go alone; part of me went with you, the day God called you home. A million times I've thought of you; a million times I've cried. If loving could have saved you, you would have never died. Missing you is a heartache that will never go away. Love, Helen Jablonsky - March 30th, 2009."

Wow, in an instant, I had the one clue I needed, Helen's last name. Suddenly, my search shifted from nearly impossible to entirely possible. Finding that needle in a haystack, the woman I had been searching for all these years, was about to become a reality. At first, I was excited by the discovery yet equally frustrated with myself for not identifying this investigative clue sooner. It was a reminder that even the most experienced investigators can miss the most obvious clues in the search for a person of interest. I immediately returned to the Lyons Township High School yearbook and found Helen Jablonsky's senior photo. Now, armed with both her last name and a high school picture, I reached out to the LTHS alumni staff, hoping they could help me locate the former Helen Jablonsky.

More than 50 years had passed since Danny was killed in Vietnam; surely Helen Jablonsky had moved on and likely married. But to whom? What was her married name? Now that I was certain I had found the right Helen, I intensified my search, determined to hear her story and the memories she shared with Danny. In my heart, I knew Helen would have much to say about Danny and the tremendous loss she had endured. I was eager to find her, hoping to preserve her story and memories within these pages. But there was one big piece of the puzzle missing that complicated my search. I didn't know if Helen had married, divorced, or remarried multiple times. How could I figure out if her last name had changed? If her last name had changed, especially if it had changed multiple times, that would make my search even more challenging. I also had a troubling thought: what if Helen had passed away and I never got the chance to speak with her about Danny? The possibility of her being gone weighed heavily on my mind, especially since when I finally located Danny's Vietnam roommate, Howard Decker, I was eight months too late when I found his obituary. I couldn't let these doubts distract me from my search. I vowed not to give up on finding Helen, my uncle's childhood sweetheart.

Helen's Yearbook Photo from 1969 & Helen's Photo from Facebook 2024

I knew I had to go back to square one, back to Danny's high school and the LTHS alumni staff. I received several responses from the alumni office and a few former classmates, but none were very helpful. However, one alumni member, Patti Cowhey, responded right away to my request for help. Patti explained that she handles all the student roster searches for the class of 1969 in support of the LTHS alumni and reunion directories. Without hesitation, Patti assured me that she would check the alumni records to see if she could find any information on Helen Jablonsky. I responded with great enthusiasm and excitement, expressing how thankful I was for her help in this endeavor. That very same day, Patti emailed me, saying she might have found the Helen I had been searching for. She noted that this particular Helen graduated from LTHS in 1970. Patti's directory information included several phone numbers and email addresses for a former Helen Jablonsky, who was now listed as Helen Jablonsky-Freid, with Freid being her new married name. My needle-in-the-haystack dilemma had just transformed from an impossible task to a very probable one. So, I took the new leads and immediately drafted the following email and sent it out to the three different email addresses, in hopes of finding Helen Jablonsky-Fried:

> *"Subject: Looking for people who might have known Daniel Joseph Davidenko*
>
> *Dear Helen, I am looking for Helen Jablonsky, a graduate of LTHS class of 1970 that may have known Daniel Joseph Davidenko class of 1968. Daniel Joseph Davidenko died on 10th of November 1969 in Vietnam. Daniel was a member of CBMU-302 stationed in Cam Ranh Bay Vietnam in 1969. I have been trying to locate a women named Helen Jablonsky, who left this heart felt message on Daniel's Vietnam Memorial Page. If I found the*

right Helen, I would love to talk to you. I have been in the process of writing this book now over the past few years and its completion hinges on stories from people who knew him. Lastly, if this is the correct Helen, then I look forward to communicating with you, if it is not, would you please let me know either way. Thank you and I look forward to hearing from you."

Unfortunately, I didn't receive a single response from my email attempts to locate Helen, and I felt like I was back at square one. However, as luck would have it, in late January 2023, I received new contact information from Patti Cowhey, including Helen's updated email address. Patti confirmed that the email was active and valid, and she mentioned that she had been in contact with a Helen Freid at this address, whose maiden name was Jablonsky. Patti understood how passionate I was about finding Helen, and in an effort to help, she reached out to Helen directly. Patti informed her that I was looking for her and that I was writing a book about Danny. Patti shared with me that she had spoken with Helen and provided her with all my contact information. Patti also mentioned that Helen had told her, *"She loved Danny with all her heart."* During my conversation with Patti, she mentioned that Helen would be reaching out to me in the upcoming weeks.

Things were really starting to come together in terms of important contacts for my book. On that same day, I also obtained numerous names and contact information for several CBMU-302 members who had been stationed in Cam Ranh Bay, Vietnam, at the same time Danny was assigned to the unit. I can't tell you how thrilled I was, knowing that I was about to hear Helen's story and possibly gain more information and insight into the events surrounding Danny's death from some of his former shipmates. Two weeks had passed since Patti spoke with

Helen, and I started to grow concerned that she hadn't reached out to me. So, I decided to email Helen at the new email address Patti gave me, hoping to get Helen's story for my book. I sent two emails initially but received no response. Helen's memories and stories about Danny were critical to this story, and I even considered jumping on a plane to speak with her in person. I knew I had to talk with her. Before hearing Helen's story in her own words, I could only reconstruct her story through the memories of friends and family who remembered her, and those memories were scarce at best.

Another two weeks had passed, and I still had not heard from Helen. As time passed by, I began to think that she may be nervous about telling her story. Did Helen feel deep down inside that everyone believed the rumors about her writing a 'Dear John' letter to Danny? Was Helen afraid that I would ask her specifically about this rumored 'Dear John' letter? I just knew that if I could talk with her on the phone, or in person, that I could tell her about all the multiple stories floating around about Danny's death. I wanted her to understand that I had uncovered enough information to challenge, if not discredit, the long-standing rumors surrounding Danny's death. I knew in my heart that if I shared some of my findings with Helen, she would be more inclined to share her story about Danny and the love and friendship they shared. I wanted Helen to know how important her story was to the book.

Out of desperation, I reached out to Patti once again to let her know that I still hadn't heard from Helen. Patti told me that Helen had communicated with her through Facebook Messenger. Patti said, *"Let me try to reach her through Facebook Messenger."* I told Patti that if you do contact her, please let her know that I would like to talk with her on the phone first. This way, I can tell her more about the book, and she can decide what she's comfortable sharing. Within a few days, Patti emailed me to say that she had shared my message with Helen and hoped I would hear from her soon. I was so thankful for Patti's efforts; I truly believe I wouldn't have found the right Helen without her outstanding research and support. So, I must pause here to say thank you to Patti Cowhey for all her help and perseverance.

On March 7th, 2023, Patti emailed me to let me know that she had received a response from Helen stating that Helen would be reaching out to me soon. Patti ended her email by saying, *"I believe she will get a hold of you. Please keep me posted."* She also attached Helen's Messenger response from Facebook (see FB screenshot below).

Helen's FB Message to Patti

Eventually, I contacted Helen via Facebook Messenger, and she agreed to share her story about Danny. We spoke a few times on the phone, and I explained the book to her, the discrepancies in the reporting surrounding Danny's death, and how much I was looking forward to hearing her story. In those first few brief conversations, Helen began to share small snippets of their love story. During each conversation, she mentioned that she was tired and feeling sick and needed to stop. She never shared what medical condition she was suffering from, and I felt it wasn't my place to ask about her personal health. Instead, I simply told her that I hoped she would be feeling better soon. She told me that once she was feeling better, she would reach out and continue sharing her story. I remember a huge wave of excitement washing over me after I hung up the phone following our initial conversation. I thought to myself, wow I hope Helen will open up her heart and be candid with me as she walks down memory lane to tell the story of their love together, a story only Helen could tell.

In my excitement, I emailed Patti to let her know that I had successfully contacted Helen and had begun writing her chapter. In that email, I told Patti that my conversation with Helen was fruitful, but she wasn't feeling well, so she had to stop. However, Helen assured me that she would reach out soon to share her full story when she was feeling better. While eagerly awaiting Helen's call, I continued working on other chapters of my book. The clock seemed to stand still as I patiently waited for Helen to reach out to me. As each day passed by, I decided that if I didn't hear from her by mid-summer, I would reach out again. Finally, on August 17th, 2023, I had my first long and detailed conversation with Helen. Our conversation was very pleasant, and I explained to her the objective behind the book and how her input was crucial. Prior to this conversation, I had sent Helen a list of questions via email to help her prepare her answers. In particular, I asked if she would be willing to share photos and letters that she received from Danny. However, Helen mentioned that in an attempt to get closure, she had disposed of all the pictures and letters that reminded her of Danny years ago. However, she did agree to tell me her memories of her short time with Danny and, hopefully, share what she remembers from their letters while he was in the U.S. Navy.

It was so nice to hear Helen's voice again as we engaged in a short conversation about how she met Danny, as well as other brief stories about their time together. Helen sounded great, and I sensed that she was excited to tell her story. I told Helen that their short time together and the story of their love for each other would provide insight into who Danny was and who he could have been. After we ended our conversation, I told Helen that I would email her the first draft of the book's introduction, which would give her some insight into the theme of the book. Additionally, I attached another list of questions I had about her relationship with Danny and the memories they made together. The story and memories to follow are what Helen herself shared with me. Both Danny and Helen believed that they would one day get married. They both believed that one day they would also have children together. Helen stated that, *"They both envisioned spending the rest of their lives together."*

Memories from Helen Jablonsky (Danny's Girlfriend and First Love)

Helen began her story by recalling that by the end of the summer of 1966, she and Danny were back in high school. In December of that year, Danny had just turned 16, while Helen, only a few weeks shy of 15, awaited her birthday in January. A few months into the school term, Helen met Chuck, Danny's older brother, at the local Woolworth's diner, a popular hangout spot for high school teenagers. Chuck was hanging out at the local diner with his high school friends and his younger brother, Danny. Helen was there with a bunch of her friends, enjoying some French fries and a soda. As the two groups mingled together, they began engaging in multiple conversations. Chuck, with his natural gift of gab, struck up a conversation with Helen. She didn't recall exactly what they talked about, but remembered it being a typical teenage exchange, lighthearted and full of teen topics.

As their conversation came to a close, Chuck casually said, *"Hey Helen, let me introduce you to my brother, Danny."* Helen remembers that Danny was a bit shy at first, but eventually, he loosened up and became more comfortable talking to her. As their conversation flowed, she noticed his warm smile and the way his eyes lit up when he spoke. There was an instant, unspoken connection between them, something gentle and sincere that set the foundation for what would become their first real love. Helen remembers their casual conversations and how she found Danny to be very pleasant. His easygoing nature and genuine kindness made her feel at ease, and she quickly became comfortable sitting next to him. There was something effortless about their connection, as if they had known each other for a time. According to Helen, from that very first meeting at the diner, she and Danny *"hit it off great."* Their conversation flowed naturally, and there was an undeniable spark between them. In fact, they both expressed interest in seeing each other again. As young teenagers do, they both exchanged phone numbers and agreed to call each other. As they left the diner that afternoon, Danny smiled at Helen and said, *"I'm already looking forward to seeing you again."*

The very next evening, Helen picked up the phone and called Danny, a bold move for the late '60s when girls typically waited for boys to call first. But Helen didn't care about outdated customs. She really liked Danny, and she wasn't about to let a silly social norm stand in the way of getting to know him better. They talked for hours, sharing their likes and dislikes, their dreams, and the places they longed to see. Conversation flowed effortlessly, as if they had known each other forever. From their first encounter at the diner, Helen learned that Danny also attended LTHS South Campus. From that day forward, they spent most of their free time together, growing closer with each passing moment. Their connection deepened as they shared their lives, dreams, and the simple joys of being young and in love.

Helen went on to say that when she first met Danny, she naturally drifted away from her regular clique of friends and began spending more time with his group of friends. It wasn't a conscious decision, but rather something that happened as she and Danny grew closer. She enjoyed the energy and camaraderie of this new group, but most of all, she loved being by Danny's side. Helen said that they both felt something special from that very first meeting, and from that moment on, they were inseparable. Whether it was spending time at school, hanging out at the diner, or simply talking on the phone, they couldn't get enough of each other. Their connection was effortless, and being together just felt right.

Helen remembers that they became boyfriend and girlfriend right away and began seeing each other exclusively. She recalled how Danny was always sweet, gentle, and full of happiness. He had a way of making everything feel brighter, and she never saw him in a bad mood. No matter the situation, Danny was always upbeat, positive, and full of energy, which made their time together even more special. Helen said that *"Danny was the sweetest guy anyone could ever meet."* He treated her with such kindness and respect, making her feel valued and loved from the very beginning. She felt like they were made for each other, and that sense of connection was immediate. Helen couldn't help but feel incredibly lucky to be with someone as thoughtful and caring as Danny. He made her feel special, and she felt that their bond was truly something extraordinary. During the winter months, it was difficult for Helen and Danny to see each other as often as they wanted, mainly due to the long walk between their homes. Helen lived in LaGrange Park, and Danny was in Brookfield. Despite the distance and cold weather, Helen would often make the long trek to Brookfield to meet Danny, since there weren't many places to hang out near her home. One of the spots they would often meet was at a local laundromat situated between their two towns, a modest but familiar place where they could spend time together. The laundromat became their cozy halfway point, a warm refuge from the winter cold. The heat allowed them to linger, and even today, Helen stated, *"The familiar scent of laundry detergent brings back memories of her and Danny in the comfort and warmth of that place."* It wasn't much, but that simple spot provided a welcoming escape where they could escape the snow and enjoy each other's company.

As the spring weather began to warm, Helen would meet Danny and their friends at the local Woolworth's diner. They had been dating for over six months now, and as they settled into their favorite corner booth, sipping on sodas across from each other, they celebrated the joy of their six-month anniversary together. While sitting across from Danny, Helen caught a glimpse of something unusual on the back of Danny's left hand. It looked like dirt smudged across his knuckles, but there was something odd about the way it looked. She gently asked him about it, but Danny hesitated, almost nervously, like he didn't want her to see what was on his hand. Helen's curiosity piqued, and she gently asked him again to show her his hand. Helen's eyes widened in surprise as she inspected Danny's hand more closely. It wasn't dirt at all. Instead, she saw something that took her breath away, a homemade tattoo, with each letter of her name carefully inked on his fingers, spelling out H-E-L-E-N. She couldn't believe it. In disbelief, she asked him softly, *"What did you do?"* Danny just looked into her eyes with a mischievous grin and gave her a big, heartwarming smile. He didn't need to say a word; his expression said it all. It was his way of showing just how much he cared for her; in the most unique and heartfelt way he could think of. Helen was both touched and amazed. At that moment, she knew that Danny's feelings for her ran deep, and his small yet significant gesture was something she would never forget. Over the last six months, both Danny and Helen's relationship continued to grow; however, the relationship had not gone beyond just kissing and physical touch. They were just happy being together and holding hands.

By the following December of 1967, Helen and Danny had been dating for one year. Everything seemed to be looking up for Danny. The winter of 1967 marked a season of new beginnings for Danny and Helen. Danny had just turned 17 and earned his driver's license, while Helen was about to turn sweet sixteen and eagerly awaited her birthday in the coming weeks. Both were ready to embrace the changes the new year would bring. With the money Danny had carefully saved from his job at Knight's Auto Body Shop, he proudly bought his first car. With his own transportation, meeting up with Helen became much easier. With Danny's new driver's license and car, a newfound sense of freedom opened up for both him and Helen. No more long walks to see each other, no more lingering in the laundromat just to stay warm.

In the days and months after gaining this newfound freedom, Danny and Helen spent even more time together. As young teenagers with a reliable way of getting around, Danny took Helen wherever she wanted to go. They went to drive-in movies, the quarry, and the forest preserve, where all their friends hung out. They often visited these places without their friends, enjoying the time alone together and each other's company. On several occasions, Danny and Helen found themselves parked at one of their favorite local hangouts, with the radio playing their favorite songs. As their relationship deepened, so did their love of music, time with friends, and ultimately time shared together. They were experiencing their first taste of love, making every moment together count. One evening, Danny and Helen met their friends at the forest preserve, where they hung out and shared a few beers. After a few hours, the group decided to grab something to eat at Pete and Tessie's restaurant, or the *"burger joint"* as they liked to call it. Everyone agreed to meet there, but as the others drove off, Danny and Helen chose to stay behind.

Danny parked the car in a secluded spot and put on some music. It was still a bit chilly outside, so they held each other close to stay warm as they listened to their favorite songs on the radio. Helen loved the way Danny held her and the warmth of his embrace. The seats were cramped and uncomfortable, but neither of them noticed. Lost in the moment, they focused only on holding each other close. That evening, it felt like time stood still, and it was a wonderful feeling, just the two of them relaxing under the stars. However, that memorable moment was interrupted by the radio DJ's voice announcing that it was 1:00 AM, one hour past Helen's assigned curfew. In an instant, Helen pulled away from Danny, her eyes widened with the realization that they had lost track of time. Helen yelled out, *"Danny, I was supposed to be home an hour ago."* Helen's parents had set her weekend curfew at 12 midnight. Both of them had dozed off and had lost track of time. In a rush, Danny drove Helen home, his heart racing as he hoped her parents wouldn't be pacing anxiously in the kitchen, waiting for her to return. As luck would have it, Helen's parents were fast asleep when she quietly slipped into the house and made her way to her room. She didn't bother showering or brushing her teeth; she simply climbed into bed, careful not to wake her parents. Despite feeling exhausted, Helen couldn't fall asleep. She lay there wide awake and thinking about how amazing it felt to be with Danny.

A few weeks had passed since that evening with Danny at the forest preserve. Due to their work and school schedules, they didn't have many opportunities to see each other over that 3-week period, but they spoke on the phone each day, and each evening to say goodnight before bedtime. On a cold Sunday morning, January 14th, 1968, Helen awoke with excitement knowing that today was her sixteenth birthday. Her parents threw her a small party early that day, but all she really wanted was to spend time with Danny. Around 3:00 PM, Danny picked Helen up for their first 'formal' date. Danny had planned an evening with fancy attire and fine dining, in an attempt to make it a night neither of them would ever forget. Danny went all out on Helen's birthday; he wanted her 16th birthday to be very special. Danny gave Helen a bouquet of flowers, took her to a movie, and afterward, he treated her to dinner at a local restaurant in La Grange, Illinois. Helen remembers that it was a nice place with a charming old-world feel, adorned with nice furniture and antiques. Danny thought it was a great place to impress Helen. He couldn't help but smile at Helen, hoping the timeless elegance of the setting would impress her as much as she impressed him.

After dinner, they decided to head to the Lamont quarry, where they sat side by side, gazing at the stars. They didn't talk much due to the heaviness of the meal lingering in both their stomachs. Instead, they simply held hands, savoring the peaceful stillness and the heavenly view above. As they sat there beneath the starlit sky, Danny reached into his pocket and pulled out a small jewelry box. Inside was a delicate silver chain with a tiny heart pendant. He had wanted to get her something more, but after buying his car, money was tight. Helen carefully fastened the necklace around her neck and whispered, *"I love it… thank you."* Even though it was Sunday night and Helen still had two hours before her curfew, Danny looked at her and said, *"Let me get you home on time."* Danny drove her home, arriving just before 11:00 PM. As he was about to leave, he leaned in and gave her one long, deep kiss. Pulling away slowly, Danny whispered, *"Sweet dreams, Birthday Girl. I'll see you tomorrow."*

By early spring of 1968, Danny and Helen were inseparable. Each afternoon, they eagerly anticipated the moment she returned home from school, and he finished his shift at Knight's Auto Shop. But it was the weekends they lived for. Two days where they could spend every waking moment together, lost in their own little world. On a cool Saturday morning in April of 1968, the thought of being alone in their favorite secluded spot at the Arie Crown Forest Preserve was on their minds. Both Danny and Helen looked forward to their weekends and spending time with their friends. One weekend, a friend casually mentioned to Danny that his parents would be out of town and that he was planning a party at his house. Danny was excited about the party and rushed off to tell Helen. As he walked up to Helen, she could see his beautiful smile and a glow in his eyes. She loved the way he looked at her, but it wasn't just in the way he looked at her; it was in the way he made her feel, a warmth that wrapped around her heart. That evening, Danny leaned in close to Helen, his voice barely above a whisper, and he said, *"I love you."* They both said the words, *"I love you"* to each other numerous times before; but this time, a single tear slipped down Helen's cheek as she held him close, her heart was full. Helen whispered back, *"I love you too".* After the party, Danny drove Helen home. She clung to his arm, resting her head gently on his shoulder, wishing she could stay with him all night, never having to let go.

As the Vietnam War raged on, Uncle Sam's efforts to draft America's youth into the U.S. Army were moving at an alarming pace, with more young men being called up each day. Helen knew that Danny wasn't happy with high school, and early that year (1968), he dropped out. Danny knew he would be eligible for the draft and would most likely be drafted into the U.S. Army like his brothers. So, before dropping out of high school, Danny met with U.S. Navy recruiters to explore the option of joining the U.S. Navy in an effort to avoid the draft. Knowing that high school wasn't working out for him, Danny somehow convinced his parents to sign the paperwork that would allow him to join the U.S. Navy. Weeks later, after Mary and Elmer signed his enlistment papers, Danny decided that high school wasn't for him and he dropped out and began working full time at Knight's auto shop. Helen knew Danny was seriously considering enlisting in the U.S. Navy, as he had often mentioned his worries about being drafted after high school during their conversations. Shortly after enlisting, Danny told Helen that he was dropping out of high school. In May of that year, Danny's focus shifted to full time work and spending as much time as possible with Helen, while he awaited his enlistment date. Danny's decision meant they would only have seven months left together before he shipped off to U.S. Navy boot camp. The next seven months seemed to fly by, and before they knew it, Danny's enlistment date was fast approaching.

On Saturday, December 7th, 1968, one week before Danny's 18th birthday, Mary and Elmer (Danny's parents) threw Danny a going-away party at the American Legion. Everyone was invited, including Helen. The party was a great success, but it left both Danny's parents and Helen with a growing emptiness in their hearts, knowing that soon Danny would be leaving for boot camp. With Danny's enlistment date rapidly approaching, Danny and Helen made sure that they spent every available moment together. Helen remembered that just a week before Danny was set to leave for boot camp, he dropped to one knee, without a ring, and asked her to marry him. Without a moment's hesitation, she said yes. They talked about getting engaged and married sometime after he was stationed at a base in the United States.

After Helen said yes, Danny knew he had to ask for her father's permission to marry Helen. The next evening, he went to Helen's home and asked to speak with her father. With his nerves under control, Danny firmly and respectfully asked for permission to marry his daughter. Helen recalled how devastated both she and Danny were when her father said, *"No, I'm sorry, Danny, but you two are much too young to get married."* Danny kept his cool, remaining silent for a moment before saying, *"Well, maybe in a few years then"*, trying to appease her father. Helen's father replied, *"Sure, Danny, maybe in a few years."* Despite her father's decision, Helen confided that it didn't stop her and Danny from visiting the jewelry store, where they picked out an engagement ring. It was a simple gold band, perfectly sized for Helen's finger. For now, the ring, the proposal, and the promise of a future together remained their little secret, an intimate moment they would hold close, just between the two of them.

On December 16th, 1968, the day arrived for Danny to begin his enlistment. The day before, he and Helen spent the entire day together, coming to terms with the fact that he would be shipping out the following morning. Helen stated that the day before Danny left for boot camp, they spent every waking moment together, reluctant to let the time slip away. Danny didn't get to see Helen the day he left for boot camp. Both Danny and Helen agreed that it would be best not to see each other that day; to avoid the emotional strain it would place on them. Helen also felt it was important for Danny to spend that time with his family, especially his mother, Mary. Helen wasn't sure but believes that only Mary and Elmer were there when Danny was picked up by the U.S. Navy and taken to Great Lakes Naval Base. Both Helen and Danny knew that his U.S. Navy training would last for two months and that they would be unable to see each other during this period of time. So, they both agreed to write to each other until they were back together in a few months.

Bootcamp graduation day could not come soon enough. Both Helen and Danny longed for the moment they would finally be together again. On Danny's bootcamp graduation day, Helen was in the crowd, as were Danny's parents and other family members. After the graduation ceremony, Danny was granted approximately two weeks of leave before being ordered to report to his next duty station. Determined to make the most of his limited time, he was hell-bent on spending every waking moment with Helen. This didn't sit well with his mother, Mary, who also longed to spend time with her son before he left. But as often happens with young love, Helen took priority. Despite his mother's wishes, Danny's heart was fully set on being with her, and in the end, his devotion to Helen won out over everything else.

Helen remembered that when Danny was scheduled to come home on leave, her family had already made vacation plans. They were busy packing suitcases in preparation for their trip to New Buffalo, Michigan, to visit relatives. The timing couldn't have been more complicated, and yet, despite the chaos of her family's plans, Helen's heart was set on the few precious days she would have with Danny. Helen recalls how she was beside herself, hysterical and upset, knowing that she and Danny wouldn't get to see each other. It seemed like a cruel twist of fate, but as it often goes with love, things worked out in the end. Danny, determined to be with her, followed Helen and her family to Michigan. It wasn't ideal, but it was enough for them to be together, even if it meant making the best of an unexpected situation. Helen didn't go into detail, but she mentioned that they had worked out a plan for Danny to stay at her aunt's house, which was just a mile away from where Helen and her parents were staying. It was an exciting arrangement for both of them, a small slice of time to be together despite the odds. However, Helen couldn't help but think that Danny's family might have been a bit disappointed, especially his mother, Mary, knowing that she wouldn't get to spend much time with him before he had to report back to duty.

The pulling forces between love and duty were always a delicate balance, and this time, Helen and Danny were making the most of it. Helen was on the verge of sharing all the details about that vacation and the time she had spent with Danny, but her voice softened, and she paused, stopping the conversation. She was feeling a bit run-down, too tired to continue. With a promise to pick it up when she was feeling better, she apologized and suggested we pause for now. Shortly after my last communication with Helen, I received the following Facebook message from her:

Helen's Facebook Message Dated 01 March 2024

I agreed with Helen and told her that once she was feeling better, we could set up a day to speak over the phone. I reassured her that we could likely capture her story in just one or two more conversations, giving her plenty of time to rest before we picked up where we left off. I closed my message by asking her to let me know what day and time worked best for her. However, it had been nearly two months, and I still hadn't heard back from Helen. She had mentioned before that she was dealing with some medical issues and was feeling very ill, so I could only assume that was the reason for the silence. I hoped she was doing better and that whatever kept her from reaching out would pass soon. A few days later, I decided to send Helen a follow-up text to see if she was feeling better and if she was ready to resume our conversation to finish telling her story. She replied through Facebook Chat, explaining that she had been very sick lately and was undergoing a series of medical tests. She mentioned having biopsies done but didn't specify what health issues she was facing. When she mentioned needing a biopsy, I knew she was probably facing some form of cancer. I could sense that she was going through a lot, and my thoughts were with her as she navigated this challenging time. Helen mentioned that she had wanted to reach out, but due to her illness, she simply didn't have the energy. Most days, she found herself sleeping for hours on end. Despite this, she promised once again that she would contact me when she was able, asking me to be patient with her as she focused on getting better.

A few days later, Helen and I resumed our conversations. I began by asking her if she remembered the last time she saw Danny. Her response was quiet but filled with emotion as she recalled, *"It was just before he deployed to Vietnam."* Helen's memory was understandably faded, as more than fifty years had passed. My research shows that immediately after boot camp, Danny went on leave and, according to Helen, spent time with her and her family during a vacation in Michigan. After that brief visit to Michigan, Helen remembers the last time she saw Danny was when they traveled together by train to Chicago. After reviewing my timeline of Danny's military movements, she agreed that the last time she saw him alive was when he boarded a train to Rhode Island. To Helen, the destination did not matter, whether Chicago or Vietnam, because in her mind, that moment marked the last time they saw each other before he shipped off to war. I later explained that after completing his training in Rhode Island, Danny flew directly to his next duty assignment with CBMU-302 in California. It was there, for the first time, that he learned his unit was already in Vietnam and that he would be heading there within just a few weeks. Helen went on to say that when the family vacation in Michigan was over, they only had two days together back in Brookfield before he had to report for duty. Before Danny left, they agreed that Helen would ride with him on the train to Chicago. She remembered the two of them boarding the train in town, tickets in hand, unaware that this would be their final journey together. It was a bittersweet moment that was exciting yet filled with uncertainty.

It was Helen's first time on a train. The thought of traveling alone on the return train trip lingered in her mind, adding a layer of unease to an already emotional goodbye. Throughout the entire train ride, Helen kept her hand firmly clasped around Danny's arm, their fingers entwined, holding on to each other as if the connection itself was the only thing that could ease the tension. The train ride took just over an hour as it pulled into the Chicago Union Station. As they stepped off the train, Helen's eyes widened in awe, taking in the breathtaking beauty of Union Station's grand architecture. The soaring ceilings, intricate details, and bustling energy of the station left her momentarily speechless, a sharp contrast to the small train station back in Brookfield. Helen stated that they shared some intimate, heartfelt moments, just looking into each other's eyes on the train ride to Chicago. Helen recalled waiting at the train station with Danny until it was time for him to board the next train, the one that would take him from Chicago to Rhode Island. The time seemed to fly by so fast, each passing minute went by like seconds, as they hugged and kissed, trying to hold on to every last moment together before Danny's departure.

According to Danny's military records, Danny was heading to the Naval Construction Battalion Center (NCBC) Davisville in Rhode Island to receive mechanics training and qualifications. The time finally came for Danny to board the next train. In that moment, both Helen and Danny shared a deep, lingering hug, their goodbye kisses filled with unspoken emotions. They held on to each other tightly, not wanting to let go, until the sharp sound of the final boarding call was announced. All of a sudden, time appeared to be in slow motion as their hands reluctantly pulled apart, and Danny stepped on that train. In their hearts, both Helen and Danny resented the uncertainty of not knowing when they would see each other again. But Helen would never forget the one long, lingering kiss Danny gave her before he boarded the train. From the doorway, she heard his voice, steady yet filled with longing: *"I love you, Helen. I'll write you and send you pictures as soon as I can."* His words, full of promise and love, stayed with her long after the train had pulled away. As the train slowly pulled away, Helen could see Danny peering out of the window, his gaze locking with hers. Within moments, he was out of sight.

Sitting on a bench at the train station, Helen felt empty and alone, and the thought of not seeing Danny for a while felt unbearable. While waiting for her own train to take her home, Helen sat quietly, her sadness overwhelming, crying as she knew she would miss him, her first true love. To Helen, the wait for the next train back to Brookfield felt like an eternity. The minutes felt like hours as the train slowly pulled into Brookfield Station. Helen kept repeating to herself, *"Stay strong, stop crying, stay strong."* She clung to the hope that soon they would be together and that one day they would be husband and wife. Helen's friend was waiting at the train station to pick her up upon her return. Once home, she went straight to her room and cried herself to sleep. Unbeknownst to Helen, that final kiss, that final hug, and that last look at Danny's face as he boarded the train would be seared into her soul forever. At that moment, a lifelong memory, unknowingly at that time, would be the last time she ever saw Danny alive. At the end of our long conversation, as Helen shared her story, I could tell she was growing tired and perhaps a bit emotional. We decided to pause and agreed to resume her story soon, when she was feeling better.

As the weeks turned into months, I could only assume Helen's health had worsened; she never reached out to finish her story. I sent numerous text messages; each met with silence. I called her several times, but the phone just rang and rang, unanswered. In a last desperate attempt to reach Helen, I went to her Facebook page to send a message. But the moment the page appeared; I gasped in horror. I had stumbled upon the unimaginable. My heart began to ache for a woman I barely knew. My vision began to blur as a few tears spilled down my face. There it was, staring me right in the face, an obituary, announcing that Helen M. (Jablonsky) Fried had passed away. I stared at the words again and again, refusing to believe them. This couldn't be real. In my mind, all I could think about was that I had lost Helen before we could finish her story. It had to be a dream. We were so close to finishing her story. Below is what was posted on her Facebook page:

Helen M. Freid Obituary –

"Helen M. Freid, age 72, of Channahon, IL passed away on Saturday, ***December 14, 2024****, at St. Joseph Medical Center in Joliet, surrounded by family. She was born on January 14, 1952, in Berwyn, IL to Joseph J. and Helen M.A. (Henkel) Jablonsky. As the eldest of five children, Helen quickly embraced the importance of family. Helen was a loving wife, patient, dedicated, and caring mother, grandmother, and sibling, as well as a dedicated friend to many. She was an avid animal lover and enjoyed fostering several fur babies over the years. Helen found great joy in spending time with family and friends. Some of her fondest moments were spending time with loved ones and traveling to many warm destinations including multiple trips to Disney World with her children, grandchildren, sisters, and their families. She delighted in hosting family and friends for parties and holidays. These gatherings were filled with laughter and love and were highly anticipated by all. Memorials may*

be made in Helen's name to the animal rescue of the donor's choice."

I was numb, crushed by the news of Helen's passing. I felt terrible for not checking in on her health sooner. Recognizing that Helen needed a break, I decided to step away from the book as well. I took a short break in November, promising myself I'd return to it in January after the holidays. But now, that holiday pause felt like a lifetime ago. Helen's death only added to the growing list of missed opportunities, as countless friends and family of Danny, those I had hoped to interview, have all transitioned to the afterlife. I missed countless opportunities while searching for friends and family with stories about Danny, only to learn that I was a few months too late. All the people who knew Danny best had already passed away. With Helen's passing, I now face the challenge of finishing her story on my own. The hardest part for me to accept is that I was only a few questions away from the answers I needed to finish her chapter. To Helen, thank you for sharing your story. Rest in heavenly peace, and may you be reunited with all your family, friends, and with Danny, your first love. The rest of this chapter will be dedicated to the last few questions I had for Helen.

After my last conversation with Helen, I felt I had enough information to capture the strength and brevity of the love she shared with Danny. As I reviewed everything we had discussed, I began outlining her story to make it a stand-alone chapter in this book. As I combed through my questions and Helen's answers, I knew I had enough to tell their story. However, a few lingering questions remained, especially about the rumored 'Dear John' letter. I narrowed my final list of questions for Helen down to these four: 1) What did you and Daniel write about to each other in your letters to and from Vietnam? 2) How did you first learn about Danny's death? 3) Do you still have the engagement ring Danny gave you? And lastly, 4) To put the rumors to rest, did you write a 'Dear John' letter to Danny? We may never know the answers to these questions, but if I had to guess, this is how I believe she would have responded to these questions: I believe she would have said their letters were filled with small talk about daily life and how they kept busy during the long separation. I'm sure they expressed their love for each other and how much they missed one another. I'd even go so far as to say they talked about getting married after she graduated, and when he returned to the States.

Danny wrote to Helen more often than he wrote home to his parents and family, and Helen most likely responded to each and every letter she received from him. As for learning about Danny's death, I can only imagine that she was first notified by Danny's mother, Mary. According to Helen, they both liked each other, and she stayed in touch with Mary after Danny left for Vietnam. Helen told me that over time, she discarded all the photos and letters from Danny in an effort to find closure. She didn't elaborate on when she let go of these mementos, but she did mention that she had held onto them for many years. As for the gold ring, I have to believe that the engagement ring wasn't discarded and that she kept it as the sole reminder of her first love. After all, who would know the difference if she wore that ring until her last breath? We may never know for sure, but my intuition tells me she still had the ring. If I had to guess, after her passing, most of Helen's jewelry was passed down to a family member, most likely one of her grandchildren. Right now, I imagine that ring is being proudly worn by a family member, though sadly, they may never know its true meaning or significance.

The dreaded 'Dear John' letter rumor was, and still is, the most important question I wanted to ask Helen. I had purposefully planned for this to be the last question I would ask her. I didn't want to upset Helen, but I needed to hear the answer from her own lips. Throughout all of our conversations, I never once got the sense that she had written such a letter. She spoke of Danny as if she were still deeply in love with him. I knew in my heart that this would be the toughest question for Helen to answer, but I eagerly needed to know the truth. I also knew it would be uncomfortable for her, and I feared she might be offended by the very notion of it. I feared that by asking the 'Dear John' letter question early in our conversations, I might upset Helen and that she would discontinue having these conversations. I didn't want Helen to think that I believed any of the many rumors that had circulated over the years. Although Helen never specifically told me, I suspect she was aware of the rumors about a "Dear John" letter and the speculation that Danny's death was a suicide. From all of my conversations with Helen, I never came away with the feeling that she wrote a "Dear John' letter to Danny. I feel that if Helen did in fact write this rumored 'Dear John' letter, that she would never have agreed to tell me her story about their lives and future plans together.

Just ten days after his passing, several rumors were circulating among friends and family about the cause of his death, everything from a friendly fire incident to an accidental shooting, or even that he had been killed in combat. In the back of my mind, I suspected that Helen may have anticipated this question, given all the rumors that had been swirling around. I'm grateful that I had the chance to share with Helen most of the information I uncovered during my research into Danny's death. If Helen didn't write this letter, then we can put the rumor to rest. However, since we don't know her answer for certain, the only way to prove otherwise would be to locate the actual letter. I've interviewed family, friends, and former CBMU-302-unit members, and not a single one has ever seen the rumored letter. According to Howard Decker's family, Danny's barracks roommate never saw the letter; he just repeated the rumor.

The only person who ever claimed to have seen 'a letter' from Danny about his emotional state was his brother, Chuck. According to Chuck, it wasn't the rumored 'Dear John' letter from Helen but a personal letter to him, where Danny expressed some form of sadness and depression. Chuck believed that Danny may have been suicidal, so he gave the letter to their mother. To this day, the letter Chuck remembers giving his mother, Mary, has never been found among the family memorabilia saved after Danny's death, and no one in the family recalls ever seeing it. During the early stages of my interviews with Helen, I shared my plausible conclusions on what I believed led to Danny's death, and suicide was not one of them. So, until I see the actual letter itself, or if someone comes forward stating that they actually saw and read the letter, then I will continue to believe that no such letter existed until proven otherwise.

On November 20th, 1969, Helen saw Danny for the last time. She attended his Catholic mass and funeral that cold November morning. With rumors still swirling about his death, kneeling beside his casket under the watchful eyes of his family must have been incredibly difficult for her. Just shy of eighteen, Helen knelt at the casket of her first love. He lay before her in his Navy uniform, medals pinned neatly to his chest, ready for his final resting place. From that first meeting at Woolworth's diner to their last kiss at the train station, Helen's heart had been filled with love for Danny. Her final glimpse of him came moments before they closed his casket and covered it with the American flag. Danny came into the world on December 14th, 1950. Exactly seventy-four years later, on Danny's birthday, Helen departed this world after a long battle with a serious illness. She left this world on the very day he was born, as though their lives were forever bound together.

Chapter 7

Danny's Unit and his Vietnam Experience

Living in Brookfield, attending school, working part-time, and maintaining a vibrant social life with his friends kept Danny on his toes. Yet, despite his hectic schedule, he and his brothers also took responsibility for household chores, always fulfilling their duties under the watchful eye of their father Elmer. Attending high school meant that Danny was juggling a full course load on top of his part-time job at Knight's Auto Body Shop. For any high school student, balancing a full schedule of classes is a formidable challenge. While the idea of working part-time and going to school full-time may seem appealing, most students quickly fall victim to sleepless nights, limited study time, and slipping classroom participation. The end result often leads to poor grades, frustrated teachers, disappointed parents, and overwhelming stress and anxiety. Through research and conversations, I have learned that in 1968, Danny dropped out of his senior year of high school but continued to work at the Knight's body shop. Dropping out of high school during the Vietnam War was like putting a 'PICK ME FIRST" sign on your back for Uncle Sam's draft board. Danny likely knew that dropping out of high school meant the military draft would soon loom over him. To avoid Uncle Sam's draft board, Danny knew he had to enlist in the U.S. Navy. Before dropping out of school, and with his parents' help, he signed his enlistment contract, ensuring that he would sidestep the draft and the near certainty of being sent to Vietnam as a U.S. Army soldier. In Danny's mind, joining the U.S. Navy meant serving aboard a ship at sea, safely distant from the fighting in Vietnam.

There were numerous military recruitment stations near Brookfield, Illinois. I'm not sure when Danny first visited the recruiting office in La Grange, Illinois, but I do know that upon arrival, he quickly discovered he was too young to enlist. The recruiter told Danny that he would need his parents' consent to enlist and that he would have to take them to the recruiting office to sign the paperwork. I have not determined exactly how Danny came to the idea of joining the U.S. Navy, but from friends and family conversations, it's apparent that Danny was conducting his own research about the U.S. Navy. Danny had talked with his brother Jim and his father Elmer, who were both former U.S. Navy men, and I am sure they told him about all the pros and cons that come with joining the U.S. Navy. He had several opportunities to talk to his brother-in-law Fred Stanko, who was on active duty in the U.S. Navy. Another possibility is that Danny might have had a friend who also wanted to join the U.S. Navy with him. Whichever the case, he made the decision to join, and with the help of his parents, they both signed Danny's enlistment contract.

As I write this story about Danny, I often wonder how he determined which U.S. Navy job was right for him and how he decided on the U.S. Navy Construction Mechanic (CM) rating, which is part of the U.S. Navy Construction Battalions (CBs), also known as the SEABEEs. I can only assume that the U.S. Navy recruiter asked Danny what kind of work he enjoyed, and Danny likely replied that he liked working on cars. With that bit of information, I'm sure the recruiter steered him toward the Construction Mechanic (CM) rating. I don't think for a moment that it ever crossed Danny's mind that a job in the U.S. Navy would land him on the ground in South Vietnam. We know that Danny talked to his sister Joan's husband, Fred Stanko, before joining the U.S. Navy. I found it interesting, and a bit of a crazy coincidence, that Danny would seek advice from Fred, because I also spoke with Fred before enlisting in the U.S. Navy in 1981. My Uncle Fred, Danny's brother-in-law (RMC Fred Stanko, USN, Ret.), gave me some of the best advice and guidance before I joined the U.S. Navy, and I will never forget it. I can still hear Uncle Fred's voice telling me about life at sea and all the great ports of call he visited. Whether Danny received information from his brother-in-law Fred, his father Elmer, his brother Jim, the U.S. Navy Recruiter, or if he just went with his gut, Danny joined the U.S. Navy with an enlistment date to start on December 16th, 1968.

A few days before Danny shipped off to boot camp at Great Lakes, Illinois, his parents threw him a going-away party with all his friends and family in attendance. One small detail that probably didn't cross Danny's mind after receiving his enlistment date was that he would be away from home for Christmas and New Year's Eve. Both of these holidays were a big deal within our family and often led to big family get-togethers. Christmas Day and the big family dinner in 1968 went on without Danny at the table. That year, Mary, Elmer, and the whole family would celebrate the holiday and share Christmas dinner together, but without Ray, Chuck, Alby, and Danny. All of Mary's boys would be deployed during the 1968 Christmas holiday season.

On February 23rd, 1969, Danny graduated from boot camp. I learned that several family members, along with his girlfriend Helen, attended his boot camp graduation ceremony. Upon completion of the U.S. Navy's boot camp, Danny was given orders to report to the U.S. Navy's Construction Mechanic School (NAVSCOLCONST) located in Davisville, Rhode Island. In 1968, the Naval Construction Battalion Center (NCBC) Davisville became the homeport of the Atlantic Fleet SEABEES. During the Vietnam War, NCBC Davisville served as a major training facility for SEABEES, and the base was used to send supplies and equipment to Vietnam. Danny was ordered to report to NCBC on February 28th, 1969, located at Quonset Point, a small peninsula in Narragansett Bay, Rhode Island. This is also the home of the SEABEE Museum and Memorial Park, located in North Kingstown, Rhode Island. The SEABEE Museum is dedicated to the men and women who served in the CB Battalions. The museum highlights the accomplishments of the SEABEE Construction Battalions during both times of war and peace. During the Vietnam War, eight Naval Mobile Construction Battalions (SEABEES) were homeported in Davisville. Approximately one year before Danny reported for duty at NCBC, on January 16th, 1968, the main body of Naval Mobile Construction Battalion (NMCB) departed Davisville, Rhode Island, for Da Nang, Vietnam.

While assigned to NCBC, Danny learned to maintain and repair automotive and heavy construction equipment for the U.S. Navy. His technical training included hands-on operations of backhoes, tactical vehicles, pile drivers, cranes, rollers, bulldozers, dump trucks, and buses. Danny also learned how to repair and maintain vehicle frames, bodies, and chassis, as well as automotive repair, consisting of automotive maintenance with vehicle batteries, brakes, valves, and tires. After 14 weeks of CM training, Danny graduated from NCBC on June 13th, 1969, and received a third set of follow-on orders to the 31st Naval Construction Regiment (NCR) located in Port Hueneme, California. Danny was excited about being stationed in California. He knew that once he settled into his new command, he would bring Helen out to join him so they could marry and live together, just like his sister Joan and brother-in-law Fred were doing in Connecticut.

Orders to Naval Construction Regiment (NCR)

On June 30th, 1969, Danny found himself in Port Hueneme, California, attending more military training. According to Danny's military records, he was going through an undisclosed (unidentified) course of instruction, and a review of his records did not indicate what type of training he received. His training likely included rifle and pistol proficiency, hand-to-hand combat drills, and basic survival, evasion, resistance, and escape techniques. Prior to the start of his training, Danny learned that his assigned unit was already deployed to South Vietnam and upon completing the course, he would be joining them. The news devastated him; his orders meant Helen wouldn't be coming to California. Danny's dreams and plans to marry Helen and to live together in California came to a screeching halt. Now they both would have to wait until his tour in Vietnam was over. Danny's fourth and final set of official military orders had him reporting to the U.S. Navy's Construction Battalion Maintenance Unit THREE ZERO TWO (CBMU-302), which was currently deployed to Cam Ranh Bay, Vietnam. He was now faced with the fact that he would be on the ground in Vietnam during the height of the war.

Orders to CBMU-302

Danny had his last bit of military training and final medical screening on August 3rd, 1969, before his deployment to Vietnam. For the U.S. Navy, it's standard operating procedure (SOP) for sailors to be granted a short leave period to visit family before deployment. However, Danny was stationed in California, and with only three days of leave, he didn't have time, or the money, for a plane ticket home to see Helen or his family. Danny's orders directed him to report for duty with his new command on August 8th, 1969. One day earlier, on August 7th, 1969, Danny boarded a U.S. Military plane and was enroute from California to his first combat assignment in Cam Ranh Bay, Vietnam.

On August 8th, 1969, Danny's plane landed in South Vietnam, along with other Sailors, Soldiers, Airmen, and Marines. After the plane pulled up to the terminal in Vietnam, all the men were placed on a bus that took them to their new assignments. Once Danny arrived at Cam Ranh Bay, he checked into his new command and was assigned a barracks room. While unpacking his gear in the barracks, Danny met his new roommate, CMCN Howard Decker, a shipmate he already knew from CM training school. He and Howard were both enlisted construction mechanics and would be working together in the CM shop. The next morning, Danny was introduced to his new commanding officer (CO), executive officer (XO), company commander (CC), and his new leading petty officer (LPO). That same day, Danny received his job assignment to the CMBU-302 auto body shop. Danny was assigned to the Mechanics Division, which handled all aspects of repair to CBMU-302 vehicles as well as the occasional U.S. Army vehicle. In one letter home to his parents, Danny told them he would be in charge of the body shop the following week and wanted to put up a sign above the entrance that read, *"Danny's Body Shop."* Prior to his first step into the auto shop, I am sure his company commander and LPO introduced Danny to the CMBU-302 Creed: *"The man who says it can't be done, usually finds himself being interrupted by a CMBU-302 SEABEE doing it."*

The auto shop was in a makeshift building that contained all the usual auto repair lifts and tools required to repair the various vehicles and equipment used by CBMU-302. Danny's first job assignment was at the auto body shop, where he was responsible for repairing exterior damage to a multitude of vehicles and heavy equipment. Danny must have been ecstatic about being in charge of this body shop. Back at Knight's auto body shop, he was a jack of all trades and never in charge of his own automotive division. The body shop was located at the southeast corner of the CBMU-302 compound and was surrounded by the ALPHA company's main building, the tire shed, and the supply department's laydown area with multiple rows of supply crates stacked nine feet high. A little further to the east and a short walk from the body shop was the base medical facility (clinic), the mess hall, the barracks, and the base enlisted club. The base was relatively small and almost everything was within walking distance. The storage area for vehicles and heavy equipment was located on the western side of the base and a short distance from the body shop. According to Danny's letters home, he gained a lot of experience driving the various vehicles and heavy equipment he worked on. He quickly learned that, as the most junior sailor assigned to the mechanic shop, he would need to work his way up the ladder and improve his skills before being considered for a transfer to other sections within the shop or for a CBMU-302 detachment assignment in the country.

CBMU-302 Construction Mechanics (CM) Auto Shop Photos

Danny's parents received a letter from LCDR Harrell, Danny's Commanding Officer (CO), informing them that their son arrived safely in Vietnam on August 8th, 1969, to begin is tour of duty with CBMU-302. In this letter, the CO explains that the unit is responsible for maintenance and minor construction support for U.S. Navy bases in Vietnam. The letter also expands the quality-of-life Danny will experience, like good living conditions, good food, great medical care, and religious services. The CO goes on to say that Danny will also continue with his professional training while serving in the unit. I'm sure this letter meant a lot to both Mary and Elmer, likely giving them a sense that Danny was in good hands and working in a safe environment. However, within a few months, they would receive another letter from Danny's CO, and this time, the feeling would shift to pain, sadness, and grief.

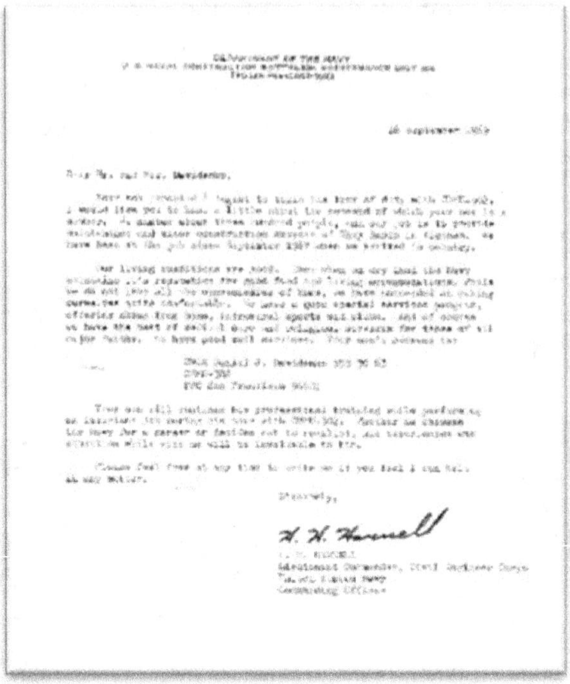

First Letter from Daniel's Commanding Officer

CRB and Operation MARKET TIME

Unbeknownst to Danny, the naval base at Cam Ranh Bay (CRB) served as the nerve center of the U.S. Navy's MARKET TIME anti-infiltration operations during the war. Cam Ranh Bay was one of the largest natural harbors on the coast of South Vietnam. In February 1965, long before Danny's tour of duty in Vietnam, the U.S. Navy discovered a 100-ton enemy trawler in Vung Ro Bay full of enemy supplies. Shortly after this discovery, the U.S. Navy and Coast Guard established an anti-infiltration program along the coast of South Vietnam. The code name for this anti-infiltration program was called Operation MARKET TIME, also known as Task Force 115, and its sole mission was to halt the sea-borne infiltration of supplies to Communist troops [25].

In Danny's small maintenance world, he was not aware that CRB was the center of coastal air patrol operations with the establishment of the U.S. Naval Air Facility. This Naval Air base was the home of SP-2 Neptune and P-3 Orion patrol aircraft squadrons. Danny's command, CBMU-302, shared Cam Ranh Bay with the Naval Air Facility, Naval Support Facility, and Coastal Division 14 that operated fourteen Swift Boats in support of Operation MARKET TIME. Additionally, platoons from SEAL TEAM ONE and SEAL TEAM TWO were also stationed at CRB and assigned to a specific operating area in Vietnam, and for the most part operated autonomously. CBMU-302 was deployed to CRB to provide public works assistance to these many dispersed naval support activities and detachments [25]. Most of these detachments were located throughout the I, II, III, and IV Corps. With only one mess hall and enlisted club on base, Danny may have crossed paths and had a few drinks with other Sailors, Airmen, Marines, as well as the U.S. Navy SEALs and Swift Boat operators.

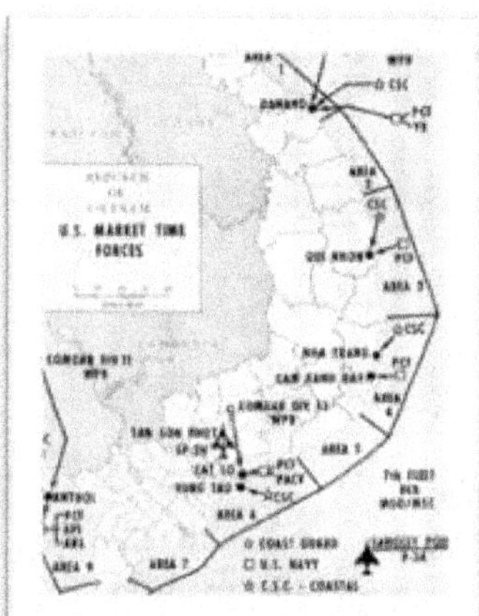

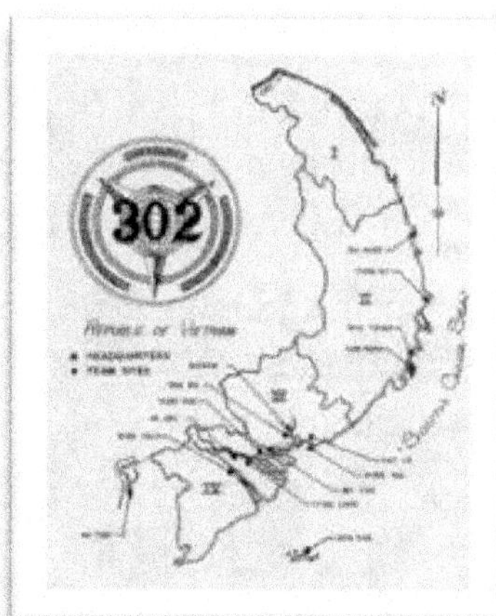

Operation MARKET TIME & Corps I, II, III, and IV Maps

The History of Danny's First Command, CBMU-THREE ZERO TWO

Construction Battalion Maintenance Unit THREE ZERO TWO (CBMU-302) was commissioned on March 31st, 1967, and the Commander Construction Battalion Pacific (COMCBPAC) directed the deployment of CBMU-302's main body to Cam Ranh Bay, Vietnam, on or about September 15th, 1967. Upon the Main Body's arrival in Vietnam, two Details were formed and were prepared to deploy. One ten-man detail was to support Inshore Underwater Group One (IUWG-1). The second detail was tasked with maintenance support of Naval Support Activity (NSA) Signal Detachment, Coastal Division 13, and Coastal Surveillance Center (GSG). Danny's command experienced a significant event on July 4th, 1967, when eight CBMU-302 men aboard a Military Airlift Command (MAC) flight enroute to Vietnam were forced down by Soviet jets over a Communist controlled island near Japan. However, two days later, they were released and reported to the command. On August 12th, 1967, COMCBPAC directed the deployment of fifty men from CBMU-302 to the Naval Support Activity Detachment located at Cam Ranh Bay in the II Corps Tactical Zone [26].

The main mission of CBMU-302 was to perform base maintenance and public works functions at the various U.S. Navy commands, as well as the Vietnamese Navy sites in the Southern portion of the country. CBMU-302 also sent out short-term special mission teams to perform specific tasks. These small groups were called Special Teams. In September 1969, the battalion's mission was expanded to include the Dependent Shelter Program (DSP). This DSP effort involved the construction of permanent structures to house Vietnamese Sailors and their families at naval facilities in II, III, and IV Corps [26]. By the end of CBMU-302's deployment to Vietnam, they had over 1400 men assigned, making it the largest SEABEE

Battalion since World War II. The unit was the last battalion-sized unit to leave Vietnam [26].

Aerial Photo CMBU-302 Base Camp

LCDR Harrell relieved LCDR Filbry as Commanding Officer of CBMU-302 in a change of command ceremony held on August 16th, 1969, at the Naval Support Facility at Cam Ranh Bay. Danny arrived on August 8th, 1969, so he would have been present at this ceremony. The picture below (from the CBMU-302, 1969-70 yearbook) was taken during the change of command ceremony. It is a grainy picture, and difficult to determine where Danny is standing among the ranks. From the photo above, you can see the base camp with four large buildings to the right (Alpha CO, Bravo CO, Supply, and the HQ buildings). To the right of those buildings, you will see the CBMU-302 Vehicle storage yard, and to the left of the buildings, you will see the supply laydown area. Though this photo is cropped down, you can clearly see that the base is relatively small in size, the equivalent of maybe 2-4 football fields.

Change of Command Ceremony (16 August 1969)

Danny's First Day in Vietnam

The first letter Daniel wrote home from Vietnam was on August 8th, 1969, and it started with this:

> *"Well, what have I got myself into? It's not so bad, really. The first night I got here, the VC blew up our hospital, that's been it. It's pretty safe here. The VC bomb the army on the other side of the hill every night, you can see the flashes all night and hear them, but they never bother us (except?)."*

On the evening of August 8th, 1969, a group of Viet Cong (VC) terrorists (also referred to as Communist Guerrillas or Sappers) cut through the perimeter barbed wire fencing of the 6th Convalescent Center at Cam Ranh Bay and proceeded to toss satchel charge explosives into the recovery wards that housed injured personnel who were recovering from their injuries. The only way to get into the convalescent compound was either by sea or through the perimeter fencing [27]. The U.S. Army released a statement that 2 soldiers were killed and 54 wounded in the attack [28]. In Danny's letter to his parents, he told them that the hospital was attacked, when in fact, it was the convalescent center and not the hospital. That night, there was security in the towers and at security posts; however, there was no security patrolling the perimeter of the base. No Viet Cong were killed or captured before, during, or after this attack. It is believed that some of the Vietnamese personnel who worked on the base provided information to the terrorists about the base and its security practices. Despite extensive changes to security and procedures, just two weeks later, the 6th Convalescent Center was attacked for the second time. Before these incidents, Cam Ranh Bay was considered to be the safest place to be in Vietnam [28].

Ward Building, Base Chapel, & Water Towers

Working at the Mechanic Shop

Throughout September and October, Danny learned to work on nearly every type of military vehicle. He especially enjoyed the body shop, repairing damaged truck doors, hoods, and fenders. Danny quickly realized that vehicles don't drive themselves and that he would have to teach himself how to operate each one safely. Damaged vehicles arriving at the base were parked in a designated section of the vehicle storage yard. Each morning, Danny received an assignment to repair a specific vehicle. He would grab his paperwork and head out to find the vehicle he was scheduled to work on. Most were easy to drive, like the smaller trucks and jeeps. The larger vehicles, however, posed a challenge for Danny. Without manuals on how to start or drive them, he had to figure it out on his own. In one of his letters, Danny mentioned that he was a quick learner and enjoyed mastering the operation of larger equipment. One of his favorite vehicles was the 5-ton truck. After teaching himself to drive it, he even dreamed of becoming a truck driver (like his father Elmer) after leaving the U.S. Navy.

After being late for work a few times, Danny's boss assigned him to the lube rack for a few weeks as punishment. Changing oil and filters was part of the mechanic's job, but Danny found it the least desirable task in the shop. By the end of the day, he was covered in grease and oil. Each day, there was a constant line of vehicles needing grease, oil changes, transmission fluid replacements, or radiator flushes. Danny considered it a good day when all that was required was tire changes and minor body work. As the weeks went by, Danny found himself back in the auto body shop, doing what he loved, fixing damaged door panels and fenders. His experience working at Knight's Auto Body back home had given him valuable skills in body repair and painting. In fact, he knew more about bodywork than most of the mechanics in his command, which is why they put him in charge of that section.

The wet season in CRB runs from September to December, with October and November being the heaviest months. During this time, the base sees frequent heavy rainfall from tropical storms. The hardest part of Danny's job was heading out into the rain to find the next vehicle to repair, trudging through mud while being blasted by wind and rain. Once Danny got into the vehicle, the drive to the repair bays was easy, and he could stay dry while working. Sunny, dry days were always welcomed, as they allowed maintenance to be done outside the bays. But regardless of the weather, repairs still needed to be made; the rain just made the job more difficult and uncomfortable. While Danny mainly worked in the body shop, he was occasionally pulled into other sections to assist with engine repairs, transmission work, brake repairs, and front-end alignments. He was gaining experience in all areas of auto repair. In a letter home, he mentioned that after a few years of truck driving, he planned to open his own auto repair shop.

By November 1st, Danny was informed that his request for a CBMU-302 detachment had been approved. While excited, he was also apprehensive about the new orders. He had just started to enjoy his work in the mechanic shop but was eager to leave CRB and join his buddies on detachment at CAT LO. Danny would spend the next few days working hard and preparing for his upcoming detachment.

A Goodwill Event

On November 4th, 1969, a detail of men from CBMU-302 conducted a goodwill event between the civilian population of Cam Ranh Village and U.S. forces in the area, by recovering the remains of a whale for interment in Cam Ranh Village's Whale Temple. The whale is a sign of good fortune in Vietnamese fishing villages, and the village was alarmed when a dead whale washed up on the beach. In an effort to support the local Vietnamese people, the command sent one bulldozer, two dump trucks, two front loaders, and other emergency equipment to the site. A group of men from CBMU-302 and some of the locals from the Cam Ranh Bay village lashed ropes around the whale to keep it from washing out to sea [29] [30]. According to Danny's executive officer, LT Jacobs, Danny was involved in this goodwill event, and he was one of the guys who operated one of the 5-ton trucks. The whale was too long and heavy to be lifted into a truck, so with the permission of the Village Chief, the whale was cut in half to make it easier to transport. The two halves of the whale were loaded into two separate trucks. I do not know which half of the whale was in the truck that Danny drove, but he managed to deliver his half of the whale carcass to the local villagers. The village people decorated both trucks with red silk to honor the occasion. The smell emanating from the whale carcass was awful, but it mattered little to the locals who were determined to honor this sacred beast. Music from drums, horns, singing, and incense filled the air from the local Vietnamese people. Danny, along with many other CMBU-302 personnel, delivered the dead whale to an area immediately in front of the Temple to Ong (whale). The whale was buried in a mound of sand about 6 feet high and braced on the sides with boards. The carcass will remain there for two years until it is time to dig up the bones for placement in their temple. Three days later, on November 7th, 1969, the village chief arranged for the village to give a feast for all the men who recovered the whale. The feast took place on the patio in front of the temple to Ong Nam Hai (The Whale Spirit of the Fishermen) [28].

The Actual Whale burial Site (side view) & The Actual Whale burial Site (front view)

From Danny's letters, we know he worked on 5-ton trucks and enjoyed driving them. However, I haven't come across any letters where Danny mentions hauling half a dead whale to a local village during a goodwill event. I believe there are more letters from Danny, hidden in a box in a family member's attic, that haven't been found yet. It's possible that Danny did write home about this whale event but never got the chance to mail them home. One of those letters might have described his involvement in this event. These events took place on November 4th and 7th, 1969, just days before Danny was found dead on base.

News Reports Indicate War Ending Soon

In 1969, when then Vice Admiral Zumwalt, the Commander of Naval Forces in Vietnam, implemented the U.S. Navy's Vietnamization Program, he gave a top priority to the construction of dependent housing for the families of Vietnamese servicemen who would command and ultimately take over all the naval bases in South Vietnam once the U.S. and Allies exited the country [31]. Shortly thereafter, it was announced that between late October and early November 1969, the first U.S. personnel drawdown would involve U.S. Naval personnel who had not completed a significant portion of their year-long tours in Vietnam and were eligible for reassignment.

In one of Danny's last letters to his parents, he states the following:

> *"I'm gonna extend here for one more year, so I'll get an early out. So, in two long years, I'll be out for good. It's only a hope right now cuz the way things are going, we might not be here for two more years. I'll try anyway. There's no use in coming back to the states when all you learn to do is rake leaves and clean barracks."*

Danny may have been referring to the news reports of a potential drawdown schedule that was set to take place in the upcoming months. Sadly, Danny would not be around to get that extension he desired so that he could end his active duty service early. Even if Danny was still alive, his hopes of ending his tour a year early may not have happened. According to CBMU-302 historical records, the command held a turnover ceremony, transferring the base to the U.S. Army and beginning redeployment to Port Hueneme, California, on January 22nd, 1972. Danny's one-year tour in Vietnam was scheduled to end in August of 1970; a 1 year extension in-country meant that he would get out of the U.S. Navy and leave Vietnam in August of 1971. In that same letter, Danny mentioned that if he were reassigned back to the United States, he would be disappointed because all he'd be doing was raking leaves and cleaning barracks. Danny's limited time in the U.S. Navy had him convinced that boot camp life was what all sailors' lives were like back home in the States. It's a bit sad that he didn't get the opportunity to see what the U.S. Navy had to offer. Who knows, he might have become a "U.S. Navy Lifer" like me and spent over 20 years in the U.S. Navy.

Drug Abuse in Vietnam

In this section, I am NOT insinuating that CBMU-302 was involved in any form of illegal drug activities. I just found out through my research that no command was immune from drug use among the ranks, according to official records. Almost every command serving in Vietnam has had some level of drug abuse within the ranks. Statistics obtained from my research and from actual CBMU-302 command documents show that drug abuse was highest among 19-21-year-old service members. The command found marijuana in barracks areas, and empty heroin vials were found on the grounds in and around the bunkers. In order to discourage drug abuse among members of the command, violators were prosecuted to the maximum extent possible. The main reason behind adding this section about drug use in Vietnam was to highlight the fact that drug use led to poor performance, poor judgement and worse, an overdose that could potentially end one's life. Also, my research revealed that numerous service men from all branches of the military were engaged in drug smuggling back to the states, and these service members were at the mercy of the drug cartels. I believe that drug use and drug smuggling present possible (plausible) scenarios that could lead to a criminal act like homicide or murder. According to a report by the Department of Defense, 51% of the armed forces had smoked marijuana, 31% had used psychedelics, and at least 28% had taken hard drugs, such as cocaine and heroin. Additionally, the House Select Committee on Crime reported that the armed forces used 225 million stimulants (pills) between 1966 and 1969, which were used to boost endurance on long missions. Sedatives were also prescribed to help relieve anxiety and prevent mental breakdowns [66].

Smuggling Drugs into the U.S.

If drug abuse in the military wasn't alarming enough, there were reports of service members smuggling drugs back to the States. Back in the late 60s and early 70s, service members departing Vietnam on U.S. military flights would have an easy opportunity to smuggle a duffel bag full of drugs back into the

United States. Back then, service members typically took off from and landed on military bases, and often, they didn't have to go through U.S. customs checkpoints, and even if they did, baggage screening was not automated or even sophisticated. So, it was easy to bring back the drugs. The difficult part was not getting caught with the drugs in your possession once you left the base. I couldn't find official records to verify these reports; however, I did find numerous news articles that report the U.S. military's significant drug problem in Vietnam at the time.

Knowing that these news reports back in the United States brought negative attention upon the U.S. military, it's not unreasonable to believe that a military command might hide or cover up these stories to avoid the negative publicity from the press. The only drug smuggling reporting that I could find suggested that the drug cartels in Vietnam somehow used military servicemen to smuggle drugs out of Vietnam on U.S. military cargo planes and ships. I did not find any information, nor did I have any interviews that suggested that Danny or CBMU-302 personnel were involved in drug smuggling activity. However, I wanted to leave no stone unturned as I looked for reasons why Danny was shot and killed in Vietnam. I did not uncover any indications that Danny participated in any form of drug use, but I have learned that he consumed various forms of alcohol, like beer and whiskey, at the enlisted club on base.

CMCN Daniel Joseph Davidenko spent 1 year, 1 month, and 17 days in the United States Navy (USN), and of this time, he spent 95 days overseas in South Vietnam. Danny, like a few of his brothers, was not that fond of military service. However, from his letters, there were times when he was excited about the U.S. Navy and times when he hated the U.S. Navy. Danny was a hard worker, and he achieved the rank of E-3. I also learned from his letters that he was working hard to gain early advancement to Petty Officer Third Class (E-4). During his time in the U.S. Navy, Danny received the Vietnam Service

Medal with One Service Star, the Vietnam Campaign Medal, and the National Defense Service Medal.

CMCN Daniel Joseph Davidenko, USN

Chapter 8

Danny's Letters Home from Vietnam

In this chapter, I've transcribed the few letters that Danny's family managed to save. Many of these letters were written by Danny to his parents (Mary and Elmer). Danny did not date his letters, and over the years, many of the envelopes were discarded after the stamps were saved by family stamp collectors. Based on the letters themselves and my research, I've done my best to organize them in chronological order, estimating the dates when Danny likely wrote and sent them home. Danny arrived in Vietnam on August 8th, 1969, and spent approximately 3 months in country before his death. These letters are transcribed exactly as he wrote them.

LETTER 1 – Danny's Letter Home (Approximate Date: 08 – 16 August 1969)

> *"Dear Ma & Dad, Well, what have I got myself into? It's not so bad, really. The first night I got here the VC blew up our hospital, that's been it. It's pretty safe here. The VC bombs the army on the other side of the hill every night, you can see the flashes all night and hear them, but they never bother us (except?) I'm in charge of the body shop here as of next week. In a week I'll be the only one here in the body shop. I've got just about every tool I can think of. I'm gonna put up a sign saying Danny's Body Shop. I'll send ya a picture when I take over. Well, I only have 1 year and 3 weeks left here. Hey, what's Ray's address? I hope he's all right. In three weeks, I'll be E-3, after that I'm gonna try for a field promotion to E-4. It sure is lonely here, especially when*

you're on watch. You walk along the outer perimeter; it gets kind of scary, but ya hack it. It was a fourteen-hour ride here. I sure was glad to touch the ground. You should see the Vietnamese people here. If you could see them, it would answer all your questions about why we're here. We have two (Vietnamese men) in the body shop, and I'm gonna teach one how to do body work. This one guy (gook) has an old set of greens that someone threw out, and his shoes are about size twelve; you can tell by the way he walks in them that he has about a size seven foot. Most of the guys treat the Vietnamese like animals. I act friendly, maybe cuz I'm soft-hearted, I guess. Well, I'd better get going, there's another day ahead of me. Love to everyone, Danny.

PS. I feel number ten."

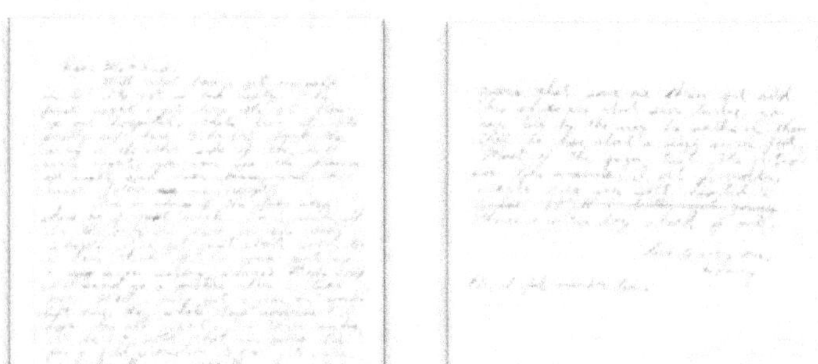

Danny's Letter Home to his Mother and Father (Letter 1)

LETTER 1 - Author Notes

In Danny's first letter home from Vietnam, he mentions that on his first night in the country, the VC (Viet Cong) attacked the hospital (6th Convalescent Facility). This was Danny's first letter home from Vietnam, and you can sense his concern about the attacks on the convalescent facility that occurred shortly after his arrival. In his mind, he must have been thinking, "What have I gotten myself into? Were these attacks a daily occurrence? Would I have to worry every night before going to bed? When would the next attack happen, and would it be my unit under attack?" However, in the letter, Danny reassures his family that he is safe. I'm sure one of the first questions he asked his shipmates and superiors after the attack was how often these incidents occurred and whether their base had ever been targeted before. I also sense that Danny felt the need to let his mother know he was safe; he didn't want her to worry or lose sleep wondering if he was being attacked by the Viet Cong. Danny's tone in this letter is upbeat, and he expresses excitement about the opportunity to be in charge of the body shop. In fact, he's already made plans to create a sign that reads, *"Danny's Body Shop."* Danny is also excited about his upcoming promotion to E-3 and the possibility of a field promotion to E-4. As an E-2, Danny's monthly pay was $127, which would increase to $155 once he was promoted to E-3. He knew that if he worked hard for the field promotion to E-4, his pay would rise to $214 per month. For an 18-year-old sailor, this was a significant amount of money, and Danny was focused on achieving his goals. In addition to his regular monthly pay, Danny also received special monthly allowances. He earned $60 a month for overseas duty and $30 a month for combat pay.

Halfway through this letter (Letter 1), Danny mentions that he's lonely in Vietnam and that it's scary when on duty walking the perimeter while on security watch. The only thing separating Danny and the enemy was a barbed wire perimeter fence and his standard U.S. military-issued M-16A1 rifle. During the Vietnam War, many Sailors, Soldiers, Airmen, and Marines were issued flak vests designed to stop shrapnel but not bullets. In the oppressive Vietnamese climate, the vests were hot, uncomfortable, heavy, and bulky, leading many to choose not to wear them. Additionally, the standard M-1 helmets issued during the Vietnam War were nearly identical to those used in World War II, with only slight design differences. By today's military standards, both the Vietnam-era flak vests and helmets offered limited protection for our service members. At the end of this letter (Letter 1), Danny refers to a Vietnamese man as a "gook." I found this troubling, as he had only been in the country for a short time and had already adopted the careless and demeaning terms many American service members used to describe the local population. However, in the same sentence, Danny also mentions how most of the guys treat the Vietnamese people like animals. It was wonderful to hear him say that he is friendly to the Vietnamese people because he is soft-hearted. I would agree with Danny's self-assessment and add that his family raised him to be a good soul, always seeing the best in people, even in the worst of situations. Danny closes his letter with, *"Love to everyone,"* and then adds a postscript: *"I feel number ten."* I have to assume that Danny was rating how he felt on a scale from 1 to 10, and on this day, he was feeling like a solid 10.

LETTER 2 – Danny's Letter Home (Approximate Date: Late August 1969)

> *"Dear Ma & Dad & Everyone, Well, today was another one of those days. It has its good points, and it has its bad points. The guy* (Howard Decker) *I share a cube* (Barracks Room) *with bought a TV today, which is a good point. On the other hand, I have the 3-7* (0300-0700 Perimeter Security Watch) *watch tonight, I don't really mind that cuz I haven't had a watch in a few days. I have another request to ask, would you send me a bottle of iodine? Working as a mechanic, I get many scratches and cuts on my hands. Payday, I'll send*

you some money for everything. Thanks! I'm saving my money for a tape recorder; I'll send some tapes home every once in a while. It's still raining here as usual. I'm pretty proud of myself for all the equipment I've taught myself to drive. I'm no expert, but I learn something new just about every day. I've been thinking I'll be a mechanic when I get out, but for the first year or so, I want to drive a truck. It feels good sitting behind the wheel of a 5-ton dump truck. Time seems to be flying by now, mostly the days, while I'm working. At night, I can't wait for quarters so I can start working again. I try to get as much as I can from the shop. I wrote Sophie and told her I wasn't a lifer, reminded her all right, kinda pounded it into her head (Sophie is Danny's sister-in-law). *Well, it's time to go. I'll see you in 13 months minus 1 month and four days. Love Danny.* (OVER- On the Back of the 2nd Page) *Ma, you said you were going to send me a St. Christopher medal. If you did, I hope it's insured."*

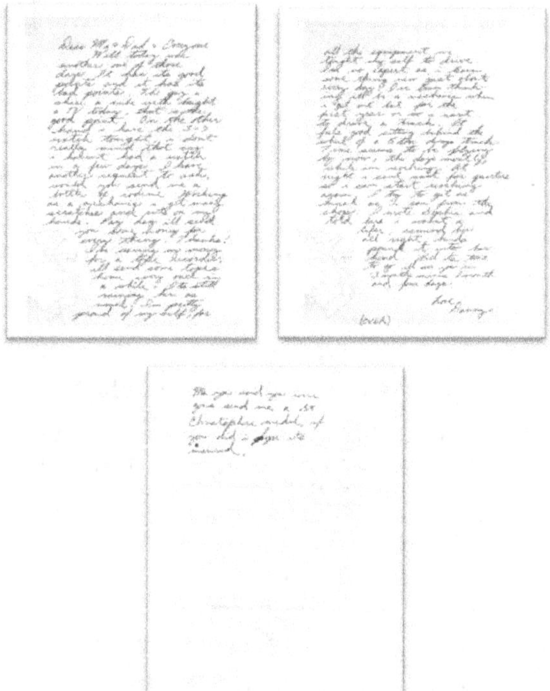

Danny's Letter Home to his Mother and Father (Letter 2)

LETTER 2 - Author Notes

Once again, Danny's letter comes across as upbeat and positive. He mentions both good and bad points, but never directly addresses the negatives. For example, he notes that the new TV is a good point, then mentions having the 0300 to 0700 (3:00 AM – 7:00 AM) perimeter security watch. It's my estimation that the bad point he refers to is likely the fact that he probably wanted to get some sleep before his watch, but his cubemate, Howard Decker, would be up all night watching the new TV, which would have distracted him from his sleep. I'm also a bit concerned about Danny asking for iodine for the cuts on his hands. This should have been a simple trip to the base medical clinic for minor treatment, or at the very least, there should have been a first aid kit in his shop. Two scenarios come to mind when I read this request. First, it's possible that one of his supervisors or buddies teased him about the cuts and told him to 'deal with it like a man,' since they were minor scratches. The second possibility is that many supervisors believed going to the medical clinic for minor treatment was just a way to get out of work. Regardless of the scenario, cuts to the hands in Vietnam could be very serious, as the chances of infection were extremely high.

In this letter (Letter 2), Danny expresses his excitement about saving up for a tape recorder so he can send home recordings to his family. In fact, there is a picture of Danny in his barracks room in Vietnam, and in this photograph (see below), you can see his M-16 rifle, his large 6-inch wooden crucifix, and his brand-new tape recorder.

Author's Note: Over the years, I have asked family members to search for any photos, letters, or memorabilia from Danny. Yet the cassette tape(s) he promised his mother he would send has never surfaced. They may still exist in a forgotten storage box somewhere, but to this day, no recordings have been found.

Danny is also very proud and excited about learning new skills and gaining experience driving various military vehicles. I can sense that things are going well for him, and he looks forward to using his military training to help him pursue civilian employment as a truck driver and auto mechanic once he leaves the U.S. Navy. He mentions that time seems to fly by and that he can't wait to get back to work each morning. I can feel Danny's excitement in this letter, especially considering that before joining the military, he was likely a mechanic's assistant at Knight's Auto Body Shop. Now, he's in charge of the auto body shop for CBMU-302, and all of this is happening in a war zone. In Danny's eyes, this was quite an accomplishment for an 18-year-old kid from Brookfield, Illinois. He must have been thinking to himself, wow, in less than a year, I've gone from being just a kid in Brookfield learning how to work on cars to an active-duty U.S. Navy man, with an M-16 rifle by my side, ready to fight when under attack or when given the order. Danny understands that during this tour of duty, he will continue to develop and advance his automotive skills for his future civilian career. Throughout it all, he doesn't seem particularly concerned that he's learning his trade in such a hostile, war-torn environment.

Danny in his Barracks Room with his New Tape Recorder, M16 Rifle, and his Crucifix

LETTER 3 – Danny's Letter Home (Approximate Date: 8 September 1969)

"Dear Ma & Dad. Things are all right here except for the weather. It's been drizzling for about four days now. It also gets up to about 110-115 degrees during the day. Other than that, everything is all right. I got a letter from Alby and one from Chuck. It was nice to hear from them. I am no longer running the body shop. I didn't go to the shop one morning till one o'clock, and they were a little mad. They put me on the dirtiest, worst job there is. I'm on the lube rack. I get to grease and oil all the

vehicles. It doesn't really bother me tho. Today makes it exactly one month I've been here, only twelve more months until I get home. The U.S. Navy gives five days of R&R; it's really nice of them, isn't it? Boy, do I hate Cam Ranh Bay. I can't wait to go on detachment. I bought a switchblade the other day. Is it cool? I can't wait to use it on someone. Not really, I just bought it cuz it's fun to play with. I've been driving just about everything here. I drive mostly 2 ½ ton dumps. I've driven a WHITE tank truck, a Mach tractor, weapons carriers, stake trucks, bull dozers, cherry pickers, you name it, I'll drive it. Oh yeah, I drove one of the army's 5-ton dumps. We do a lot of the army work cuz there mechanics aren't too bright. Well, I gotta go, so take care. Love Danny"

Danny's Letter Home to his Mother and Father (Letter 3)

LETTER 3 - Author Notes

In this letter (Letter 3), Danny mentions that he's been in Cam Ranh Bay for one month. Since he arrived in Vietnam on August 8th, 1969, this letter was likely written around September 8th, 1969. Danny begins this letter by saying that everything is all right, except for the weather. Unbeknownst to him, the rainy season in Vietnam lasts from May to November, with the heaviest rains falling between September and October. The combination of heat and rain creates intense humidity, making for a miserable environment to live and work in. Danny is excited to receive letters from his brothers, Alby and Chuck. I believe these letters were especially important to him because both brothers are in the U.S. Army and can relate to what he's experiencing in the U.S. Navy. While I didn't have access to those letters, I can only surmise that they shared their own likes and dislikes about the U.S. Army and U.S. Navy, as well as advice on what to do and what not to do, to get through their respective tours of duty.

Danny clearly states that he is no longer in charge of the body shop, which, in a previous letter, he described as his dream job. He was excited to name it *"Danny's Body Shop."* In this letter, Danny explains why he lost that job and was reassigned to the lube rack (*the dirty old grease rack*). It turns out that Danny was more than a few hours late for work. A few minutes late might be forgiven but being late for more than half the workday will get you in serious trouble with your Leading Petty Officer (LPO) and Company Commander (CC). Work begins after morning quarters at approximately 0700 (7:00 AM), but Danny didn't show up until 1300 (1:00 PM), meaning he was AWOL for six hours. This wasn't just a little tardiness; Danny was missing more than half a day's work. Such an absence is unacceptable and aligns with some of LT Jacobs' (Danny's Executive Officer) findings that Danny may have been staying up late each night, possibly overdrinking at the club on base. It's important to note that Danny was only 18 years old and likely not accustomed to drinking excessive amounts of alcohol. He was probably learning the hard way what excessive nightly drinking can do to the body. In a previous letter, Danny expressed how excited he was to start each workday and learn all there is to know about auto mechanics. However, it appears that he didn't fully understand the effects of alcohol on the body and most likely got a crash course in hangover management 101.

This letter (Letter 3) marks Danny's one-month anniversary in-country, and he alludes to the fact that he only has 12 months remaining on his tour of duty in Vietnam before he gets to go home. Danny expresses his dislike for Cam Ranh Bay (CRB) and his excitement for the opportunity to get a detachment assignment. From this letter, it's clear to me that Danny must have made a few mistakes and is hoping for a do-over, a fresh start, or a clean slate to get back on track. Danny believes that a new assignment (a detachment) would give him the chance to leave CRB and get that fresh start he's hoping for, a chance to rebuild his reputation, particularly the one damaged by being late for work. In the military, being late is not tolerated, and I must assume that Danny was late more than once, which is likely why he's so eager for a detachment assignment with new bosses. Danny knew these detachments were small, typically consisting of 5-10 men, and were located in isolated areas, away from CRB and the enlisted club where alcohol was a temptation. As Danny ends his letter, you can sense his pride and excitement about teaching himself to drive a variety of heavy equipment and vehicles. He even brags about driving a 5-ton U.S. Army dump truck.

LETTER 4 – Danny's Letter Home (Approximate Date: Late September 1969)

> *"Dear Ma & Dad, I want to say I'm sorry for not writing. I know you're mad, but I had a reason. Everything has been rotten here. All I've been doing is getting into trouble. I didn't want anyone to worry the way I've been feeling I couldn't make anything up, so I just didn't write. Everything is all right here now, so don't worry. Boy, do I hate the U.S. Navy; it's done nothing but bring problems. There's no one to talk to. I used to have Helen* (Helen was Danny's Girlfriend) *to talk to, but it's impossible in a letter, so now everything is kept*

to myself. 23 more months here and I'll be home for good. I got a letter from Chuck, and I saw on the return address that he's spec/4. I'm gonna take the E-4 test in November. Here are some pictures so you'll remember what I look like. I'm still on the dirty ol' grease rack. I keep trying to get off it, but no such luck. It's good to hear that Jan & Bill are getting along so well. Well, it's 6:50 in the morning where you are, and it's 6:50 at night. It seems like morning comes faster here, so I'll go for now. I'll be up at seven at night, wherever you are, in case you want to say good morning. Love Danny"

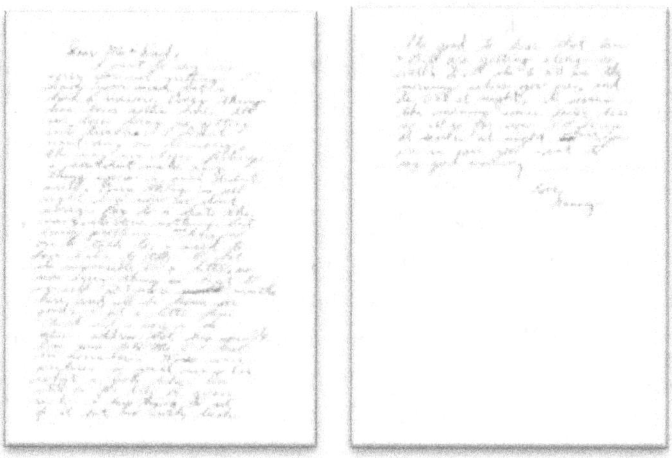

Danny's Letter Home to his Mother and Father (Letter 4)

LETTER 4 - Author Notes

In this letter home, Danny emerges from a few undisclosed events that got him in trouble with his superiors. Based on previous letters, I must assume that Danny has been late several times for work and for duty, including Perimeter Security Watch and Patrol. Oversleeping or forgetting to be at his assigned place of duty seems to be an ongoing issue for Danny, one he has yet to correct. Being late once or twice might be forgivable, but being habitually late only draws negative attention from co-workers and Leading Petty Officers (LPOs). If Danny doesn't address this problem soon, he will find himself in even more trouble.

To understand what might be contributing to the situations that keep landing Danny in trouble, I need to put myself in his shoes and consider the factors affecting his ability to self-regulate his schedule and meet his military responsibilities. My first thought was that Danny is only 18 years old, and in many respects, still a kid. However, after boot camp, the U.S. Navy doesn't see him as a kid anymore. In fact, during U.S. Navy boot camp, he was reprogrammed and purposely trained to transition from a civilian (a kid) to a well-trained U.S. Navy sailor (a man), ready to work and fight for his country. Then, I thought that Danny might be stressed over the constant threat of the enemy fighting around him. On top of that, he has to cope with the terrible weather, extreme heat and rain, and the limited physical environment of the small base in Cam Ranh Bay, Vietnam. However, Danny isn't the only one on base facing these distractions. Everyone has to deal with the worries and stress that come with the constant possibility of an attack on the base. He had shipmates, leaders, and the base chaplain to talk to if these were, in fact, some of the things causing his distress and tardiness.

The most likely underlying factors contributing to his problems could have been regular visits to the enlisted club each night, where he may have had a few too many drinks. Alternatively, he might have stayed up all night hanging out with friends, playing card games, watching TV, or writing letters home. The fact that he can't just pick up a phone and talk to Helen and is limited to just letters alone, weighs heavily on his mind. Each of these explanations is a viable reason why I believe Danny keeps getting in trouble. However, there is no acceptable excuse for his failure to meet his military work-related obligations. Danny quickly learns that his seniors have had enough of his tardiness and are not pleased with his lack of military bearing.

In this letter (Letter 4), Danny never specifically explains what keeps getting him into trouble or how these issues have affected his ability to write home. He mentions that everything is alright now and reassures his parents not to worry. Yet, in the very next sentence, he states that he hates the U.S. Navy because it brings him nothing but problems. It's hard to interpret Danny's true emotional state, as the letter reveals that he is dealing with many unidentified issues, which seem to send him on an emotional roller coaster with periods of highs and lows. Danny seems excited that his brother Chuck has been promoted to SPEC/4 (Specialist/E-4/Corporal) in the U.S. Army. From the letter, it's clear that Danny looks up to Chuck and is also excited about the possibility of passing his E-4 (Petty Officer 3rd Class) exam in November and getting promoted in the near future. Additionally, Danny expresses his dislike for working the *"dirty ol' grease rack"* and how he keeps trying to get out of that particular job. Deep down, I believe Danny knew that it was his own actions and lack of military bearing that had led to him being assigned to that undesirable task. He also knew that the only way to get out of that position was to improve his abilities and address the issues that led him to his current situation.

Danny expresses his hope for an assignment to one of CBMU-302's detachments within Vietnam, possibly to work with friends and to establish a better reputation, one where he would be timelier and more responsible. As a retired Lieutenant Commander in the U.S. Navy (the same rank as Danny's Commanding Officer), I have seen firsthand young sailors who exhibited similar tardiness issues. Some young sailors found it difficult to get to work on time. They were often late for work and showed a lack of responsibility while carrying out their military duties. Typically, 90% of these young sailors correct the situation quickly and have a successful tour or career in the military. However, 10% seem to never figure it out and struggle through their tour, never get promoted, or worse, end up in the Brig (U.S. Navy Jail) or are kicked out of the Navy on a bad conduct discharge (BCD). For Danny, being assigned to a CBMU-302 detachment was more than just orders, it was a second chance to prove to the command that he had matured, understood his responsibilities, and wanted nothing more than to succeed.

LETTER 5 – Danny's Letter Home (Approximate Date: Mid-October 1969)

"Dear Ma & Dad, I got your letter today. It's good hearing from someone. I'm gona extend here for one more year so I'll get an early out. So, in two long years I'll be out for good. It's only a hope right now cuz the way things are going we might not be here for two more years. I'll try any ways. There's no use in coming back to the states when all you learn to do is rake leaves, and clean barracks. I'm glad you liked hearing my voice. It was nice hearing yours too. I guess we all need practice on saying, "over". I'll call again on Christmas or whenever I can close to it. It's Sunday and almost time to eat. Nothing has changed here. Our Battalion had a party yesterday, that's where I got these pictures. I tried to get some color film, but whenever our exchange gets color film, it only lasts about an hour. I'll get some next week. I finally started saving, I put fifty-five in last payday. I'll get 10% interest on it. I guess I'll buy a car when I get back home. How's everyone doing back in the world? Well, I better go now so take care. Love Danny"

Danny's Letter Home to his Mother and Father (Letter 5)

LETTER 5 - Author Notes

In Letter 5, Danny's mood shifts from lonely and sad to upbeat and positive again, and he starts contemplating extending his time in Vietnam for another year. He was told that a one-year extension in Vietnam would grant him an early release from his initial 4-year service commitment with the U.S. Navy. In an earlier letter, Danny was getting in trouble and expressing his dislike for the U.S. Navy. So, what transpired during such a short period that made him want to extend his time in Vietnam for an additional year? Danny also alludes to the fact that CBMU-302 may not be in Vietnam for the next two years. From this statement, I can only assume that during morning quarters, when leadership discusses the day's plan, they were also keeping the unit informed about the political situation and the potential for a quicker-than-expected withdrawal from Vietnam.

According to CBMU-302 historical records, CBMU-302's mission commenced on August 16th, 1967, when CBMU-302 sent fifty men to Cam Ranh Bay, Vietnam. The battalion relocated from Cam Ranh Bay to the large U.S. Army base at

Bien Hoa in April 1971. Additionally, Danny's unit, CBMU-302, was the last SEABEE battalion to leave the Vietnam war zone. CBMU-302 folded its colors at its basecamp in Bien Hoa, Vietnam, on January 22nd, 1972, and the unit redeployed back to Port Hueneme, CA. Therefore, based on the command's history and the context Danny provides in the letter, it seems that if he had accepted the extension, he would have likely remained in Vietnam for two years. The extension would have allowed him to fulfill his time in Vietnam while reducing his overall 4-year enlistment contract, possibly leading to an earlier release from his service obligation. This would also align with the military practice of offering incentives like extensions for sailors who agreed to stay longer in-country, especially during times of uncertainty like the Vietnam War era.

Danny mentions getting color film from the "Exchange." In the U.S. Navy, the Exchange refers to a store or shopping center where sailors can purchase clothing, food, and other personal items. These stores are officially called U.S. Navy Exchanges. In contrast, the U.S. Army uses the term PX, which stands for Post Exchange, for their equivalent facilities. Danny mentions a battalion party that took place on base, where he had a few pictures taken. The U.S. Navy often holds quarterly events like this to give the troops a break and allow them to enjoy the camaraderie of their unit. When deployed at sea, the U.S. Navy refers to similar events as "Steel Beach Picnics." During these occasions, the command brings beer aboard, allowing sailors who are not on duty to enjoy up to two beers, along with food, music, and a day off from regular work. A key difference between sea duty and shore duty is access to alcohol. Ships do not have enlisted clubs or allow daily alcohol consumption, while U.S. Navy shore bases, including CBMU-302's base in Vietnam, operate enlisted and officer clubs where alcohol is available during non-working hours.

Danny references a time when he was talking to his mother and father via radio transmission from his base in Vietnam. Calling home during the Vietnam War was not an option afforded to every Soldier, Sailor, Airmen, or Marine. There were no personal cell phones or landlines available for service members to contact family members back home. To bridge this communication gap, the United States Military Affiliate Radio Service (MARS) established stations throughout Vietnam, supporting all branches of the military. The MARS system offered military service members a way to personally communicate with loved ones back home through a "phone-patch" connection over shortwave radio. Occasionally, MARS stations would provide each service member with a five-minute personal radio-telephone call to the United States, offering a rare and cherished opportunity to speak directly with family during their deployment. All they needed was a stateside phone number and parents willing to accept a collect call. The only concern was the location of the MARS radio station in the United States. From the MARS station in Vietnam, the operator would contact an available stateside station, which would then place the collect call to the number provided by the service member. If that radio station were located in California, then it would result in an expensive long-distance 'collect call' from California to Brookfield, Illinois. Once the collect call was accepted by a family member, the MARS operator would instruct the family member that only one person could talk at a time, and they had to say the word "OVER" when they were finished talking and not start talking again until after the service member said "OVER." When using this system, service members were required to follow a certain protocol. Since this was a non-secure means of communication, the service member was not allowed to mention anything about where they were or what they were doing in Vietnam. The time limit was also strictly enforced. The MARS operator would hold up his hands and count down from 10 before terminating the call. The MARS station in Cam Ranh

Bay, callsign AB8AC, was operated by C Company, 41st Signal Group.

LETTER 6 - Danny's Letter Home (Approximate Date: Late-October 1969)

"Dear Ma & Dad, I received two letters from you all right. I haven't been in a very good state of mind. It seems like everything has been going wrong for the last few weeks. I have someone on my back every minute, and the more I'm pressured, the more I try to do wrong. I guess things will work out. Your son made E-3 on the ninth of Oct. On the tenth of October, I almost lost it cuz I got written up again. I had the 3-7 watch, and I was 2 ½ hours late. I got written up for being AWOL. I went to the mast and pleaded guilty, and he dismissed the charges. Boy, is this letter hard to write. I feel like a stranger. How's everyone back home? Tell everyone I said hi! You've probably found the three stamps by now. Give them to Dickie Roberts. Tell him they cost 4 boxes of soap, cuz that's how much it costs to get a damn gook to get them for me. I've got to go, at least you know I'm still alive and can write. Love Danny"

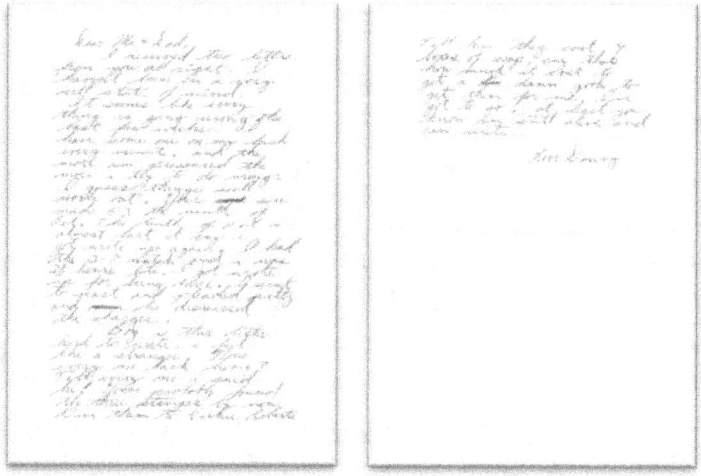

Danny's Letter Home to his Mother and Father (Letter 6)

LETTER 6 - Author Notes

In this letter, Danny clearly acknowledges that he has been making mistakes and getting into trouble. You can sense from his tone that he's upset with himself. From experience, I can tell you that being in the military is challenging for any young person, and especially for an 18-year-old kid stationed in a war zone. Conforming to strict military regulations and following orders without fail from superiors can be overwhelming. In such circumstances, emotional distress at that age is completely understandable. During the Vietnam War and previous conflicts, the branches of the U.S. military did not provide training on how to cope with personal stress, or the emotional challenges service members might face during their tours of duty. The military's primary focus was (and still is today) on training personnel in their specific fields or trades and emphasizing the importance of following orders, especially in times of war. A Soldier's, Sailor's, Airman's, or Marine's personal feelings were never considered part of that training. After expressing that he had not been in a good state of mind and that things were going wrong, Danny quickly shifts to excitement upon mentioning his promotion to E-3 on October 9th, 1969. However, that excitement is short-lived. In the very next sentence, he expresses obvious frustration, explaining that the following day, October 10th, he was written up for being Absent Without Official Leave (AWOL) for 2 ½ hours.

Though a minor infraction, Danny's tardiness couldn't go unpunished. His command had to make an example of him to reinforce that such behavior wouldn't be tolerated. For the average civilian, 2 ½ hours late might seem trivial, but in a war zone, when others are depending on you to be at your assigned post on time, being late is far more than just an inconvenience; it can jeopardize the mission and the safety of everyone involved. In fact, the military trains you to be early. In this letter, Danny states that he was taken to Captain's Mast to face his AWOL charges. Captain's Mast is not a courtroom trial; it is simply a term used by the U.S. Navy to impose "non-judicial punishment" for minor offenses. The Captain's Mast is a disciplinary process conducted by a commanding officer for relatively minor infractions, without the need for criminal charges or formal proceedings. It's used to address issues like tardiness for duty or mandatory formations, improper use of equipment, or failing to show proper respect to supervisors or superiors. Traditionally (historically), the proceeding would take place at the main mast of the ship, and all sailors who are not on duty are expected to be present. What follows is effectively a process of reading the offending sailor's actions, followed by a determination as to disciplinary measures. These measures often involve extra duty and a temporary loss of rank.

The term "Mast" may also refer to when a captain of the ship or a commanding officer makes him/herself available to hear concerns, complaints, or requests from their crew. From Danny's letter, I can sense that he must have had someone give him good advice on what to do and not to do while attending his Captain's Mast. The commanding officer knew it was a minor infraction and dismissed the charges. Danny admitted his mistake and made no excuses for his lack of military bearing. This may have been beneficial to Danny because his Commanding Officer dismissed the charges and assigned extra duties. Dismissing charges in a Captain's Mast is very important, because that means it would not go in his military record. I'm sure that Danny was thrilled that he did not get a reduction in rank, especially when he just got promoted to E-3. Danny had to perform extra duties as payback for his errors in judgment and ultimately for embarrassing his immediate supervisors, who had to deal with the whole mess.

Danny also talks about three stamps he obtained through the trading of four bars of soap. In Vietnam at the time, trading items with the local population was frowned upon; however, it was often overlooked and rarely punishable. In the States, back in 1969, Ivory soap cost twenty-five cents for four bars of soap. The base exchange sold soap to the Sailors a bit cheaper than that, so Danny probably bought the soap for half that cost, so he could get some stamps for his nephew, Rich Roberts (aka Dickie Roberts). Back in 1969, a first-class postage stamp cost only six cents. Today, that same stamp would cost about fifty-three cents. Lastly, the final sentence of this letter, he stated, *"I've got to go, at least you know I'm still alive and can write"*. This was written just weeks before they found Danny's lifeless body on base. Danny mentions October dates regarding some of the things that happened to him, but he does not mention any details as to what actually happened. Based on the dates mentioned in this letter, I can surmise that Danny wrote this letter sometime after October 10th, 1969. During the Vietnam War, the mail coming and going from the States to Vietnam and from Vietnam to the States, on average, would take about seven to ten days. Mail could take much longer because military personnel and supplies take precedent of the U.S. mail system.

LETTER 7 - Danny's Letter Home (Approximate Date: 03 November 1969)

"Dear Ma & Dad, How ya been? I'm doing all right. I got my wish; I'm leaving Cam Ranh Bay. I'll be going on detachment. It will either be BEN LUC, CAT LO, *or* BEN THUY. *I'm hoping to go to* CAT LO, *cuz I know a few guys there. It's good to hear that the house is coming along so well. If you can, send me some pictures. I guess it's pretty cold or getting there back in the world. It's still hot here, as if ya didn't know. I've got duty today, so I can't do anything. In the pictures you have in the envelope, the guy is me. That's what the cube I stay in looks like. My hair is short again. I had it pretty long, but the lifers got on my ass. The tape recorder is mine; I paid $125.00 for it. I'm one of the lucky ones who gets to have an M-16. They're taking all the guns now. I don't know what's next. The TV is the guy* (Howard Decker) *who sleeps under me. We get only one station here, but they usually have pretty good stuff here. In the picture of me sitting in the chair. The real small fan was given to me, the second biggest fan I found, and the real big one I stole from the snack bar last night. I had a buddy help me carry the big one. As you can see, I'm pretty cool at night. I'll send some color pictures home next pay day. Well, gotta go for now. Tell everyone I said hi and I miss you all, OK. (TRICK or TREAT) Love Danny"*

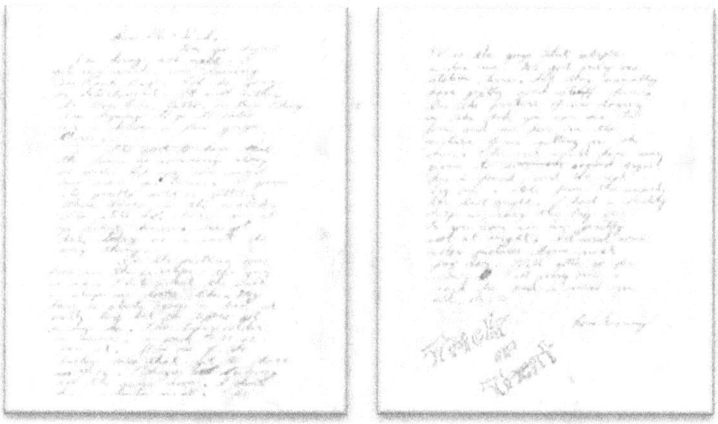

Danny's Letter Home to his Mother and Father (Letter 7)

LETTER 7 - Author Notes

In this letter, Danny really sounds excited again about the opportunities of leaving the CRB base. He mentions getting his wish to go on assignment with one of CBMU-302's detachments located in the country. From this letter, Danny's request to go on a detachment in support of CMBU-302 operations was approved. Going on a detachment meant that he would be leaving Cam Ranh Bay (CRB) for a different location within Southern Vietnam. Danny mentions that he really would like to be assigned to the CAT LO detachment, where he has a few friends. Danny does not mention where he made these friends, but I would assess that he made them while stationed in Vietnam, or maybe some buddies from either CM school or boot camp, who also received orders to CBMU-302 in Vietnam.

In this letter, Danny's upbeat and positive attitude toward his work and his future assignment in Vietnam is evident. However, it's important to keep things in perspective. Danny is still a kid at heart, having leapt headfirst into military life, just beginning the transition from boyhood to manhood. Military service is stressful, and in an instant, reality can strike, harshly reminding you of life's fragility. In that moment, you either grow up fast or crumble under pressure. Like most Vietnam veterans, Danny had to become a man the moment he set foot on Vietnamese soil. Unfortunately, not every young man makes that transition.

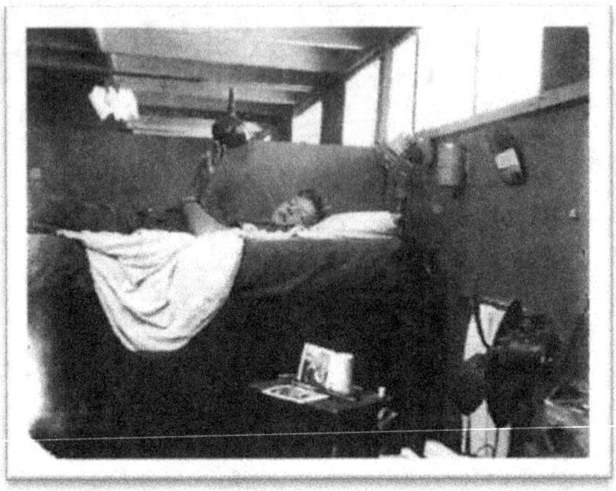

Danny in His Bunkbed (Barracks Room)

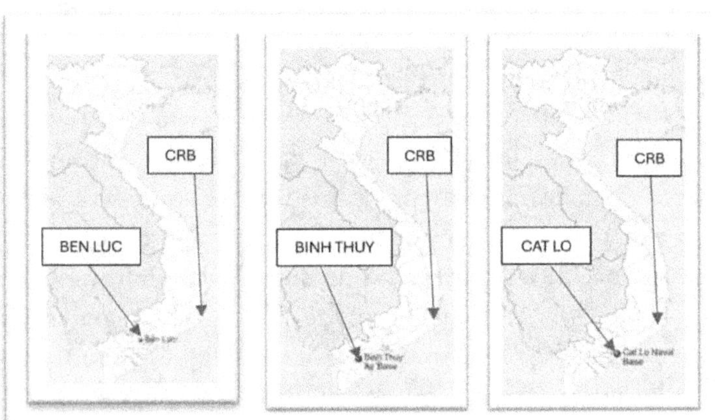

BEN LUC / BINH THUY / CAT LO

For a better understanding of what Danny's unit (CBMU-302) did during the Vietnam War, it would be easier to look at them like a 'public works department' similar to those found in most cities. However, CBMU-302 had more extreme responsibilities that included constructing and maintaining military bases throughout Vietnam. CBMU-302 built roads, runways, parking aprons, buildings, ammo bunkers, and most structures you would find on a military compound or base. CBMU-302 had multiple detachments (i.e., Detachment Alpha, Bravo, Charlie, Delta, etc.), which were located in 10 different locations that were spread across II, III, and IV Corps. Each detachment consisted of approximately 7 to 12 men led by a Chief Petty Officer (E-7) or a First-Class Petty Officer (E-6). Each detachment had at least one builder, steel worker, utilities man, construction electrician, equipment operator, and construction mechanic. Danny would have filled the construction mechanic position on the detachment team.

If Danny had been selected for the BEN LUC detachment, he would have been supporting the U.S. Navy's "GIANT SLINGSHOT" operations. BEN LUC was a U.S. Naval Support Base (from 1968-1970) located close to the junction of South Vietnam's Vam Co Tay and Vam Co Dong rivers. BEN LUC served as the main support base for the U.S. Navy operations during the Vietnam War. The main objective of the GIANT SLINGSHOT operations was to interdict communist infiltration from the nearby 'Parrot's Beak' border area of Cambodia using river patrol, riverine assault, and helicopter units. Parrot's Beak was a base and rest area for the People's Army of Vietnam (PAVN) and the Viet Cong, and one of the terminating points of the Ho Chi Minh Trail. Beginning in December of 1968, American Naval Forces deployed to forward staging positions along both rivers [15, 17]. The U.S. Navy sent the SEABEES in to set up advanced base defenses, fuel and ammunition storage, a helicopter pad, sleeping and messing facilities, and other essentials.

If Danny had been selected for the BINH THUY detachment, he would have been supporting the U.S. Navy's regional operations. BINH THUY, a U.S. Naval Support

Activity (from 1966-1972), was the hub of U.S. Naval Operations in that region. The site was chosen because of its location on the Bassac River, making BINH THUY ideal as a base from which U.S. Naval Forces could operate against Viet Cong supply traffic on surrounding waterways. To enhance support from BINH THUY to naval combat units throughout the delta, a great effort was made during 1968 to improve the airstrip at BINH THUY and to complete a 1,500-foot airstrip, hangars, aircraft repair shops, and berthing and messing facilities. Between 1969 and 1972, American naval personnel trained their Vietnamese counterparts and turned over to them river craft, equipment, and the installation, and then commenced the U.S. withdrawal [16].

If Danny got his wish and was selected for the CAT LO detachment, his preferred location, he would have been supporting the Naval Base and surveillance and river patrol operations. CAT LO was on the northern shore of South Vietnam's Cape Vung Tau, serving as a naval base and logistics base during the Vietnam War. U.S. Naval leaders chose the site for several reasons: It had ready access to the South China Sea, the river approaches to Saigon, and the Mekong Delta. Three distinct operational forces were based at CAT LO: Components of the Coastal Surveillance Force, the Saigon Minesweeping Force, and the River Patrol Force. CAT LO also acted as a staging and logistics base for the move deep into the Mekong Delta of the U.S. Navy's river force. To support the various combat forces at CAT LO, a CBMU-302 detachment would provide construction support to ship berthing, messing, supply, repair, transportation, communications, and other logistic facilities. The U.S. Navy selected CAT LO as the base for Swift Boat Squadron 1, Division 103, with 14 Swift Boats to be based there with berthing, messing, and basic repair facilities [18]. Construction Battalion Maintenance Unit 302 Detachment C was based at CAT LO during the base construction.

Sadly, Danny never got to go on a detachment. As I write this book, I can't help but wonder: What if? What if that assignment had started on November 1st, 1969? What if Danny had gotten his wish and joined the CAT LO Detachment with his friends? Would I still be writing about the loss of my Uncle Danny? Would he be alive today? Would my family and I have had the chance to know him better? Would he have married Helen, settled down in Brookfield, and raised a family? What if? Well, that detachment assignment came a little too late for Danny, because just days after he got this news about his selection for assignment to a CBMU-302 detachment, Danny's lifeless body would be found in the supply department's laydown area on base.

Chapter 9

What Happened on November 10th, 1969?

On November 10th, 1969, a new children's show, Sesame Street, premiered on television, captivating young viewers across the country. The airwaves were filled with chart-topping hits; 'Wedding Bell Blues' by the 5th Dimension dominated in the U.S., while the song 'Sugar-Sugar' by the Archies led the UK charts. Two days later, our family no longer cared about the new music on the radio or the latest children's shows on television. All of it was overshadowed by somber news reports on TV and in the headlines in the newspapers reporting the death of a young boy from Brookfield, Illinois. The local news stated that a U.S. Navy sailor had been killed in Cam Ranh Bay, Vietnam on November 10th, 1969. By the morning of November 12th, 1969, the city of Brookfield was stunned, another one of its sons had been lost to the Vietnam War. Like many local papers across America, the 'Brookfield Citizen' newspaper kept its community informed with world events, hometown news, store coupons, and an ever-growing list of young men who would never return home. On this cold November morning, the front page of the local papers carried yet another name, CMCN Daniel Joseph Davidenko, another son of Brookfield lost to a distant war. Danny's name was now and forever etched into the town's collective memory.

Newspaper Article #1 (Unknown Publication / November 12th, 1969)

BROOKFIELD GI IS KILLED AT CAM RANH BAY - The Defense Department announced yesterday the death of a Brookfield man in Vietnam. He is U.S. Navy Construction Mechanic Daniel J. Davidenko, 18, son of Mr. and Mrs. Elmer Davidenko, 9527 Congress Park St., Brookfield. Davidenko, a member of Construction Battalion 302, was killed on November 10th at Cam Ranh Bay. He attended Lyons Township High School, La Grange, before enlisting in the U.S. Navy in Dec. 1968. Davidenko had been in Vietnam for three months before he was killed. Besides his parents, he is survived by four brothers, James Nichols, 26, Raymond, 24, Charles, 20, and Albert, 20; and three sisters, Mrs. Joan Stanko, Mrs. Jean Roberts, and Mrs. Jan Stockdale. Services for Davidenko will be held today and tomorrow at 6:00 PM in the chapel at 3452 Grand Ave., Brookfield.

Author's Note: Our family saved this paper-clipped article. It is in poor condition, and it is unknown when or by which newspaper it was originally printed.

Newspaper Article #2 (TIMES Publication / November 12th, 1969)

Danny Davidenko, 18, of Brookfield, killed in Vietnam - Danny Davidenko, 18, son of Mr. and Mrs. Elmer Davidenko, 9527 Congress Park, became the third Brookfield boy to lose his life in the defense of Vietnam when he was killed Sunday, November 9th, 1969, at Cam Ranh Bay. He would have been 19 next month. Danny, a member of the SEABEEs, was spending six months on shore patrol and had been in Vietnam one day less than three months. He had been in the service less than a year, enlisted in the U.S. Navy on December 16th, 1968. A resident of Brookfield for 14 years, he had graduated from St. Barbara's elementary school and attended Lyons Township High School before going to work at Knight's body shop in Lyons. He has two brothers stationed in Korea and has just received his orders to Vietnam. All three were in the U.S. Army, and ***Mrs. Davidenko told Tuesday evening, after news of her son's death had been reported by telegram and a visit by U.S. Navy personnel, how relieved she had been that he had joined the U.S. Navy because she thought he would "be safe" there. Ironically, he had been in more danger than his U.S. Army brothers. "I guess it was just meant to be," sighed Mrs. Davidenko in her grief.*** **(Reported by The Times, an Enterprise Publication- no date/location).**

Author's Note: Our family saved this paper-clipped article. The reporter of this article did not report Mary's story correctly when it came to the list of sons in the U.S. Army. Mary had three boys in the U.S. Army: Ray and Chuck were stationed in Korea, and Alby was stationed in Japan. The dates and descriptions are inaccurate as well, but it was published by the newspaper without confirming the accuracy of the information.

Newspaper Article #3 (Unknown Publication / November 12th, 1969)

Area Sailor Is Killed in Vietnam – U.S. Navy construction mechanic Daniel J. Davidenko, 18, son of Mr. and Mrs. Elmer Davidenko, 9527 Congress Park, Brookfield, was killed November 10th in Cam Ranh Bay, Vietnam. Funeral services will be held today at the Paloucek Funeral Home, 3452 Grand Brookfield. A requiem mass will be offered at 10:00 AM at St. Barbara's Catholic Church. Interment will be at St. Mary's Cemetery. Mr. Davidenko, a member of construction battalion 302, enlisted in the U.S. Navy last December. He attended Lyons Township High School. He is survived by four brothers, all also in the armed Services. James Nichols, 26; Raymond, 24; Charles, 20; Albert, 20; three sisters. Mrs. Joan Stanko, Mrs. Jean Roberts, and Mrs. Janice Stockdale, in addition to his parents.

Author's Note: These three newspaper articles were carefully preserved within our family's memorabilia. Unfortunately, the name of the newspaper that published two of the articles is unknown.

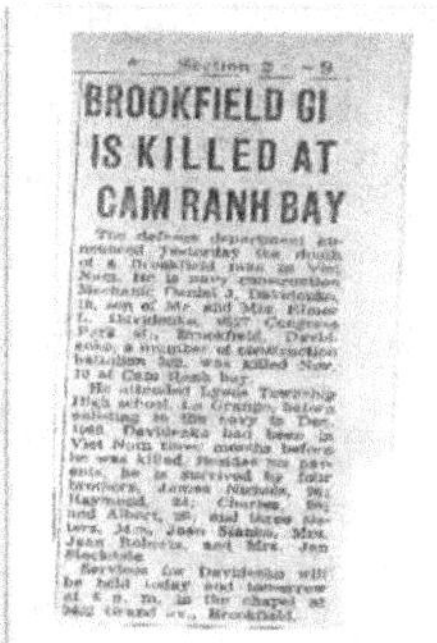

BROOKFIELD GI IS KILLED AT CAM RANH BAY

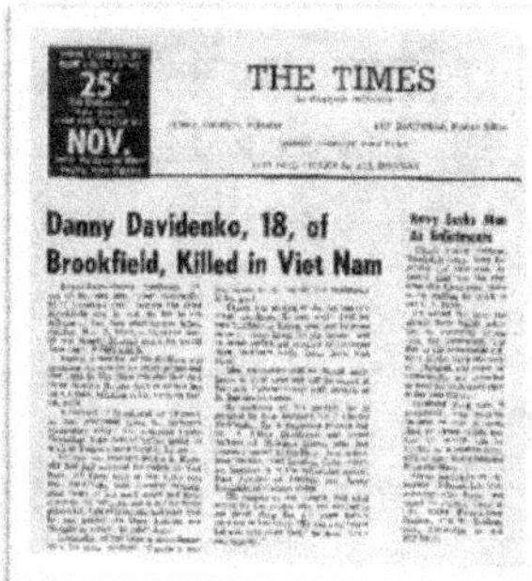

THE TIMES

25¢

NOV.

Danny Davidenko, 18, of Brookfield, Killed in Viet Nam

Area Sailor Is Killed In Vietnam

News Article #1, News Article #2, & New Article #3

Multiple articles were printed in the various local community newspapers announcing Danny's death in Vietnam and providing information on the time, date, and location of the funeral service. In Brookfield, they held a Mass for Danny at Saint Barbara's Catholic Church, and the Interment was held in Saint Mary's Cemetery. It took extensive research to find documents and personnel from CBMU-302 to help me piece together the events that took place in early November 1969. I personally interviewed numerous CBMU-302 members to learn more about the events that took place between the 7th and 11th of November 1969. The most critical interviews were held with LT Paul Jacobs (Executive Officer), LT Dennis Casey (Base Chaplin / Catholic Priest), LTJG William Underwood (Supply Officer), ENS Dave Kelly (Administrative Officer), CM3 Ted Mottashed (Unit Member and friend of Howard Decker), CM3 Sterling Ramsey (Unit Member), CM3 Ken Wilson (Unit Member), and CM3 Tom Vandenburg (Unit Member).

In addition to these CBMU-302 members, I reached out to over 100 other team members through emails and phone calls, hoping to gather their stories and memories of that fateful day. During these phone calls, several members hung up on me, presumably assuming that my phone call was from a telemarketer or an automated spam call center. Many other members that I reached out to couldn't recall the events from over fifty years ago. As for the emails, I received a handful of responses; however, I never heard back from more than forty recipients (most likely the email addresses were old, or they were all deceased). Of the team members who have passed away, I had the opportunity to speak with a few surviving wives, hoping they could recall any stories their husbands had shared about their time with CBMU-302. However, only one wife, Mrs. Lee Decker, the widow of CMCN Howard Decker (Danny's roommate in Vietnam) recalled that Howard often spoke of Danny's death. She shared that whenever he visited the Vietnam Memorial Wall, Howard would always touch Danny's name. Lee also sent me a few photos from Howard's Vietnam scrapbook, offering a glimpse into their shared memories. She also shared the story of Danny's death, as it was told to her by Howard.

The main source of information in this chapter came from LT Jacobs, Danny's executive officer. LT Jacobs was assigned as the sole Judge Advocate General (JAG) investigation officer into the death of CMCN Dainel Joseph Davidenko. Through numerous phone conversations, letters, and emails, LT Jacobs relied on his faded memory to help me reconstruct the events that unfolded at the Cam Ranh Bay base during those first few weeks of November 1969. It was evident to me, that LT Jacobs made a sincere effort to accurately recall all events that took place before, during, and after conducting his JAG investigation. In addition to my research, interviews, and correspondence with family, friends, and CBMU-302 members, I submitted multiple Freedom of Information Act (FOIA) requests to obtain all of Danny's military records, including the JAG investigation, Naval Investigative Service (NIS) investigation, Military Police (MP) reports, and the official autopsy report. I received some of Danny's military records from the National Archives; however, no NIS investigative report or the official autopsy was included. Despite numerous calls to the Naval Criminal Investigative Service (NCIS) Norfolk, JAG offices in Washington, D.C., the

Mid-Atlantic Region offices, and the Armed Forces Medical Examiner at Dover AFB, I have been unsuccessful in obtaining these documents. Therefore, the following information is based on my personal interviews, email conversations with LT Jacobs, Danny's friends, family, and former CBMU-302 members, as well as the limited documents I received through my FOIA requests. If and when I receive the official JAG investigation, NIS investigation, MP reports, and the official autopsy report, I will assess these findings and report on them in a separate chapter of this book.

November 7th and 8th, 1969 (Friday / Saturday), CBMU-302 Compound at CRB

Based on LT Jacobs' limited recollection of his JAG investigation and his investigation interviews with CBMU-302 personnel, he stated that, *"no one remembered seeing Danny after work on Friday or throughout the weekend."* For Danny, Friday marked the end of the workweek, and according to his last letter home, he was filled with excitement and positive emotions. He had just received word of his assignment to one of CBMU-302's detachments stationed throughout Vietnam. Based on Danny's letter, we know that earlier that week, he had been informed of his assignment for a new detachment and was most likely instructed to begin preparations for his departure. Once he received the news, Danny likely started immediately on the necessary tasks for his upcoming detachment. These preparations would have included finishing projects in the auto shop, getting any required medical shots or checkups, having gear issued for the detachment, and packing up his belongings while cleaning out his barracks room. In addition to packing, Danny would have needed to secure and store any personal items he wasn't authorized to bring on detachment. The one person who could have confirmed his preparations was his roommate, Howard Decker. Unfortunately, Howard passed away before I had the chance to interview him.

As Danny focused on his new assignment, he likely began preparations during the workweek and continued over the following weekend. It's hard to believe that no one remembers seeing him that weekend (as stated by LT Jacobs). Danny was ecstatic about his new orders, and it's likely he celebrated with a few drinks at the enlisted club on base. Based on my interview with LT Jacobs, he couldn't remember if anyone had seen Danny at all that weekend. According to LT Bakers' (Medical Officer) medical report, the time of Danny's death was estimated at approximately 0500 (5:00 AM) on Monday, November 10th, 1969. This estimated time of death was thirteen hours before Danny's body was found by LTJG Underwood (CBMU-302 Supply Officer) at approximately 1800 hours (6:00 PM). The time of death assessment was based on the level of postmortem issues observed by LT Baker that he recorded in his certificate of death report. In the summary block of the certificate of death, LT Baker pronounced Danny dead at 1815 hours (6:15 PM). This was his preliminary time of death reporting at the scene, which was approximately 15 minutes after Danny's body was discovered.

The official time of death will be determined during the official autopsy. What remains unclear is the exact date and time of Danny's official autopsy and where the autopsy was performed (In Vietnam or at Dover AFB). There is a significant gap in my data between Thursday, November 13th, 1969, when Danny's body was present at a memorial service at the base chapel in CRB Vietnam, and Monday, November 17th, 1969, when he finally arrived home aboard a plane from Dover AFB. Based on the postmortem details identified in the death report, Danny had been dead for at least 13 hours, which suggests that he was shot in the early morning hours of Monday, November 10th, 1969.

If LT Jacobs' memory is correct, how is it that during this entire weekend, no one remembers seeing Danny near his barracks room, the chow hall, the enlisted club, or even on the ball field, where the whole command was engaged in a softball game, enjoying some R&R (Rest and Relaxation)? If LT Jacobs' memory of events is accurate, no one reported seeing Danny between 1800 (6:00 PM) on Friday and when his body was found around 1800 (6:00 PM) on Monday evening. If Danny was shot and killed around 0500 (5:00 AM) Monday morning, surely someone should have reported hearing gunshot(s) early that morning? If the recorded time of death is accurate, multiple people, especially those personnel on duty, should have heard the gunshot(s) that Monday morning. As an intelligence officer with an education in criminal investigation and crime scene analysis, I had to ask myself, how could Danny go unseen or missing for such a long period of time? At a minimum, you would think that Danny was at least seen at the chow hall over the weekend, which serves at least four meals a day (breakfast, lunch, dinner, and MIDRATS (MIDRATS is the fourth meal prepared each evening). Additionally, Danny was excited about his new detachment assignment. Why wasn't he out at the enlisted club, having a few beers with his peers, or bragging in the barracks about his new orders?

Something about this JAG investigation report, as verbally relayed to me by LT Jacobs, doesn't make sense. How is it possible that not one member of a command comprised of several hundred personnel failed to see Danny anywhere on base over a 72-hour period? How many command members did LT Jacobs actually interview? To fully understand how these conclusions were reached, I need to obtain a copy of the official JAG investigation to see who was interviewed, how many personnel were questioned, and how it was determined that no one remembered seeing Danny that weekend. Once I have the JAG investigation, I will be able to review it with LT Jacobs and help him to recall from his memory the events that unfolded during this tragic event. His completed JAG investigation will be critical to helping me reconstruct a timeline of events.

In the meantime, we have to ask the question: What factors could have explained Danny's disappearance? Could he have established some new friends at the U.S. Navy Exchange, maybe some Army, Marine, or Air Force guys, and decided to travel with them to their respective bases to hang out for the weekend? Is that possible? Knowing what events happened over 50 years ago and the fact that most unit members' memories have faded, I am concerned that we may never truly know Danny's whereabouts before his lifeless body was found on base. The only recorded documentation I have at this time is the medical officer's time of death assessment (0500 / 5:00 AM), which would suggest to investigators that Danny was killed on CBMU-302's base sometime early Monday morning before his death… **or does it?**

November 9th, 1969 (Sunday), CBMU-302 Compound

According to LT Jacobs (XO of CBMU-302), Sunday, November 9th, 1969, began like any other typical Sunday at Cam Ranh Bay. The construction battalion followed a Sunday work routine, meaning only personnel in the duty section would be working or on duty. The chaplain would have held church services that morning, and after church, the men would enjoy a day off, with many playing at the ballfield. With Sunday being largely a day off, the command would often use trucks to shuttle men back and forth to the U.S. Navy Exchange, so they could shop for what they needed or wanted. For Danny, Sunday should have been a restful day with no work scheduled unless he was assigned to the duty section. LT Jacobs mentioned during one of our interviews that if Danny had any duty responsibilities that weekend, his absence would have been noticed much sooner, especially if he hadn't shown up for duty. Based on this, we can conclude that Danny did not have weekend duty. Therefore, Sunday was likely a time for Danny to enjoy church services and have a relaxing day before his busy work week began on Monday morning. It's also reasonable to assume that over the weekend, Danny could have visited the Base Exchange, the Chow Hall, the Enlisted Club, attended Sunday church services, and ultimately, he should have been packing for is new duty assignment. So why didn't anyone see Danny? Perhaps LT Jacobs' recollection of events is not as precise as it once was; after all, it has been well over 50 years since Danny's death.

On base that Sunday evening, the Enlisted Club was likely filled with sailors enjoying a few drinks and hanging out with friends and coworkers. It's equally likely that the Officers' Club was filled with officers relaxing and unwinding. At both clubs, there would have been music playing for entertainment as the men talked, played cards, or watched a movie. It's probably safe to assume that no one in the clubs that night heard the reported single gunshot that rang out at approximately 2000 (8:00 PM) Sunday evening. That evening, the tower watch reported hearing a gunshot to his dispatcher, who should have documented the report in the official logbook. According to LT Jacobs, it seems the dispatcher didn't think much of the single gunshot report and assumed that the base armory, located just across the street from the compound, was testing a weapon. The dispatcher's lack of concern over the gunshot reported by the tower watch stander and his failure to log it were alarming. Standard operating procedures clearly require that it be recorded. LT Jacobs explained that the Gunner's Mate in charge of all base pistols and rifles had a habit of testing weapons by shooting them out the back door of the armory. The single gunshot heard by the tower watch that evening did not raise any particular alarm. There is no confirming evidence that the gunshot heard from the watch tower was one of the shots that took Danny's life.

With the armory close by, there could have been numerous gunshots, which were likely ignored or dismissed as the armory discharging ammo from jammed pistols and rifles. The timing of the reported gunshot at 2000 (8:00 PM) does not match the initial estimated time of death of 0500 (5:00 AM) provided by LT Baker. There is a nine-hour discrepancy between the shot heard and the recorded time of death, which suggests that the shot may have simply been from the armory. Based on this basic timing analysis, the investigator (LT Jacobs) should have interviewed the armory personnel on duty that evening to determine if they in fact were clearing jammed weapons around 2000 (8:00 PM). Securing statements from armory personnel should have been a priority. Their testimony would have established whether the shot reported that evening came from the armory itself or from a different location on or off base.

By 2200 (10:00 PM) that evening, the clubs had closed, and the men returned to their barracks to get some sleep before the 0600 (6:00 AM) reveille on Monday morning. With the clubs closed and the men back in their barracks, Danny's roommate, Howard Decker, should have noticed if Danny was in his bed sleeping or sitting at his desk. The next morning, before muster, as Howard was getting out of bed, he should have noticed whether Danny was or was not in his bed. Why didn't Howard and the other sailors in the barracks see Danny heading to the Head (U.S. Navy restroom/shower) to shave, shower, and prepare for the day? Why didn't anyone from CBMU-302 notice whether Danny was or was not at morning chow? None of this makes any sense, and it leads me to believe that maybe LT Jacobs does not accurately remember or recall the details of his JAG investigation. Again, I am not pointing fingers at LT Jacobs, hopefully the answer to these questions will be revealed in his JAG investigation. I know I would struggle to recall details and memories from any event that happened more than 50 years ago. One fact we do know is that the first time Danny was "reported" missing was when he didn't show up for morning quarters at 0700 (7:00 AM). The answers to all of the questions mentioned above should be in LT Jacobs' JAG investigation. However, when I started the development of this chapter, I had not received a copy of his official JAG investigation after submitting multiple FOIA requests.

November 10th, 1969 (Monday), CBMU-302 Morning Quarters

On November 10th, 1969, like clockwork, the daily routine began with reveille at 0600 (6:00 AM). The men would have about an hour to grab a shower, head to the chow hall for breakfast, and then report to morning quarters next to the headquarters (HQ) building. By 0700 (7:00 AM), all hands would fall into formation, standing at attention with their company commanders (CC). During morning formation, each company conducts a roll call to ensure everyone is present before the Commanding Officer (CO), LCDR Harrell, and the Executive Officer (XO), LT Jacobs, come out from the HQ building to pass pertinent information and to issue the orders of the day. Promptly at 0730 (7:30 AM), the XO would read the plan-of-the-day, followed by the CO providing his comments and closing remarks. Once all pertinent information was passed, the companies would be dismissed from quarters to begin the workday. The workday would continue until 1200 (Noon), when the men would head to the chow hall for lunch. At 1300 (1:00 PM), lunch would end, and it was back to work until 1730 (5:30 PM), when the command would stop all work for the day. Afterward, the men would go to the chow hall for supper. After supper, the men had on-base liberty to do as they pleased until the next morning's quarters.

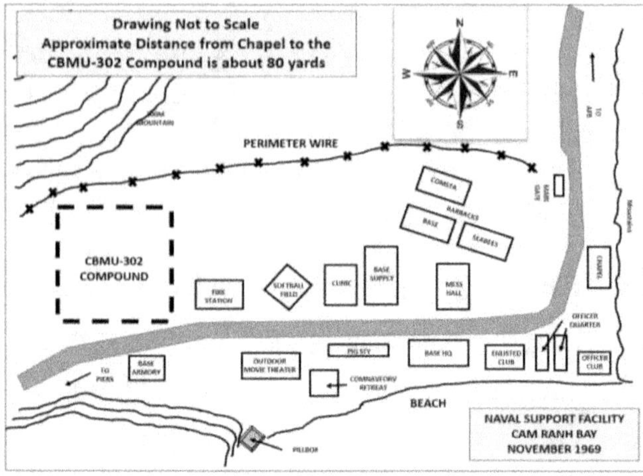

Overview of CRB Base Map

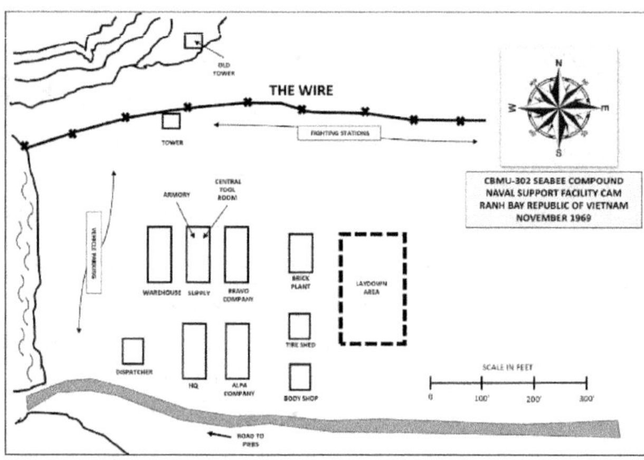

Overview of CBMU-302 Compound Map

During Monday morning quarters on November 10th, 1969, Danny did not show up to say present when his name was called out during morning roll call. It was not uncommon for the occasional sailor to oversleep and miss morning quarters or to be late for work. Not being at your assigned place of duty, especially at morning quarters, would get you into trouble with your leading petty officers (LPOs), and these absences were not taken lightly by the Company Commanders (CC). In one of

Danny's letters to his parents, he mentions getting into trouble for oversleeping and being late for work. Given this, it's likely that missing morning quarters on this particular Monday wasn't Danny's first time. Presumably, there was no sense of worry or urgency from Danny's superiors, especially if it had happened once or twice before. At a minimum, Danny's company commander (CC) or his leading petty officer (LPO), should have questioned CMCN Howard Decker (Danny's roommate), as to Danny's whereabouts.

With Danny not being present, his Company Commander immediately sent a sailor to the barracks to see if he was still in his bed sleeping or if he was feeling sick. While the sailor went looking for Danny, the Company Commander started asking the men if anyone had seen him that morning at the barracks or chow hall. According to LT Jacobs, none of his barracks mates or other company personnel remembered seeing him that morning. When the sailor returned to the Company Commander, he reported that he had searched the barracks rooms and the head (restroom/shower), with no luck in finding Danny. The Company Commander was not happy about the situation, as it was his responsibility to know where his men were at all times. By now, it was approximately 0800 (8:00 AM), and Danny was still nowhere to be found. The Company Commander, not yet thinking there was an immediate cause for alarm, reported to the CO and XO that one of his men was missing from muster and currently unlocated.

According to LT Jacobs, since no one could remember the last time they had seen Danny, the CO instructed the Company Commander to assemble a six-man search party. He ordered them to conduct a thorough search of the CBMU-302 compound and the rest of the CRB base. In addition to the six-man search party, the Command Chief Master at Arms also joined the search. During my interview, LT Jacobs stated that the search team was instructed to cover every structure, the hill, the beach, and even an old WWII pillbox overlooking the beach. The search continued with no results, and by 1730 (5:30 PM), the end of the workday, the search was called off. The entire command grew increasingly concerned that Danny was still missing. On the CBMU-302 compound, the supply department had a 'supply laydown area' filled with crates and building supplies. Located to the east of ALFA Company's main building, the laydown area contained multiple parallel rows of crates and supplies, stacked approximately nine feet high and spaced about five feet apart. Each row was roughly twenty five feet long. According to statements made by LT Jacobs and LTJG Underwood, approximately thirty minutes after the search ended, LTJG Underwood went over to his supply laydown area to conduct an inventory of newly arriving supplies, and that's when Danny's body was discovered between two rows of crates. LT Jacobs recalls that LTJG Underwood found Danny's body, and he immediately notified LCDR Harrell (CO), LT Jacobs (XO), and the command's Administrative Officer (AO), ENS Kelly.

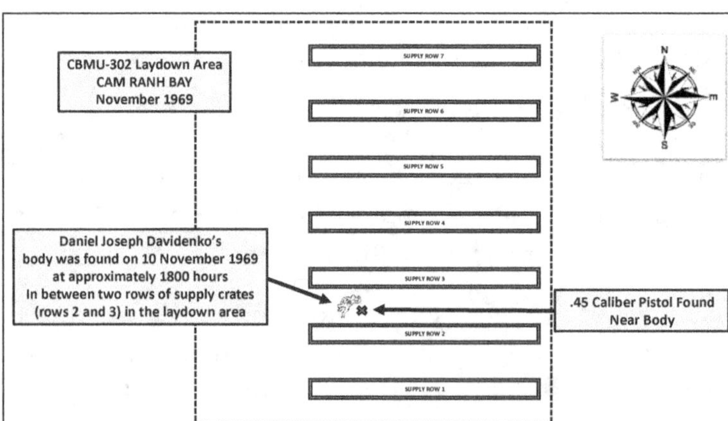

CBMU-302 Supply Laydown Area

For most of the men of CBMU-302, Monday, November 10th, 1969, was like any other workday, filled with routine military duties; however, this was not the case for Danny's company and the assembled search party. By the end of the day on Monday, everything came to a screeching halt when LTJG Underwood discovered the body of CMCN Daniel Joseph Davidenko between some crates in the command's supply laydown area. What should have been a typical evening took a drastic turn when LTJG Underwood reported finding the body of the command's missing sailor. The mood and tempo of the base shifted immediately. The situation created a lot of confusion; however, both LTJG Underwood and ENS Kelly remained calm enough to inform the chain of command and to call for medical staff to provide emergency care, despite the fact that those present knew Danny appeared to be dead. Following orders, the Medical Officer and a U.S. Navy Corpsman were dispatched to the scene to render first aid assistance. Upon their arrival, the medical staff quickly determined that Danny was dead, based on the lack of a pulse and the color of his skin. The response shifted from first aid to preparing a medical death report. Once Danny was pronounced dead, the command promptly notified CBMU-302's higher headquarters via secure message traffic. The details regarding the handling of Danny's body by the medical staff remain unclear because LT Jacobs left the scene to report the death to higher headquarters. It wasn't until after Danny's body was removed that LT Jacob was informed that he would be the sole JAG investigator assigned to the case. As a result, he was not present to document these events. It is unknown whether anyone in the command had knowledge of how to conduct a death investigation. However, what is clear is that no one in the command treated Danny's death or the location of the scene as a potential crime scene that needed to be preserved.

Danny's Body Found

CBMU-302's supply laydown area consisted of several rows of crates arranged in an East-West direction. According to LT Jacobs, Danny's body was discovered lying on his back between these crates, with his head facing east. He was only dressed in his standard military-issued green SEABEE uniform and his Navy "Boon-Docker" boots. His green military shirt was unbuttoned, and he was not wearing his standard military undershirt. Additionally, Danny was without his hat. These uniform discrepancies may seem trivial; however, all military personnel are required to wear their complete uniforms with hats at all times, unless the CO, for various reasons, relaxes the Navy's standard dress code. At the scene, approximately four to five feet to the east of Danny's body, there was a non-military issued pistol, possibly a Sig Sauer M11A1, a .45 caliber semi-automatic. LT Jacobs referred to the pistol as an "orphan," a term he used to describe weapons that were not issued by the U.S. Army, Air Force, Marine Corps, or Navy. The XO also recalled that Danny was clutching a large wooden crucifix in his right hand, which was pressed against his chest, covering his embroidered name tag. Danny's stiff hand had to be moved in order to read his name tag. It wasn't until I learned about the crucifix from LT Jacobs that I realized I had previously seen a crucifix in a photo Danny had sent home to his mother. In that photo, Danny was sitting at his desk, and you can clearly see a crucifix hanging from the end of his M-16, which is propped up on his desk inside his barracks room. In that black and white photo, the crucifix appears to be about 6 to 8 inches long, not the typical size carried by most service members. It remains unclear how Danny obtained such a large crucifix. Did he bring it from home? Was it something he bought at the Navy Exchange, or did his mother send it to him via the mail? An even more important question is whether Danny carried the crucifix around with him, or if it stayed displayed in his barracks room. I asked LT Jacob if he found it odd that Danny had this crucifix clutched in his right hand, pressed up against his chest, and covering his name tag? He didn't really give it much thought. Something about the way the crucifix was positioned against his chest, covering his name tag, raises investigative questions, particularly when evidence is being evaluated to determine the manner of death (MOD).

As part of my research, I attempted to locate LT Baker, a Medical Officer who may have been assigned to CBMU-302. After reviewing the historical records of the unit, I found no records indicating that a medical officer by the name of LT Baker was assigned to CBMU-302, nor did I come across any officers with this last name. Based on this, I must conclude that LT Baker was a U.S. Navy doctor assigned to the Air Force Hospital at Cam Ranh Bay. LT Baker's examination of Danny's body revealed a 1.5 centimeter penetrating wound in the mid-sternum (entrance wound) and a 2 centimeter wound in the mid-back (exit wound). Further examinations showed acrocyanosis, a bluish or purple discoloration of the hands and feet. Danny's eyes were dilated with fixed pupils, and he showed no signs of respiratory or cardiac function. Additionally, there was segmented hemostasis of the retinal veins, indicating a stoppage of blood flow. At the scene, LT Baker pronounced Danny dead at 1815 (6:15 PM) on November 10th, 1969. By the time the on-site examination began, Father Denis Casey (LT, USNR) arrived and administered Last Rites to Danny at 1830 (6:30 PM). Since LT Jacobs did not remain on scene and no one that I interviewed recalled the details of how Danny's body was removed, I must assume that Danny was placed in a base ambulance and transported to the hospital at Cam Ranh Bay. According to LT Jacobs, after Danny's body was removed, the command had no further knowledge of where his body was located. As it turns out, Danny's body was taken to the Air Force Hospital at CRB. Danny's body was taken into the medical examiner's room, where the medical staff conducted a preliminary death examination and completed the official death report. It is important to note that this examination was not the official autopsy, nor did this examination include any form of forensic analysis to determine Danny's manner of death (MOD). Three days after Danny's death, Father Casey requested that Danny's body be present during the Mass held on Thursday, November 13th, 1969, at the base chapel. At the conclusion of Danny's memorial service, he was honored with an 18-gun salute before his body was returned to the hospital morgue for processing and transportation back to the United States.

LT Jacobs, the Executive Officer of CBMU-302, was given the overwhelming task of serving as the sole investigator into the death of CMCN Daniel Joseph Davidenko (Danny). According to LT Jacobs' faded memory, Danny was last seen at work on Friday at approximately 1730 (5:30 PM). He was not seen again until his body was discovered the following Monday evening, around 1800 (6:00 PM). This does not necessarily mean Danny wasn't on base during that time; people simply couldn't recall seeing him. LT Jacobs approached the investigation with sincerity and dedication, despite having no prior experience in death investigations or JAG procedures. He shouldered this significant responsibility without any formal training. What remains unclear is whether an official NIS investigation into Danny's death was ever conducted. As of this chapter's writing, I have been unable to locate the NIS investigation (if one was completed), the Military Police investigation (if one exists), or the official autopsy report and photographs. What perplexes me is LT Jacobs' statement that it took him a full ten days to complete his command JAG investigation and to certify his findings. He also mentioned that he personally certified the final report. It is important to note the time and date that LT Jacob's investigation was completed. LT Jacob stated that it took him ten days to finish his investigation. That would mean that the JAG investigation was completed on November 20th, 1969, coincidentally the same day Danny was laid to rest at Saint Mary's Cemetery in Illinois.

After learning that LT Jacob's investigation was completed 10 days after Danny's death, I began to question the veracity of his JAG investigation. If it took LT Jacobs ten days to complete the investigation into Danny's death, how did the Commanding Officer of CBMU-302 (LCDR Harrell) determine within 24 hours of finding Danny's body that his death was the result of a "self-inflicted" gunshot wound to the chest? Based on the CO's letter, we can infer that the CO reached this premature conclusion regarding the manner of death (MOD), as self-inflicted, before the investigation was complete. Stating that the gunshot (s) were self-inflicted suggests that the manner of death could have been either suicide or an accident. The medical examiner's report only indicates the cause of death (COD) as an *"Injury to Sternal Region, Penetrating Gunshot Wound"* (gunshot wound to the chest). Nowhere in the medical report does it suggest or conclude the manner of death. Therefore, we must ask: Why did the CO make this premature judgment about the manner of death? The formal letter from the CO to Mary and Elmer Davidenko, dated November 11th, 1969, approximately 24 hours after Danny's death, states that Danny shot himself. But how could the CO know the gunshot(s) wounds were self-inflicted? According to LT Jacobs, there were NO witnesses to the shooting. Did the CO assume the rumors about the 'Dear John' letter were true?

Even if he believed this rumor, it still doesn't justify using rumors or hearsay to determine the manner of death. In this letter, the commanding officer states that Danny was missing for 12 hours, which directly contradicts LT Jacobs' statement during my interview that Danny had not been seen for approximately 72 hours. An investigation must uncover evidence to support a conclusion about the manner of death. Based on all my research, there is no physical evidence or witnesses to support such a conclusion. In fact, most of the physical evidence was destroyed. According to LT Jacobs' recollection of events, he acknowledged that there were no witnesses to the shooting, yet he nonetheless concluded that Danny's death was accidental. Or at least that's what his faded memory recalls. Therefore, we must ask: How did LT Jacobs reach a conclusion about Danny's manner of death without any evidence or witnesses?

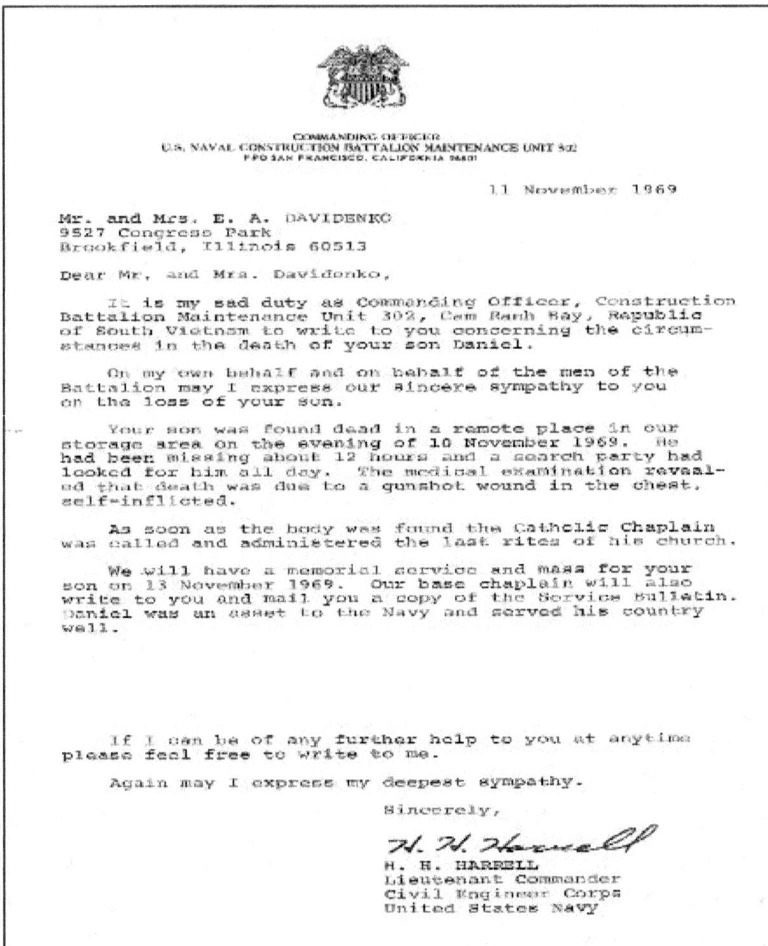

COMMANDING OFFICER
U.S. NAVAL CONSTRUCTION BATTALION MAINTENANCE UNIT 302
FPO SAN FRANCISCO, CALIFORNIA [illegible]

11 November 1969

Mr. and Mrs. E. A. DAVIDENKO
9527 Congress Park
Brookfield, Illinois 60513

Dear Mr. and Mrs. Davidonko,

It is my sad duty as Commanding Officer, Construction Battalion Maintenance Unit 302, Cam Ranh Bay, Republic of South Vietnam to write to you concerning the circumstances in the death of your son Daniel.

On my own behalf and on behalf of the men of the Battalion may I express our sincere sympathy to you on the loss of your son.

Your son was found dead in a remote place in our storage area on the evening of 10 November 1969. He had been missing about 12 hours and a search party had looked for him all day. The medical examination revealed that death was due to a gunshot wound in the chest, self-inflicted.

As soon as the body was found the Catholic Chaplain was called and administered the last rites of his church.

We will have a memorial service and mass for your son on 13 November 1969. Our base chaplain will also write to you and mail you a copy of the Service Bulletin. Daniel was an asset to the Navy and served his country well.

If I can be of any further help to you at anytime please feel free to write to me.

Again may I express my deepest sympathy.

Sincerely,

H. H. Harrell

H. H. HARRELL
Lieutenant Commander
Civil Engineer Corps
United States Navy

Official Letter Informing Mary and Elmer Davidenko of the Death of Their Son Danny

LCDR Harrell's Letter to Elmer and Mary November 11th, 1969 - Transcribed

"11 November 1969
Dear Mr. and Mrs. Davidenko,

It is my sad duty as Commanding Officer, Construction Battalion Maintenance Unit 302, Cam Ranh Bay, Republic of South Vietnam to write to you concerning the circumstances in the death of your son Daniel. On my own behalf and the behalf of the men of the Battalion, may I express our sincere sympathy to you on the loss of your son. Your son was found dead in a remote place in our storage area on the evening of 10 November 1969. He had been missing for about 12 hours, and a search party had looked for him all day. The medical examination revealed that death was due to a gunshot wound in the chest, **self-inflicted**. *As soon as the body was found the Catholic Chaplain was called and administered the last rights of his church. We will have a memorial service and mass for your son on 13 November 1969. Our base chaplain will also write to you and mail you a copy of the Service Bulletin. Daniel was an asset to the U.S. Navy and served his country well. If I can be of any further help to you at any time please feel free to write to me. Again, may I express my deepest sympathy. Sincerely, H.H. HARRELL, LT Commander, Civil Engineer Corps, USN."*

It is important to note that LT Baker's medical report did not specify how Danny was shot. He could not have made such a determination, as he was not present to witness the shooting. At a minimum, rendering that type of conclusion would have required certification as a forensic pathologist and access to detailed forensic evidence gathered during the investigation. Additionally, LT Baker's report makes no reference to Danny's wounds being "self-inflicted." Such a determination would fall outside the scope of his medical evaluation and would require specialized forensic pathology training. Because there were no witnesses to the shooting, no physical evidence collected, and no forensic photographs taken at the scene, any attempt to determine Danny's manner of death (MOD) would be nearly impossible. Therefore, you cannot say the gunshot was self-inflicted without supporting evidence or witnesses.

Somehow, the CBMU-302 Commanding Officer made this premature judgment without the required evidence and prior to the completion of the full investigation. How did the CO determine that the gunshots were self-inflicted? Based on what physical evidence? The only reliable way to confirm that the shots were self-inflicted would be eyewitness testimony and/or a thorough crime scene analysis, supported by a comprehensive autopsy performed by a forensic pathologist. The CO predetermines the manner of death nine days before the official JAG investigation is completed. In my professional military opinion, LCDR Harrell made a significant error in judgment, and his letter was detrimental to LT Jacobs' JAG investigation.

The CO's predetermination of the manner of death may have influenced the Naval Investigative Service (NIS) to forgo an official crime scene investigation. With NIS already stretched thin with counter-insurgency operations across Vietnam, if the CO of CBMU-302 informed them that Danny's death was a self-inflicted accident or suicide, they may have decided no further investigation was necessary. If the narrative mentioned above, about the CO and NIS, were true, it would represent a gross miscarriage of justice, because any non-combat death of a civilian or military member on a U.S. Military installation must be investigated by NIS (Today the NIS organization is currently called the Naval Criminal Investigative Service - NCIS). Did the

CO's letter play a role in NIS's decision not to conduct a formal death investigation? According to reports, NIS was focused on counter-insurgency operations, with all other investigations given second priority. In my subsequent research into the facts of this case, I uncovered a significant amount of potential evidence that went uncollected, unprocessed, and unpreserved.

I eventually received Danny's military records from the National Archives. When they arrived through the mail, I was extremely excited to finally review the official investigations and autopsy records. However, all I received from Danny's military record were administrative documents: 1) DD-373, Consent, Declaration of Parent or Legal Guardian, 2) Record of Discharge; 3) Release from Active Duty or Death; 4) U.S. Navy Occupation, Training, and Awards History; 5) History of Assignments; 6) Enlistment Contract; and (6) his initial medical and dental records. Among the documents obtained, Danny's medical records emerged as a critical source of information used to support my crime scene analysis and to question the veracity of the investigation. No documents pertaining to the official JAG investigation, the official NIS investigation, the official military police report, or the official autopsy were included. However, a key piece of evidence contained in Danny's medical records documented that he was right-handed (right-hand dominant), a fact that would provide a critical clue in determining his manner of death (MOD).

ENS Dave Kelly Interview (CBMU-302 Administrative Officer)

After years of searching for CBMU-302 personnel, I finally found another witness to the events leading up to the discovery of Danny's body in the supply laydown area. LT Jacobs mentioned ENS Dave Kelly as one of the unit members he recalled being present when Danny's body was found. I spent a great deal of time trying to locate ENS Kelly, but without success. However, once I found the CBMU-302 historical website, I was able to locate BUCS (SW) David W. Schill, a former member of the unit and the current CBMU-302 historian. Although Senior Chief Schill served honorably throughout Vietnam, he did not serve with Danny. Through numerous emails and conversations with David, he provided me with CBMU-302 historical links and the CBMU-302 alumni rosters. After finding ENS Kelly's name on the roster, I made several phone calls and sent numerous emails using the listed contact information, but my attempts to reach him were unsuccessful. Senior Chief Schill managed to locate ENS Kelly for me and informed him about my efforts in writing this book. As a result, ENS Kelly was somewhat prepared to listen and provide answers to my questions regarding Danny's death. ENS Dave Kelly served as the Administrative and Personnel Officer for CBMU-302, working directly under the Commanding Officer LCDR Harrell and Executive Officer LT Jacobs. I also used the unit roster to reach out to as many CBMU-302 members as possible. The following description of the events that follow was obtained from my interview with ENS Dave Kelly:

During our initial discussion and some informal dialogue about the content of this book, I asked ENS Kelly a few questions and requested that he share what he remembered from those fateful days in November 1969 when Danny went missing, and the command discovered his body in the supply laydown area. ENS Kelly explained that when Danny didn't show up for muster, the command went into a panic. As the personnel officer, he had a mountain of reports to handle because there was a sailor missing. ENS Kelly stated that, *"When Danny didn't show up, the whole day was spent scrambling to write reports."* After morning quarters, ENS Kelly remembers that the search parties were looking for Danny, but they were focused mainly on the barracks area, assuming he'd most likely be around there. ENS

Kelly also mentioned the beach that was adjacent to the base, with its secluded spots and wooded areas, all providing possible areas to hide. However, he didn't go as far as to say whether they had searched those beach areas. I asked him if the supply laydown area was searched, but ENS Kelly couldn't recall if that area was searched or not. ENS Kelly did state that, *"no one was supposed to be in that storage area… that was like… if you didn't have business back there… You weren't supposed to be there… so umm…that was someplace he wasn't supposed to be... and if he had a gun…he had something he wasn't supposed to have."* Based on that response, I redirected my question about weapons issued to the men and if they were issued .45 caliber pistols along with their M-16 rifle when on duty. ENS Kelly responded, *"No, and that all the weapons were locked up in the armory and that the enlisted guys had zero access to weapons or ammunition, or at least they weren't supposed to. However, when general quarters was sounded over the base speakers, all the sailors in the barracks would jump into a truck and head straight to LTJG Underwood's armory. There, they'd be issued helmets, flak jackets, rifles, and ammunition. From the armory, the men would run to their assigned foxholes and set up defensive positions."* Knowing that the men of CBMU-302 were only issued M-16 rifles and not .45 caliber pistols lends more credibility to the idea that the pistol found near Danny's body was an 'orphan', a non-military-issued weapon, or more importantly, not a CBMU-302 issued weapon.

I asked ENS Kelly to walk me through the events, as he remembered them, from the day Danny's body was discovered. He explained that the entire day was chaotic as they couldn't locate Danny. Later that same day, around 1700 (5:00 PM), ENS Kelly walked down to the officers' mess for supper. Upon returning to his office, he passed the ball field on base, where a baseball game was in progress. As he walked by, ENS Kelly saw LTJG Underwood standing by the ball field, watching the game with a beer in his hand and one leg resting on a short wooden fence. He noticed that Underwood appeared deep in thought.

ENS Kelly waved to him, and immediately, LTJG Underwood called out for him to come over. When ENS Kelly approached, Underwood said, *"I need your help."* ENS Kelly asked what was going on, and Underwood said, *"Come with me".* They both got into Underwood's jeep and drove toward the supply laydown area. As they were driving, LTJG Underwood asked, *"What's the name of that missing guy?"* ENS Kelly replied, *"Davidenko."* Underwood then said, *"I think I might have seen something,"* as they continued toward the supply area. ENS Kelly remembers the supply laydown area being a large lumber yard with pallets of building supplies. ENS Kelly vaguely remembers what LTJG Underwood said exactly, but it was something like, *"I saw out of the corner of my eye… that somebody was back there* (meaning the supply laydown area). *I thought he was probably just taking a nap… I didn't think anything of it… But then, when I* (Underwood) *was back at the ball game, I started thinking about it… You know, thinking, I'd better go check to see what the fuck that was that I saw… or who the hell was back there… we got a missing guy, don't we?"* ENS Kelly felt that LTJG Underwood didn't want to check it out alone, so he asked him to come with him. Since it was after dinner, and ENS Kelly stated that, *"he usually ate around 1700* (5:00 PM)*", and I believed we went to the supply area sometime after that."* Assuming ENS Kelly went to chow around 1700 (5:00 PM), and it took about 45-50 minutes to eat and walk back to the ball field, it's reasonable to conclude that they arrived at the supply laydown area at approximately 1800 (6:00 PM), which aligns with other reported times. ENS Kelly took a pause and stated, *"Now, please distinguish between stuff I heard, which is absolutely hearsay and not admissible in court, from stuff I remember that I had seen. So, with that disclaimer, here is what I recall hearing and seeing."*

Authors Note: ENS Kelly is now a practicing attorney, so his decision to include this disclaimer makes perfect sense. With my limited law experience in legal matters, I can honestly say that I would have made the same disclaimer. On a supporting note, after several years of correspondence with LT Jacobs, he eventually asked me if I had plans to pursue a wrongful death investigation in connection with my uncle's death. At that time, I assured him that I had no intention of filing a wrongful death investigation and that I just wanted answers to bring closure for my family. However, if I uncover information that could definitively point to foul play or a cover-up, then I would consider filing a wrongful death claim and request that NCIS conduct a new criminal investigation into Danny's death.

According to ENS Kelly, he and LTJG Underwood pulled the jeep up to the supply laydown area, and both walked to the spot where Underwood thought he might have seen someone sleeping. ENS Kelly recalls seeing a figure leaning back over a pallet, as if the person had just leaned back and fallen asleep. As they approached the suspected sleeping sailor, LTJG Underwood called out to him and then snapped his fingers in front of the sailor's face. There was no response, no movement, no signs of life. As ENS Kelly moved closer, he couldn't see the sailor's name tag. On a U.S. Navy SEABEE uniform, the service member's name tag is positioned above the right shirt pocket, and the word "SEABEES" is displayed above the left shirt pocket. ENS Kelly explained that he couldn't see the name tag because the sailor was clutching an object in his right hand, which was positioned over the name tag. I asked ENS Kelly if he remembered what was in the sailor's hand that blocked his name tag. He paused for a moment, searching his memory, and said to himself, *"What was it that I heard was in his hand?"* ENS Kelly did not recall the object at first, until I mentioned that others had reported seeing a 6–8 inch long wooden cross (crucifix) held firmly in Danny's right hand, obscuring his name tag. ENS Kelly replied, *"Yes, that's right."* I remember standing there in front of the body, and I could see that he was clutching something to his chest. Initially, I couldn't make out what was in his right hand because his shirt was open, obscuring the object.

I started asking ENS Kelly about LT Jacobs' JAG investigation, and before I could finish my next question about the scene, ENS Kelly said, *"You know, we didn't have a camera… and that's something... I don't know what they were thinking… but I think we were all just amateurs… nobody knew how to investigate a death* (crime scene)*."*

ENS Kelly goes on to describe the actions of LTJG Underwood. Underwood leaned over the body, grabbed the sailor's forearm, and pulled it back to check the name tag. At that point, it became clear to both Underwood and Kelly that the sailor was in fact clutching a cross in his right hand. Underwood noted that Danny had a firm grip on the cross and observed that he was already stiff, indicating that he had been dead for a while. After lifting Danny's arm, Underwood said, *"Yep, Davidenko… that's our guy."* ENS Kelly and LTJG Underwood exchanged a look, silent but understanding; both knew they had found the command's missing sailor. ENS Kelly recalled getting back in the jeep and heading to LTJG Underwood's office, where they began making phone calls. Though he didn't remember the exact order of the calls made by Underwood, Kelly recalled notifying the CO and XO, and that LTJG Underwood received guidance on who else to contact. Additional calls were made to Medical, Security, and the Chaplain's office.

To ENS Kelly, everything was happening quickly, and it seemed like everyone showed up almost instantly. Within minutes, the CO, XO, and security personnel were at the scene. Kelly believed the security personnel were from the CRB base, though he didn't recall seeing MP arm sleeves. In his opinion, the security personnel didn't seem well-trained for this type of event, though he noted that was just his personal assessment. Everyone stood around the body, and it was clear to all that Danny's right hand was firmly clutching the cross. From an investigative standpoint, ENS Kelly's account that numerous personnel were present and standing around the body demonstrates that the scene was not sealed or controlled in accordance with standard investigative procedures, thereby jeopardizing the integrity of any evidence relevant to determining the manner of death (MOD).

As ENS Kelly looked around at the growing crowd, he saw expressions of confusion, as if everyone was thinking, 'What the hell happened here?' The large wooden crucifix was clearly visible as the first responders moved Danny's right hand. Additionally, ENS Kelly and all the onlookers could see a small amount of blood on his shirt and chest. When I asked about the location of the gun, ENS Kelly stated that *"he did not see a gun on the ground near Danny's body."* I stopped him immediately and said, "Wait a minute, now that contradicts every story I've heard about a gun lying just a few feet from Danny's body." ENS Kelly immediately responded, *"I don't remember seeing a gun while I was there, but I remember hearing about a gun being found next to the body."* I suggested that maybe LTJG Underwood or someone else picked up the gun to keep it safe. ENS Kelly replied, *"Well, what I heard was that it took them a while to find the gun."*

Authors Note: It is worth noting that if a crime scene investigator had heard such a statement from any member of the unit, it would have immediately raised red flags, indicating that the scene's integrity had been compromised and that evidence may have been disturbed prior to the investigation. Such circumstances would suggest that the integrity of the investigation itself was jeopardized and could also raise the possibility of foul play.

At one point, ENS Kelly described the scene like this: *"Ten minutes seemed like forever. It was just riveting, with a lot of adrenaline pumping, everyone standing still, nowhere to go, and nothing to do. I just stood behind my superiors, watching and waiting for orders."* Returning to the question of the gun, ENS Kelly insisted he didn't recall seeing one. However, he added, *"Somebody might have found a gun while I was there, but I just don't remember. I do remember hearing later that a gun was found, several feet from the body, but I don't recall which direction."* He added, *"You know, those .45's have a hell of a kick."* ENS Kelly continued, *"I kept thinking, if he shot himself, the gun would have flown out of his hand. With that kind of recoil, it would have taken off like a rocket in one direction or another."*

I proceeded to ask ENS Kelly a few more direct investigative questions about the scene: Did he notice any blood splattered on the crates, blood on the ground, or blood on Danny's back when the body was moved? Was he there when they moved the body? ENS Kelly paused before responding with: *"Was I there when the body was moved…? That's a good question."* He hesitated, then added, *"I'm sure I didn't see any other blood, that I can answer pretty straightforward."* The fact that no one at the scene recalled seeing any additional blood on the crates, the ground, or Danny's back is deeply troubling. A 2 centimeter wound in the middle of Danny's back should have left a significant amount of blood in those areas. Yet not a single witness reported seeing any blood at the scene. The absence of blood should have led any investigator to consider the possibility that Danny was not killed at this location. Instead, it would strongly suggest (to a seasoned investigator) that his body was placed in the supply laydown area sometime after he had been killed elsewhere. The lack of blood at the scene must be investigated further to either corroborate or dismiss this possibility of Danny's death occurring at that location.

Continuing my interview with ENS Kelly, I turned my attention to questions about the JAG investigation. I mentioned how the JAG manual is to be used and how there is a small section of the manual that covers death investigations. I went on to ask him if he remembered seeing the JAG manual LT Jacobs used to assist him in his investigation. ENS Kelly replied, *"I don't think we had a JAG Manual."* I asked him if he remembered seeing any NIS investigators present at the scene. He stated that, *"the only investigation that he knew about was conducted by LT Jacobs."* How about the MPs? Did you see them conducting any type of investigation? ENS Kelly stated that, *"I don't remember any arm bands, they were just base security guys".* I went on talking about how this scene should have been treated and how an official investigation by NIS should have been conducted. In terms of the investigation, ENS Kelly replied, *"It seemed like they just wanted to get it cleaned up and get done with it internally, I mean, I just felt like the whole thing was kind of like an embarrassment and solving it as you would on a TV show...it just wasn't really the focus at all."*

Moving on, I began to ask him about the bullets and the entrance and exit wounds. Before I could get my next question out, ENS Kelly interrupted and stated, *"Here is something else I saw...I saw that bullet multiple times in Paul's* (LT Paul Jacobs) *desk drawer...because I needed to go in that drawer every once in a while to get his* (LT Jacobs) *signature stamp...I was allowed to put Paul's signature on a few types of documents...I might have even picked the bullet up and handled it...and umm, it was kind of deformed... I recall it looked like a deformed piece of led...there could have been a fragment missing off of it, but it looked to me like the whole thing."* The bullet that ENS Kelly was referring to was the bullet that was removed from Danny's chest during the initial medical death examination.

ENS Kelly continued explaining that Danny wasn't supposed to have a weapon or ammunition at all. ENS Kelly stated, *"If he'd been caught with it, he would've been in big trouble. So, he had to hide it somewhere, especially with barracks inspections happening every so often. Someone always got caught with something... usually marijuana."* He paused before adding, *"Once, we had a guy caught with C4 explosives".* He went on to say that the guy was court-martialed and kicked out of the U.S. Navy. ENS Kelly stated, *"The funny thing is, he was our photographer* (the guy caught with the C4*). But when we actually needed a photographer, for this whole thing with Danny, he wasn't around."* ENS Kelly doesn't remember if the command had military-issued cameras. I told ENS Kelly that LT Jacobs stated that he didn't have a camera, and interestingly, LT Jacobs didn't think to ask for a camera to take photos of the scene. We know for a fact that the command had access to a camera and most likely a photographer, because approximately two months before Danny's death, CBMU-302 held a change of command ceremony in which the unit took official photographs of the entire event. In fact, CBMU-302 had a dedicated team assigned to develop a semi-annual cruise book (yearbook), documenting the men's tours of duty with photos and descriptions. We also know for a fact that some command members had personal cameras because of all the CBMU-302 photos I discovered during my research for this book. So, if the command had a camera and a photographer for routine events, why did they not have a camera for LT Jacobs JAG investigation which should have been considered a potential crime scene? The assertion that no camera was available at the command is difficult to reconcile with standard operational practices and suggests that the investigation into Danny's death was incomplete and not conducted in accordance with the JAG Manual or accepted death investigation procedures.

I then asked ENS Kelly about the reported location of the gun in relation to Danny's body. I stated that, according to other unit members, the gun was found approximately four feet from Danny's left hand, while he was clutching a crucifix in his right. I told ENS Kelly that this didn't make sense to me because Danny was right-handed (right-hand dominant). As an investigator, I would have to ask two crucial questions: Why and how? Why would Danny have used his non-dominant left hand to fire a weapon into his chest twice? How could he have shot himself twice in the same entrance wound with his non-dominant left hand, all while holding a crucifix in his dominant right hand? ENS Kelly asked, *"Was he right-handed?"* I told ENS Kelly that from my research, I had learned from Danny's medical record and from family members that Danny was, in fact, right-handed. ENS Kelly didn't recall hearing anything about whether

Danny was left-handed or right-handed. However, he found the information troubling, remembering a photograph where Danny was wearing a pistol holster on his left side. This led ENS Kelly (and maybe LT Jacobs) to assume Danny was left-handed. I, too, had seen the photograph and found it concerning. My logical reasoning kicked in as I searched for an explanation for the holster's placement on Danny's left side.

Authors Note: From the photo ENS Kelly was referencing, it appears that Danny may have been attempting to capture a "Cool Navy Guy" photo of himself with the pistol, perhaps to send to his family or girlfriend back home. When I first analyzed the photograph, two possibilities came to mind: 1) the photo negative may have been inverted (reversed) during printing, or 2) the lighting and camera angle may have favored a left-side profile. If the latter is true, Danny may have positioned the holster on his left side simply to provide a clearer view of the pistol, creating a more natural-looking photo. Regardless of the explanation, the facts remained: Danny was right-handed. ENS Kelly stated, *"Well, if the medical record said he was right-handed, that would be more reliable than a photograph."*

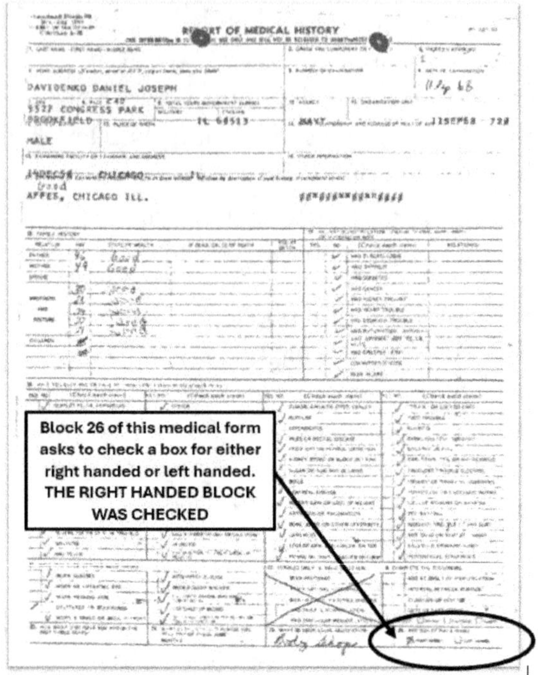

Danny's Medical File Confirming that he was Right-Handed (Right Hand Dominant)

Continuing with my questions, we moved on to the entrance and exit wounds described by LT Baker in his medical report. According to ENS Kelly, he didn't think there was an exit wound. He mentioned the one bullet that was found lodged in Danny's chest, the bullet he had seen numerous times in LT Jacobs desk. ENS Kelly's mind seemed to be churning, consumed by the possibility of my 'two-bullet' theory. Could one of the bullets have been a dud, lodging itself in Danny's chest, while the other passed completely through his body, causing the exit wound? The medical report was the single key piece of indisputable evidence in support of my 'two bullet' theory. It clearly stated that the body had both an entrance and an exit wound. The report also stated that the bullet removed from Danny's chest was completely intact and not fragmented, clearly confirming my two-bullet theory.

My next question to ENS Kelly was about any rumors he remembered hearing that were circulating immediately after Danny's death. The main rumor he recalled was about a 'Dear John' letter. I told him that no one in our family had ever seen this rumored letter, nor did anyone remember it being returned with Danny's personal effects. In fact, over the years, our family has heard numerous rumors, including the "Dear John" story, but no such letter exists among the family memorabilia. My grandmother never mentioned seeing a 'Dear John' letter, nor did she ever tell my mother or any other family members that one existed. ENS Kelly recalled hearing a rumor that *"there was a note of some kind in his pocket when they found the body... but I didn't see it."* I heard the rumor that it was the 'Dear John' letter... but once again, *"I didn't see it."* He went on to say, *"It seems to me that Paul* (LT Jacobs) *told me there was no 'Dear John' letter. If such a letter had existed, it would have been mentioned in LT Jacobs' JAG investigation."*

Over the years of interviewing LT Jacobs, not once did he mention a physical 'Dear John' letter during my interviews. Instead, LT Jacobs only acknowledged that they heard rumors about the letter. I then asked about who packed up Danny's personal belongings to be sent home. ENS Kelly believed that LTJG Underwood was responsible for packing and shipping Danny's things, as he handled the shipment of items in and out of the command. I told ENS Kelly that this contradicts what I was told by LT Jacobs. LT Jacobs stated that Danny's Alpha Command CC packed up his belongings. I told ENS Kelly that some of the things he mentioned didn't make much sense to me, based on other interview findings. Why would LTJG Underwood, the person who found the body, also be responsible for packing up Danny's personal belongings? I suggested to ENS Kelly that we should look at the situation from the perspective of an investigator: 1) LTJG Underwood finds the body but waits before telling anyone. 2) He eventually solicits investigative support from ENS Kelly before officially notifying the CO and XO. 3) LT Jacobs does not recall interviewing LTJG Underwood during his investigation; the very person who found Danny's body. 4) LTJG Underwood is then directed to pack up the deceased's personal belongings. If all of this were true, a seasoned investigator would certainly be concerned about these findings. ENS Kelly admitted that he wasn't sure and was merely speculating. He assumed that LTJG Underwood handled these tasks because it made sense to him. After all, LTJG Underwood was the supply officer, and the supply department was essentially the closest thing they had to a UPS store, responsible for shipping and receiving. ENS Kelly also mentioned hearing that whoever packed up Danny's belongings was responsible for sanitizing the contents of the package. He explained that the U.S. Navy would remove non-essential items, such as pornography. He went on to say that *"it was rumored they also disposed of the 'Dear John' letter, along with a few letters Danny had written but never mailed."* He recalled that one letter in particular was troubling. The correspondence reportedly found among Danny's possessions included language that appeared to express feelings of worthlessness and depression. However, ENS Kelly was quick to emphasize that these were all just rumors he had heard. I told ENS Kelly that this was the first time I'd heard this story, and that LT Jacobs had never mentioned these unmailed and possibly disposed of letters that were among Danny's personal possessions. Furthermore, had LT Jacobs been in possession of the actual "Dear John" letter, it would have constituted the most significant piece of physical evidence supporting a suicide manner-of-death determination. There would have been no logical reason for him to dispose of this critical piece of evidence, because it would have supported his suicide MOD determination. Again, ENS Kelly emphasized that they were just rumors and that he was trying to recall events from over 55 years ago. I responded by saying that everyone I'd interviewed mentioned the rumor about the 'Dear John' letter, but not one of them could confirm its existence or say they had actually seen or read it, to include LT Jacobs. Therefore, the 'Dear John' letter remains a rumored mystery. I ended our conversation with a heartfelt thank you and a request for follow-up questions in the future, which he agreed to.

LT Paul Jacobs Interview (CBMU-302 Executive Officer)

The following is based solely on letters, emails, documents, and interviews (via phone conversations) collected from multiple CBMU-302 unit members. According to LT Jacobs, when Danny's body was found, he remarked, *"There was just a clean single bullet hole in the chest, no pool of blood or external bleeding."* He also stated, *"The bullet didn't penetrate the body, so all bleeding was internal,"* at least to the best of his understanding and knowledge. This statement left me perplexed. If LT Jacobs didn't observe any blood, did he simply assume the bullet didn't penetrate the body? How did he come to that conclusion? Did he not fully observe and document the scene for his investigation? He mentioned that he left the scene once medical personnel arrived, so he could begin drafting a message to higher HQ regarding Danny's death. However, he never mentioned actually reading the medical officer's death report. Had LT Jacobs fully read and understood the contents of that report, he would have noted LT Baker's identification of a 2 centimeter exit wound. Could it be that LT Baker's report didn't reach LT Jacobs until his JAG investigation was almost completed?

LT Jacobs told me that he was given the full metal-jacketed (FMJ) bullet removed from Danny's chest and that he kept the bullet in his desk drawer until his last day in Vietnam. LT Baker must have known that LT Jacobs had been assigned as the sole JAG investigation officer, which is why he gave him the bullet, which he must have believed was intended to be part of the evidence collected during the death investigation. However, since LT Jacobs wasn't experienced in death investigations, he didn't fully understand why the medical officer gave him the bullet. It never occurred to LT Jacobs that the bullet was a key piece of evidence. When I asked him if he still had the bullet, LT Jacobs stated, *"No, I threw it into the South China Sea on my last day in Vietnam"*.

I am completely confused as to how LT Jacobs came to believe that Danny was shot only once. Is it because they only found one bullet, the bullet that remained inside Danny's body? Is it because they only found one bullet casing, the casing jammed in the ejection port of the gun? Did LT Jacobs arbitrarily conclude that Danny died from internal bleeding because he didn't stay at the scene to conduct a full examination of the body and the surroundings? Had LT Jacobs remained at the scene, he would have had the opportunity to check for blood on Danny's back, on the ground, or even on the crates. A 2 centimeter hole in Danny's back would likely have caused a significant amount of blood at the scene, especially if the bullet hit a vital organ like the heart or a major artery. Regardless, there still should have been a large blood stain on the ground at the scene. It is important to note that LT Baker's medical report did NOT specify how Danny was shot. He could not have made such a determination, as he was not present to witness the shooting. At a minimum, rendering that type of conclusion would have required certification as a forensic pathologist and access to detailed forensic evidence gathered during the investigation.

Additionally, LT Baker's report makes NO reference to Danny's wounds being "self-inflicted." Such a determination would fall outside the scope of his medical evaluation and would require specialized forensic pathology training. No one at the scene fully understood the complexity of conducting a crime scene investigation. However, in LT Jacob's defense, I don't believe that he was appointed as the sole JAG investigator until a few hours after Danny's body was found. This would explain why he didn't secure the scene for evidence collection and why he left before the body was removed. With a 1.5 centimeter entrance wound and a 2 centimeter exit wound, coupled with a full metal-jacketed bullet remaining in Danny's body, I am led to conclude that Danny was shot twice. One bullet penetrated his chest and exited his back, while the second bullet remained lodged in his chest. Why didn't LT Jacobs come to this conclusion of a two-bullet theory? The evidence clearly shows that Danny was shot twice, as the bullet recovered from his body was fully intact, with only a slight bend at the tip. This fact alone proves it did not fragment to create the exit wound. So why didn't LT Jacobs ask what caused the 2 centimeter exit wound?

As mentioned earlier, LT Jacobs identified the gun found next to Danny's body as an 'orphan', a non-military-issued weapon. It was sent to the SEABEE armory for inspection and tracing. Since it wasn't listed in any military inventory, it was classified as a personal firearm, which was prohibited by regulation, and subsequently destroyed. Another troubling issue is that the gun supposedly found near Danny's body, possibly the one that killed him, was never processed or treated as evidence. Why wasn't the gun treated as evidence? And why was it handed over to the Master at Arms and subsequently destroyed? In my opinion, the fact that the gun was not a military-issued weapon should have been enough to launch a criminal investigation. Why was the bullet casing found at the scene discarded instead of being treated as evidence? Why wasn't the gun, casing, and the bullet extracted from Danny's body properly processed, inventoried, and included with any other evidence collected during a JAG or NIS investigation? Why didn't LT Jacobs ensure the entire scene was photographed for the investigation?

I am not certain whether the military had the capability to conduct ballistic testing in Vietnam at the time. However, had the weapon, cartridge casing, and bullet been preserved as evidence, ballistic analysis could have determined whether that firearm was the one that killed Danny. According to LT Baker's medical report, Danny also had a 2" x 1" burn on the inner aspect of his left elbow. What caused this burn? Could it have come from the barrel of a hot .45 caliber pistol that had just been fired twice? The burn mark on Danny's inner left elbow could be a significant piece of physical forensic evidence, potentially caused by contact with the hot barrel of a gun. If the firearm had been retained as evidence, it could have undergone forensic comparison with the burn mark on Danny's body to establish whether the gun was responsible for the wound. It would have been nearly impossible for Danny to shoot himself twice in the exact same location with his non-dominant left hand, while also burning his left elbow on the barrel of the gun. As LT Jacobs continued his JAG investigation, he ordered the ALPHA Company Commander to take an inventory of Danny's personal effects and have them packed up and shipped home. Despite my multiple record requests, no military inventory report documenting Danny's belongings was ever provided. All of Danny's military-issued items (i.e., pistol, rifle, helmet, assorted gear) were removed and sent back into CBMU-302 inventory.

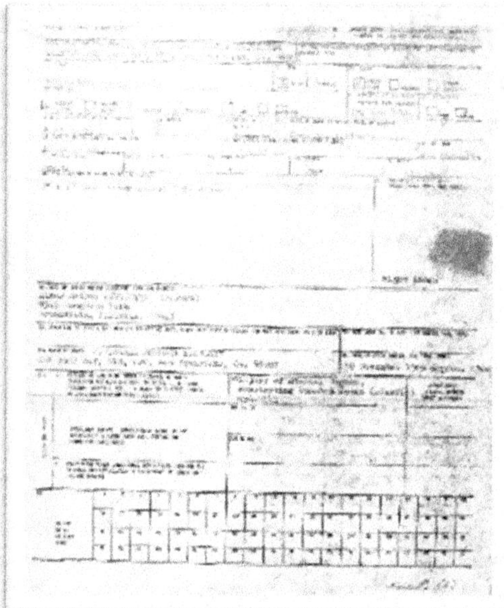

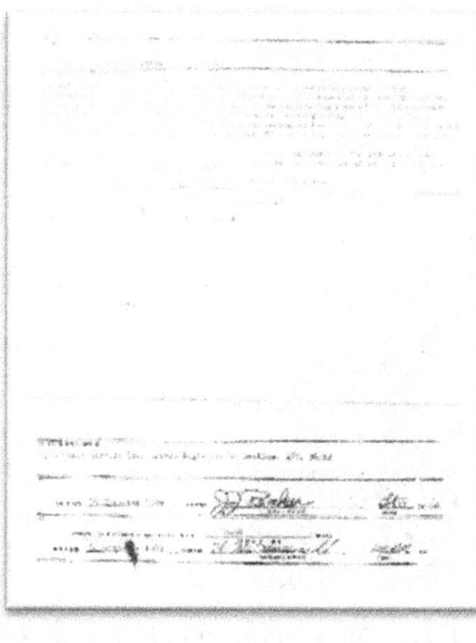

Original NAVMED M (REV. 4-58) Danny's Certificate of Death (Front and Back Pages)

Medical Officer's Certificate of Death/ Front Page - Transcribed

"CERTIFICATE OF DEATH

NAVMED M (REV. 4-58) FRONT

FROM: CBMU-302, FPO, San Francisco, CA 96601

1. NAME: DAVIDENKO, DANIEL JOSEPH

2. SEX: MALE

3. RACE: CAUCASIAN

4. STATUS: REGULAR ACTIVE

5. LENGTH OF SERVICE: 1 year 1 month 17days

6. AVIATION: NO

7. FILE OR SERVICE NO: B ######

8. RATE: CMCN

9. CORPS: (left blank)

10. BRANCH OF SERVICE: USN

11. PLACE OF BIRTH: Chicago, Illinois

12. DATE OF BIRTH: 14 December 1950

13. AGE: 18 years 11 months

14. RELIGION: Rom Catholic

15. COLOR OF EYES: Green

16. COLOR OF HAIR: Blond

17. COMPLEXION: Fair

18. HEIGHT: 71"

19. WEIGHT: 140 lbs.

20. MARKS AND SCARS: ***2"x 1" burn inter-aspect of left elbow***

21. FINGERPRINT: Right index (actual fingerprint)

22. NEXT OF KIN OR FRIEND: Elmer Arthur DAVIDENKO (FATHER) 9527 Congress Park Brookfield, Illinois, 60513

23: ADMITTED TO SICK LIST FROM: (left blank)

24. DATE ADMITTED TO SICK LIST: (left blank)

25. PLACE OF DEATH: US NAVAL SUPPORT FACILITY CAM RANH BAY, FPO San Francisco, CA. 96697

26. TIME OF DEATH: 10 November 1969, approx. 0500

27. CAUSE OF DEATH: Injury of Sternal Region, ***Penetrating Gunshot Wound (missile)*** *036-4141*

28. (left blank)

Medical Officer's Certificate of Death / Back Page - Transcribed

NAVMED N (REV. 4-58) (BACK)

29. NAME: DAVIDENKO, Daniel Joseph ######

30. SUMMARY OF FACTS RELATING TO DEATH: Subject found by LTJG William F. UNDERWOOD ###### who immediately reported the discovery to the dispensary. The Medical Officer and an HM2 immediately departed for the scene. Examination of the subject revealed a ***1 ½ centimeter penetrating wound*** *of the mid-sternum with a* ***2 centimeter wound in the mid-back.*** *Subject revealed acrocyanosis, dilated and fixed pupils, no respiratory or cardiac function, and segmented hemostasis of the retinal veins. The subject was pronounced dead by the Medical Officer at 1815 hours on 10 November 1969.* ***An anterior-posterior and a left lateral chest x-ray performed on the remains of the subject revealed a metal projectile still present in the right posterior chest cavity.*** *Last rites were performed by Father Denis Casey at 1830 hours. Autopsy will be performed and forwarded upon receipt. Identity of remains established by shipmates.*

31. DISPOSITION OF REMAINS: 59th Field Service Co., Graves Registration Section, APO, 96312

32. DATE SIGNED: 10 November 1969

SIGNATURE: J J BAKER (Medical Officer) LT, USNR

33. APPROVED: Court of inquiry or Board of Investigation WILL Be Held.

DATE SIGNED: 10 November 1969

SIGNATURE: H.H. HARRELL (Commanding Officer) LCDR, USN"

Letter From Father Denis Casey, LT, USNR - Transcribed

"Dear Mr. & Mrs. Davidenko, I would like to express my sincere sympathy to you and the other members of your family on the death of your son Daniel. While I did not personally know your son, I am informed by his Company Commander that he was an excellent young man who performed all his duties in an outstanding manner. As evidence of the high esteem in which your son was held, a very large turnout attended a memorial Service and Mass held at the base chapel of the Naval Support Facility, Cam Ranh Bay, on 13 November 1969. As LCDR Harrell, your son's Commanding Officer, mentioned in his letter, your son received the last Sacraments when his body was found. I know this is a very trying time for you and your family. I will remember you all in my Masses and prayers. I enclose a copy of our Memorial Service Bulletin. If I can be of any help at any time, please feel free to write to me. May I again express my sincere sympathy. Sincerely, DENIS CASEY, LT, CHC, USNR, Catholic Chaplain, US Naval Support Facility Cam Ranh Bay"

Letter from Father Casey - Letter dated November 17th, 1969

Original Mass Bulletin Prepared by LT Denis Casey, Catholic Priest, USNR

According to LT Jacobs, Danny received an 18-gun salute when men from CBMU-302 fired three volleys, six shots each. Immediately following the salute, the command played Taps to conclude the memorial Mass for Danny.

NIS Response – To my request for their Investigation into Danny's Death

This is incredibly frustrating, after all those years of research and FOIA requests, to be told that no records exist of an NIS investigation into Danny's death raises serious questions. If Danny's death was officially recorded as self-inflicted, there should have been a thorough investigation conducted to reach that conclusion, especially since he died on a secure military base and not in a combat-related situation. I finally received a document from the Navy pertaining to my inquiry, to see if an NIS investigation was conducted on the death of CMCN Daniel Joseph Davidenko. To my disappointment, the letter stated that no records exist in any of the Department of the U.S. Navy (DON) databases.

According to DON, a records search was conducted within the following DON databases: 1) The Defense Central Index of Investigations (DCII), 2) The Consolidated Law Enforcement Operations Center (CLEOC), 3) The NCIS Knowledge Network (K-Net) database, and 4) The Records and Information Management System (RIMS). All four databases could not locate any death investigation into Danny's death in Vietnam. The fact that none of the Departments within the U.S. Navy's databases (DCII, CLEOC, K-Net, or RIMS) contained any records, suggests one of three possibilities: 1) The investigation was never conducted. Despite standard procedures, there may have been a failure (or deliberate decision) not to document or pursue an inquiry. 2) The records were lost or destroyed. Military records, especially from Vietnam, have been known to be lost due to mismanagement, bureaucratic errors, or from a natural disaster, like a fire. 3) The records were never properly archived. If the NIS investigation into Danny's death was informal or conducted outside standard procedures (which I doubt), it might not have been properly documented in any official government system. In more than three decades of working with the U.S. Navy and multiple government agencies, I have never heard of an investigation, especially one involving a death, being conducted informally. Since my FOIA requests produced no NIS records or other investigative documents, the only remaining course of action is to appeal the findings in a final effort to identify any official investigative report, should one exist. Below is the actual letter I received from the Department of the U.S. Navy in response to my FOIA request.

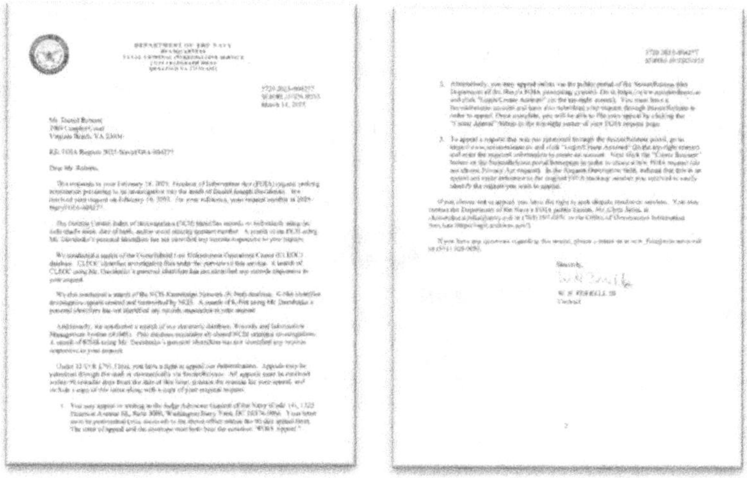

Department of the U.S. Navy FOIA Response Dated March 14th, 2025

So, what really happened to Danny? What chain of events led to his death? In this chapter, we examined the information gathered from the official documents obtained through my numerous FOIA requests, as well as firsthand accounts from several CBMU-302 members who were present at the scene. Danny's tragic death occurred over 55 years ago, and most of the men I interviewed are now in their seventies, eighties, or nineties. Understandably, their memories of that fateful day have faded over time. As for the unit members I interviewed, it was difficult for them to reach back into their collective memories to piece together the specific details surrounding Danny's death, but each man did his best to share what he remembered. After reviewing all the data collected from my interviews, it became abundantly clear that each person I spoke with had their own unique, but minor, variations in recalling specific details about Danny's death. The most significant inconsistency in their accounts was the location of the gun in relation to Danny's body. However, they were unanimous on one critical point: none of them saw any blood at the scene. After my initial communications with LT Jacobs back in 2006 he said we would look through his boxes in the attic to see if he could find his memorabilia from his tour in Vietnam with CBMU-302. Below is the cover letter dated June 10th, 2007, that he sent me with a few documents that he had kept over the years.

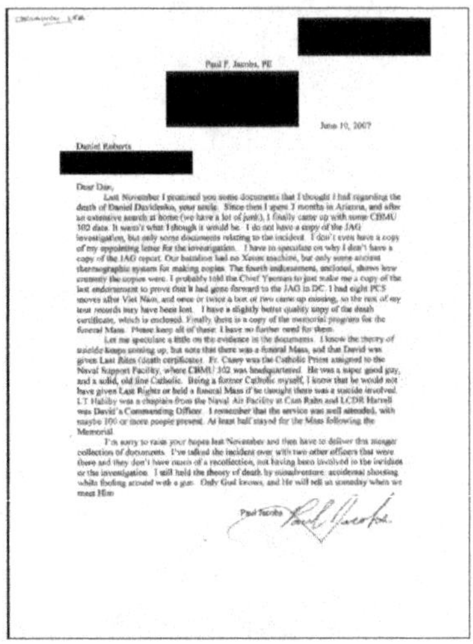

Paul F. Jacobs, PE

June 10, 2007

Daniel Roberts

Dear Dan,

Last November I promised you some documents that I thought I had regarding the death of Daniel Davidenko, your uncle. Since then I spent 3 months in Arizona, and after an extensive search at home (we have a lot of *junk*), I finally came up with some CBMU 302 data. It wasn't what I though it would be. I do not have a copy of the JAG investigation, but only some documents relating to the incident. I don't even have a copy of my appointing letter for the investigation. I have to speculate on why I don't have a copy of the JAG report. Our battalion had no Xerox machine, but only some ancient thermographic system for making copies. The fourth endorsement, enclosed, shows how crummy the copies were. I probably told the Chief Yeoman to just make me a copy of the last endorsement to prove that it had gone forward to the JAG in DC. I had eight PCS moves after Viet Nam, and once or twice a box or two came up missing, so the rest of my lost records may have been lost. I have a slightly better quality copy of the death certificate, which is enclosed. Finally there is a copy of the memorial program for the funeral Mass. Please keep all of these. I have no further need for them.

Let me speculate a little on the evidence in the documents. I know the theory of suicide keeps coming up, but note that there was a funeral Mass, and that David was given Last Rites (death certificate). Fr. Casey was the Catholic Priest assigned to the Naval Support Facility, where CBMU 302 was headquartered. He was a super good guy, and a solid, old line Catholic. Being a former Catholic myself, I know that he would not have given Last Rights or held a funeral Mass if he thought there was a suicide involved. LT Habiby was a chaplain from the Naval Air Facility at Cam Rahn and LCDR Harrell was David's Commanding Officer. I remember that the service was well attended, with maybe 100 or more people present. At least half stayed for the Mass following the Memorial.

I'm sorry to raise your hopes last November and then have to deliver this meager collection of documents. I've talked the incident over with two other officers that were there and they don't have much of a recollection, not having been involved in the incident or the investigation. I still hold the theory of death by misadventure: accidental shooting while fooling around with a gun. Only God knows, and He will tell us someday when we meet Him.

Paul Jacobs

Actual Letter from LT Paul F. Jacobs – Dated June 10th, 2007

LT Paul F. Jacobs Letter dated June 10th, 2007 – Transcribed

> *"Dear Dan, Last November I promised you some documents that I thought I had regarding the death of Daniel Davidenko, you Uncle. Since then, I spent 3 months in Arizona, and after an extensive search at home (we have a lot of junk), I finally came up with some CBMU-302 data. It wasn't what I thought it would be. I do not have a copy of the JAG investigation, but only some documents relating to the incident. I don't even have a copy of my appointing letter for the investigation. I have to speculate on why I don't have a copy of the JAG report. Our battalion had no Xerox machine, but only some*

ancient thermographic system for making copies. The fourth endorsement, enclosed, shows how crummy the copies were. I probably told the Chief Yeoman to just make me a copy of the last endorsement to prove that it had gone forward to the JAG in DC. I had eight PCS moves after Vietnam, and once or twice a box or two came up missing, so the rest of my tour records may have been lost. I have a slightly better quality copy of the death certificate, which is enclosed. Finally, there is a copy of the memorial program for the funeral mass. Please keep all of these. I have no further need for them. ***Let me speculate a little on the evidence in the documents. I know the theory of suicide keeps coming up, but note that there was a funeral Mass, and that David*** (He meant Danny) ***was given Last Rites (death certificate). Fr. Casey was a Catholic Priest*** *assigned to the Naval Support Facility, where CBMU 302 was headquartered. He was a super good guy, and a solid, old line Catholic.* ***Being a former Catholic myself, I know that he would not have given Last Rights or held a funeral Mass if he thought there was a suicide involved.*** *LT Habiby was a Chaplin from the Naval Air Facility at Cam Ranh and LCDR Harrell was David's* (Danny's) *Commanding Officer. I remember that the service was well attended, with maybe 100 or more people present. At least half stayed for the Mass following the Memorial. I'm sorry to raise your hopes last November and then have to deliver this meager collection of documents. I've talked the incident over with two other officers that were there, and they don't have much of a recollection, not having been involved in the incident or the*

> *investigation.* ***I still hold the theory of death by misadventure; accidental shooting while fooling around with a gun.*** *Only God knows, and He will tell us someday when we meet Him. Signed Paul Jacobs"*

Over the years, in my many conversations with LT Jacobs, he consistently maintained his belief that Danny accidentally shot himself. Not once did he suggest that Danny's death was an apparent suicide. This is clearly reflected in the letter above, where LT Jacobs notes that Father Casey would not have given Danny his Last Rites or held a funeral Mass if he had believed suicide was involved. All I can hope for now is to locate the official JAG investigation and the official autopsy report, so I can conduct a thorough review of the findings and, hopefully, determine Danny's true manner of death.

One last critical question I never got to ask LT Jacobs before he passed away, was the following: If Danny was the only casualty from CBMU-302 and it was determined that his death was due to an accidental shooting (or suicide for that matter), then why wasn't Danny ever mentioned in the command's annual yearbook? During CBMU-302s Vietnam tours between 1968 and 1970, CBMU-302 produced several command yearbooks documenting the unit's achievements by department and division. These publications featured imagery from each tour as well as official photos of all personnel assigned during that period. However, there is not one photo of Danny to be found in the yearbook produced during his tour of duty in 1969 with CBMU-302. There is not one mention or photo of Danny, not even within his own division, the CM shop. Why isn't Danny listed? I did, however, find Danny's name mentioned in the commands master muster list, but that was all I could find. As the only CBMU-302 member to die in Vietnam, you would think there would be some memorial page dedicated to Danny in the yearbook that was developed while he was assigned to the unit. But no, there is no mention of Danny or this tragic event… Why? In my mind, no matter the circumstances, each unit member assigned should have been mentioned in these yearbooks.

Chapter 10

Evidence Collection Procedures

On November 10th, 1969, the lifeless body of an 18-year-old sailor by the name of CMCN Daniel Joseph Davidenko (Danny) was found in the Construction Battalion Maintenance Unit THREE ZERO TWO (CBMU-302) supply laydown area located in Cam Ranh Bay (CRB), Vietnam. The cause of death (COD) was determined to be a single gunshot wound to the chest. Davidenko had been missing for several days prior to this discovery. As we look into the facts surrounding this case, one critical question of concern arises: Upon finding Danny's body, did CBMU-302, the base Military Police (MP), or the Naval Investigative Service (NIS) treat the site as a potential crime scene? Available records suggest that no criminal investigation was initiated, despite the presence of Danny's body and a non-military-issued .45 caliber weapon at the scene.

This raises further concerns about the handling of the case and whether a thorough investigation was conducted into the circumstances surrounding his death. All discoveries of deceased individuals, whether civilian or military, must be treated as potential crime scenes. An initial death investigation should proceed under the assumption that a crime has occurred until it can be definitively ruled out. Crime scene processing requires trained professionals with the appropriate education and experience in crime scene investigation and evidence collection. This process is meticulous and often lengthy, involving detailed documentation of the scene's conditions and the collection of all physical evidence that may aid in determining the circumstances surrounding the death. This evidence is crucial in identifying one of the five manners of death (MOD): Natural causes, accident, suicide, homicide, or, in the most severe cases, murder.

The Judge Advocate General (JAG) Investigation into the death of CMCN Daniel Joseph Davidenko was overseen by Danny's Executive Officer (XO), LT Paul Jacobs. LT Jacobs, a commissioned officer in the U.S. Navy and a civil engineer by trade. He was filling the executive officer (XO) billet while assigned to CBMU-302 in Cam Ranh Bay (CRB), Vietnam. As is common for many commissioned officers, LT Jacobs was assigned duties outside of his primary field of expertise during his career, including serving as the JAG investigator for this case. The Commanding Officer (CO), LCDR Harrell, appointed LT Jacobs as the sole investigator into the death of CMCN Davidenko. This decision raises questions about the thoroughness and impartiality of the investigation, considering LT Jacobs' background in civil engineering rather than as a crime scene investigator (CSI).

The vast majority of JAG investigations in the U.S. Navy are conducted by regular line officers and typically focus on mishaps, such as ship or aircraft accidents, whether or not they involve loss of life. However, the case involving CMCN Daniel Joseph Davidenko was not a mishap or an on-base accident. It warranted a full suspicious death investigation, one requiring professional crime scene investigators with the training to identify, collect, and preserve critical evidence.

In this chapter, I will discuss every aspect of a 'typical death investigation report' and look closely at the formal JAG investigation process as outlined in the Navy's JAG Manual. I will also use the standard death investigation procedures of the Naval Criminal Investigative Service (NCIS) and the Military Police (MP) to help me guide the reader through a death investigation. Lastly, if it becomes available, I will also break down the coroner's report and the completed autopsy as it relates to the manner of death (MOD) decided in this case. The intent is to examine all the evidence and facts collected to help the reader come to their own conclusion as to what might have led to the death of CMCN Daniel Joseph Davidenko, USN.

Before I go any further, it is important to understand the difference between the two terms: Cause of Death (COD) and Manner of Death (MOD). Here is a good definition of the two terms from Wikipedia: "In law, medicine, and statistics, cause of death (COD) is an official determination of the conditions resulting in a human's death. A cause of death is determined by a medical examiner. In most cases, an autopsy needs to be performed by a pathologist. The manner of death (MOD) refers to the circumstances under which a person dies, as determined by medical examiners or coroners. It is distinct from the cause of death, which identifies the specific disease or injury that led to the death [58].

In Danny's case, the COD was a gunshot wound to the chest, the MOD was determined by his commanding officer to be an "accidental death" via self-inflicted gunshot wound. From all the collected documents I uncovered during my research (minus the autopsy, NIS investigation, JAG investigation, and MP report), I have not found any physical evidence or documentation that would lead me to conclude or confirm the manner of death in which Danny died. Therefore, it is imperative that we examine all the physical evidence and testimonies to help us determine the actual manner of death in this case. It is also important to note that a manner of death determination may be ruled as "Undetermined", if there is insufficient evidence to reach a definitive conclusion, or if the evidence was improperly collected, processed, or preserved during the investigation.

During one of my interviews with LT Jacobs (JAG Investigation Officer), I asked the following question: Did you conduct a search of the scene for the collection of evidence? Paul replied, *"Base Legal did that, and probably called on Naval Investigative Service (NIS) in Saigon, or the local Army Criminal*

Investigative Department (the MPs) for help." If LT Jacob's memory is correct, and this is what actually took place, then these other agencies should have been responsible for processing all the death scene evidence. Evidence items such as the gun, the bullet casings, the bullet removed from Daniels' body, fingerprints on the gun/bullet casings, blood splatter photographs, gun powder residue on the hands, the entry and exit wound photos, pocket litter, blood-stained clothes, and many other pieces of evidence. All of this evidence should have been documented, photographed, processed, protected, and archived. Unfortunately, according to his own omission, LT Jacob did not do any evidence collection during his JAG investigation and assumed it was being done by another investigative organization. LT Jacobs stated that the gun, the bullet, and the bullet casing were either discarded or destroyed and not preserved as evidence. Once he made this statement, I immediately knew that no other investigative agency had been involved, because if they had been, all of those discarded items would have been collected and preserved as evidence. When LT Jacobs admitted that he didn't collect this physical evidence as a part of his investigation, my heart sank deep into my chest, I knew instantly that his JAG investigation was flawed from the beginning.

Let's broaden our knowledge about crime scene investigation to help us better understand what is required and necessary for a successful investigation that ultimately leads to the manner of death. Let's start with an introduction to how a crime scene investigation must be conducted and processed. At the start of an investigation, the assigned lead investigator should conduct a 'walk-through' of the scene. This must take place before the body or any items are touched or moved. This walk-through will provide the investigator with the opportunity to locate and view the body, identify valuable and fragile evidence, and determine initial investigative procedures required for a complete, systematic examination and documentation of the scene and body. Upon arrival at the scene, the investigator should assess the crime scene boundaries and adjust as needed, establish a path of entry and exit, identify, document, photograph, and collect all evidence, including a limited review of the body.

The lead investigator must minimize scene disturbance to prevent the loss and contamination of all evidence. It is essential to maintain a proper chain of custody for evidence. This way, through proper documentation, collection, and preservation, the integrity of the evidence can be assured. A properly maintained chain of custody and prompt transfer will reduce the likelihood of a challenge to the integrity of the evidence [34]. All investigators should assume, until proven otherwise, that the location where LTJG Underwood found Danny's body may not be the actual location where the injury (gunshot) or death took place. It is imperative that the investigator attempt to determine the locations of any and all injuries that may have contributed to the death. Physical evidence at all known locations the victim routinely visits (barracks room, body shop, enlisted club, etc.) may contain evidence pertinent in establishing the cause, manner, and circumstances related to the death. The physical location and positioning of Danny's body is crucial to this investigation.

In many movies, we see criminals dispose of a body in a remote area, hoping that its isolation will prevent it from being discovered, thereby aiding their efforts to evade justice. However, in this case, Danny's body was found in a location that raises significant questions. The CBMU-302 supply laydown area was a place not frequently accessed by non-supply staff. It was hardly the kind of desolate location one might expect if the intent was to conceal a crime. This location warrants a closer examination, as it may provide valuable clues about the circumstances surrounding his death. Every investigator should look for evidence to determine if the victim was killed in that location or if the victim was moved to that location after being killed [34].

LT Jacobs reported that he did not observe blood at the scene. He did not see any blood on Danny's body, clothes, or around the area where he was found. This raises more investigative questions: Why was there no blood observed at the scene? One theory is that Danny may have been shot and killed elsewhere, with his body later placed in the supply laydown area. This hypothesis introduces several important questions regarding the manner of death (MOD) and the possibility of foul play. A key piece of evidence missing from the investigation is photographic documentation. If photographs of Danny's body had been taken at the scene, it would have preserved crucial details regarding the body's position and condition. As I reflected on the two-bullet theory, one troubling fact stood out: neither the second bullet nor its casing was ever found at the scene. This is yet another red flag calling into question the credibility of the investigation surrounding Danny's death. If photographs had been taken at the scene, they would have provided insight into whether his body had been moved post-mortem.

Given the nature of the exit wound in the middle of Danny's back, there should have been significant blood at the scene. Blood would likely have been present on Danny's body, particularly on his back and on his clothing. Additionally, there should have been blood splattered on nearby objects, such as supply crates. The absence of such evidence raises serious questions about the investigation's thoroughness and the circumstances of Danny's death. I communicated via email and phone with LTJG Underwood (the supply officer who found Danny's body), with LT Casey (the base Chaplain), and ENS Kelly (the administrative officer), and neither remembers seeing any blood at the scene. Their responses (LT Jacobs, LTJG Underwood, ENS Kelly, and LT Casey) illustrate the importance of crime scene photographs of Danny's body and the surrounding area in which it was found. Photos should have been taken before and after the body was removed. We all know that memories fade over time, but photography permanently captures the scene for future research and investigation. Had photographs been taken before and after the body was moved, and had those images shown no blood present at the scene, such evidence would be consistent with foul play and support the conclusion that Danny's body was placed in the supply laydown area following his death.

There are five possible outcomes to a death investigation, and they are: natural causes, accidental, suicidal, homicidal, and murder. A natural death is defined as a death due to a natural disease or the aging process and not precipitated by an external factor. An accidental death is defined as a death caused by an unexpected event that is not intended, expected, or foreseeable. A death by suicide is defined as the act of taking one's own life voluntarily and intentionally. A death by homicide is defined by the act of one human killing another, however, with no intent to kill. A death by murder is defined as an unlawful killing, where there is the intent to kill. With five possible scenarios on the cause of death, you can see why a death investigation is necessary.

More importantly, each death investigation scene should be treated as a crime scene in an effort to preserve all evidence that will ultimately lead to cause and manner of death determinations. Without this detailed investigation, the manner of death may never truly be known. The circumstances surrounding Danny's death called for a detailed death investigation with the collection and preservation of evidence, so that the lead investigator(s) and medical examiner could assess the most probable manner of death.

In preparation for this chapter, I reviewed the JAG manual procedures that should have been used in the death investigation of CMCN Daniel Joseph Davidenko. I also compared the JAG manual procedures to, the Guiding Principles for Crime Scene Investigation and Reconstruction (by OSAC March 2020), and the U.S. Department of Justice's Death Investigation Guide for the Scene Investigator (by NIJ June 2011), to help me evaluate the limited evidence and documentation pertaining to Danny's death. Obtaining the official JAG investigation, the official autopsy report, and any other related investigative documents is crucial to understanding the scope of the inquiry into Danny's death. These records will allow for a detailed assessment of how thorough and complete the investigation was, or if there were critical aspects that were overlooked or inadequately explored. By analyzing these documents, it will be possible to determine whether the investigation was conducted with the necessary rigor and if all potential leads and evidence were properly considered. These critical documents, along with well-documented crime scene photos, will ultimately be needed to determine the manner of his death, if, in fact, it can be determined. Statistically, in many death cases, the MOD cannot be determined if there is a lack of physical evidence.

As I pointed out earlier, all death investigations should be treated as crime scene investigations. Crime scene investigations are designed and established to collect all required evidence at the scene. These investigations require that photographs be taken from a multitude of angles, both with the body and after the body is removed. The investigation must also require numerous physical measurements to be taken of the scene, so that investigators can reconstruct the scene to help them determine what happened.

From emails and phone conversations with LT Jacobs, the Executive Officer (XO) of CBMU-302, it appears that his JAG investigation was started after Danny's body was removed and taken to the base hospital for further medical examination and casualty reporting. When I asked LT Jacobs specific questions about what he remembered about the scene and about his investigation, he stated that *"he did not stay at the supply laydown area"*, the location where Danny's body was found. LT Jacobs stated that, *"after the medical team arrived, I went to my office to begin drafting the required messages to be sent to higher headquarters"*. He also stated that, *"I was not present at the site when the body was removed"*. LT Jacobs recalls that the Commanding Officer (CO), the Administrative Officer (AO), and the Supply Officer (SUPO) may have stayed on site with the body until the body was removed by the medical team. LT Jacobs stated that, *"his JAG investigation was all based on face-to-face testimony (interviews) to establish facts"*.

Based on my research, there were no eyewitnesses to Danny's fatal shooting, so these statement interviews would be categorized as circumstantial evidence. Circumstantial evidence is defined as indirect evidence that does not directly prove a fact but suggests its existence through inference. It requires reasoning or drawing logical conclusions from related facts to support a claim. This type of evidence can lead to multiple interpretations and does not provide definitive proof on its own [58]. With no witnesses, it is imperative that all physical evidence be collected to determine Danny's manner of death. According to LT Jacobs, his JAG investigation did not include photographs of the scene. By his own omission, he was not present at the scene to collect evidence before or after the body was moved from the scene. I appreciated LT Jacobs' honesty in his responses to my questions; however, I found several of his responses to be very troubling. Many of his responses indicated to me that LT Jacobs, though a highly qualified and trained civil engineer, was not qualified, nor properly trained to conduct a death investigation. Sadly, he is not alone, as most military officers do not have the prerequisite training and skills to conduct such investigations. Only the personnel assigned to the U.S. Navy's Military Police, the Naval Investigation Service (NIS), and the U.S. Navy Legal Department (U.S. Navy Lawyers) have the minimum required training and skills to conduct this type of investigation.

As for the base at Cam Ranh Bay (CRB), it was a multi-service affiliated base, with all branches of the military having similar legal organizations responsible for conducting these investigations. It is crucial to emphasize that I am in no way assigning fault or blame to LT Jacobs. On the contrary, he followed the orders of his commanding officer and conducted the JAG investigation to the best of his ability, hopefully in accordance with the guidelines set forth in the JAG Manual. However, I am deeply perplexed by the premature conclusions regarding the manner of Danny's death that were drawn by LCDR Harrell, the commanding officer (CO). These early determinations raise significant questions about the thoroughness and objectivity of the investigation. Why did the CO choose to send a letter to Danny's parents that pre-determined Danny's manner of death (MOD) as a "self-inflicted" gunshot wound? The letter was drafted within 24 hours of Danny's death and sent before the investigation(s) were completed and before the coroner rendered an official determination of the manner of death. These questions are critical to understanding whether the investigation was conducted with the necessary care and attention to detail.

As a retired military officer, I understand the importance of getting something out to the family as soon as possible after a service member has died. In fact, I applaud the CO's timeliness; however, I do find fault with the CO and the premature content of his letter. Below is an example of what Danny's commanding officer's letter SHOULD have said, pending the official JAG and other investigative findings:

It is my duty as Commanding Officer of CBMU-302 to inform you of the death of your son, Daniel, at Cam Ranh Bay, Vietnam. On behalf of the men of CBMU-302, I would like to express our sincere sympathy to you on the loss of your son. Your son was found dead in the command's supply laydown area on the evening of 10 November 1969. ***A full investigation as to the manner of death is currently ongoing. Once the investigation is complete, I will inform you of the investigation findings.*** *If I can be of any further help to you, please feel free to write to me at any time. Again, may I express my deepest sympathy.*

Until I find information to confirm that other investigations were or were not conducted, I will have to assume that the ONLY investigation conducted was the CBMU-302 JAG investigation. Should this prove to be accurate, it would suggest that CBMU-302's higher naval headquarters, along with the Naval Investigative Service (NIS) assigned to that region of Vietnam, bore ultimate responsibility for the failure to conduct an official death investigation. Both higher headquarters naval legal teams and the NIS should have been involved with this death investigation, and both should have treated it like a crime scene investigation, until proven otherwise. Danny was not killed in combat, and therefore, his death should have been handled by an agency trained and qualified in crime scene/death scene analysis and investigation. All of my requests for information through the FOIA process to the National Archives, local Naval Criminal Investigation Service (NCIS) offices, and local JAG offices have yet to yield a single document of other investigations or the official autopsy.

It is important for the reader to understand that it is a standard practice in all death investigations that when you discover a dead body, especially with no eyewitness, it would be treated as a crime investigation with evidence collected to assist in determining how that person died. Only a complete investigation with a detailed review and analysis of all the collected evidence will lead the investigator to a possible MOD conclusion. Think about all the unsolved crimes in the United States. Here are a few statistics from a 2017 FBI report: roughly 40 percent of the nation's slayings (death investigations) went unsolved that year [32]. Nearly half of all U.S. murders went unsolved in 2020. According to Thomas Hargrove, who runs the Murder Accountability Project, in the last seven months of 2020, most murders went unsolved. In 2022, almost 80 percent of all murders were committed with firearms [35]. From this random selection of unsolved crime statistics, you can clearly see how difficult it is for 'seasoned experts' to solve a crime and to determine the manner of death. To complicate the issue further, imagine these investigations being conducted by someone with no crime scene investigation experience.

In my estimation, the 2017-2022 statistics regarding unsolved murders can largely be attributed to a combination of unqualified crime scene investigators, limited evidence, and the actions of someone well-trained in covering up a murder. These failures in solving crimes are further compounded by inadequate investigation techniques, poor evidence collection, and failure to maintain proper chain of custody. Despite the advances in forensic technology and modern investigative tools, which allow for the collection and preservation of evidence, it is perplexing that a higher percentage of these crimes are not solved. Now, let's turn back to November 10th, 1969, a time when investigators lacked the sophisticated tools, we have today. However, it's crucial to note that, despite the limitations of the era, they still had the responsibility to conduct a thorough investigation, one that should have included photographic documentation and proper evidence collection.

The JAGMAN

It is important to note that the JAGMAN is a resource used by military officers to guide them through a difficult investigative process. This manual is large and complex, and it attempts to cover investigations from A through Z. However, the manual is only as good or as thorough as the person using it and can be quite overwhelming for the untrained investigator. To give you some understanding of how cumbersome the JAGMAN really is, the Judge Advocate General's Core had to develop a "JAGMAN Investigations Handbook" to assist first-time users of the manual. This JAGMAN handbook is a fraction of the size of the actual JAG manual. Here is a quote from the 2015 JAGMAN Investigations Handbook:

> *"This handbook is designed to assist commanding officers and investigating officers with the 'administrative investigation' process. There is no substitute for a working knowledge of Chapter II (Administrative Investigations) of the JAG manual; this handbook is not designed to replace reference to, and study of, the source document. Rather, this publication gives you a simplified "nuts and bolts" summary to initially orient your approach to the investigative process. The organizational approach to this handbook is to discuss how an investigating officer (IO) goes about conducting a Preliminary Inquiry, Command Investigation, or Litigation Report Investigation [36]."*

As a retired Naval Officer, I know for a fact that most officers assigned as the JAG investigation officer are not fully qualified for such an important investigative process, but the U.S. Navy sees it differently and provides these officers with the JAGMAN and the JAGMAN Handbook to walk them through the process. As a commissioned officer in the U.S. Navy, you have no choice in the matter. Once you are selected as the investigator, you accept your assigned duty and get the job done to the best of your abilities. LT Jacobs did just that, as he was tasked with the enormous responsibility of conducting an investigation into the death of CMCN Daniel Joseph Davidenko, and he executed this duty to the best of his abilities.

Typically, a majority of these military JAG investigations that involve a death usually revolve around aircraft crashes, heavy equipment malfunctions or misuse, shipboard fires, and accidents from working in a dangerous environment. None of these scenarios fit the circumstances surrounding the death of Danny. According to LT Jacobs' memory of his investigation, he stated that not one sailor could recall seeing Danny for the preceding 72 hours. Within this 72-hour period, the command conducted a full day's search of the entire base, with no luck in finding Danny. Ironically, the search effort was unsuccessful, and shortly after the search ended, somehow, miraculously, Danny's body was accidentally found by the supply officer. If LT Jacobs' memory is correct, that Danny had not been seen for more than 72 hours, and the search effort produced no results, then this alone should have indicated that a criminal investigation was necessary once his body was discovered. The unexplained 72-hour period alone would lead a seasoned investigator to presume that a crime had occurred until proven otherwise.

So, what does it take to become a Crime Scene Investigator (CSI)? CSIs and forensic science technicians undergo extensive classroom instruction and hands-on training before they are qualified to process crime scenes. Many law enforcement agencies require new CSIs to complete multiple training programs before they are even allowed to assist seasoned investigators. A newly hired CSI typically works under the supervision of an experienced investigator, gaining practical experience and honing their skills before they can operate independently or lead an investigation. Even with all this training, the CSI's investigation can be jeopardized with contamination because the CSIs and the forensic experts arrive at the scene after the first responders (Policemen, Firefighters, Emergency Medical Technicians, etc.).

CSIs and forensic experts are the ones specifically trained on how to preserve a crime scene. However, many of the first responders to a crime scene do not initially concern themselves with preserving or preventing contamination of the crime scene. This is due to a multitude of factors, like an active shooter situation in progress, walking through the crime scene to clear the area and make it safe, and searching for victims or perpetrators. On top of this, each first responder adds more foot traffic to the scene, and if the weather is bad, this will further destroy evidence. I am not trying to degrade our first responders; I just want to be clear that their initial job is to respond, protect, and get control of the situation before any form of investigation begins. We need to keep in mind that first responder situations can last for minutes, hours, or even days, depending on the situation.

The Facts as Described by LT Jacobs

During my discussions with LT Jacobs, he stated that Danny was last seen after work on Friday, November 7th, 1969, at approximately 1730 hours (5:30 PM) and that Danny's body was found on the following Monday, November 10th, 1969, at approximately 1800 hours (6:00 PM). Over this 72-hour period, no one recalls seeing Danny. According to LT Jacobs, the weather was clear, dry, and sunny during the period between Danny's last known sighting and the discovery of his body.

While the weather may seem like a minor detail, it is, in fact, a critical factor in the investigation. Why? Because inclement weather, such as rainstorms, could have hampered search efforts and, more importantly, compromised physical evidence at the scene. However, given that the weather remained dry, warm, and sunny over those three days, most, if not all, physical evidence should have remained intact. This raises an important question: If the environmental conditions were favorable for preserving evidence, why were there no apparent signs of blood or other crucial forensic indicators at the scene?

After the command determined that Danny was missing, the commanding officer dispatched a six-man search party with the responsibility of thoroughly searching every inch of the base. Despite this effort, the search team failed to locate Danny's body. This raises a critical question: How did a six-man search party fail to find his body on a relatively small base? If Danny's body was in the CBMU-302 supply laydown area, an area within the base, then why wasn't it discovered during the initial search? Was the search incomplete, improperly conducted, or hindered in some way? Or does this suggest that Danny's body was placed in the supply laydown area sometime after the search had already been conducted? Approximately one hour after the search for Danny ended, his body was discovered in a clearly visible area within the supply laydown area. According to LT Jacobs, Danny's body was found accidentally by supply officer LTJG Underwood while conducting an inventory check at 1800 hours.

LT Jacobs told me that the search party conducted an extensive search but failed to locate Danny's body, yet during their search, they did find a second illegal handgun in the Mechanics Shop on the CBMU-302 compound. This raises even more serious questions. How could a second illegal firearm, small and easily concealed, be located during the search, while

Danny's body, a significantly larger and far more visible object, went undiscovered until later in a conspicuous area? Even more troubling is the fact that the second illegal handgun was identical to the one found just a few feet away from Danny's left hand (according to LT Jacobs). This coincidence warrants further scrutiny. Was this truly an unrelated discovery, or could this other firearm be connected to Danny's death? Only a ballistics test on both weapons could have determined if either weapon was used to kill Danny. Upon later review of my interview notes with LT Jacobs, I realized I had failed to question him about the second identical pistol found in the mechanics shop where Danny worked. I must assume this gun was not preserved as evidence and was destroyed by LTJG Underwood's armory, raising further concerns about the thoroughness of the investigation.

How is it possible that a six-man search party was able to locate a small handgun hidden in an obscure location within the Mechanics Shop (Danny's assigned work area), yet failed to find the body of a full-grown adult male and a similar handgun in a relatively open area on base? This raises serious questions about the thoroughness of the search, the timeline of events, and whether Danny's body was placed in the supply laydown area after the search had already concluded. Lastly, while not entirely unusual, another question arises: Why was the head of the supply department, LTJG Underwood, conducting an inventory check after normal work hours? According to LTJG Underwood's own statement to me: *"While walking about one afternoon and informally inspecting the CBMU construction supply yard after work to ensure it was all properly stowed, I came across the body of the young Seabee in question."*

Again, I am not personally accusing LTJG Underwood of wrongdoing or questioning his motives. I am simply noting that a seasoned investigator would view his statement as concerning, prompting several important questions. Why was it necessary for the supply department head to conduct this seemingly routine after-hours inventory inspection at the end of the workday? If this inventory was necessary, then why wasn't it conducted by one of LTJG Underwood's junior supply clerks? Why did LTJG Underwood feel that he needed to conduct this inventory at that particular time, and why did a six-man search team, which had just finished an extensive search, fail to find Danny's body in this relatively open area? Once LTJG Underwood noticed what appeared to him as a "sleeping sailor" in his supply area, why didn't he immediately confront the sailor or, at a minimum, check on the sailor's health and welfare? Standard protocol requires senior enlisted personnel or officers to initiate an immediate welfare check on any sailor discovered sleeping or unconscious in an unauthorized or unusual location.

LT Jacobs, LTJG Underwood, and ENS Kelly each had slightly different recollections of the events that day. According to ENS Kelly, LTJG Underwood was at the ballpark, drinking a beer and appearing deep in thought when he noticed ENS Kelly walking by. At that moment, LTJG Underwood said he had noticed something in the supply laydown area and asked ENS Kelly to accompany him to investigate, believing they may have found the command's missing sailor. This version of events raises several questions. Why does ENS Kelly's account differ from that of LTJG Underwood and that of LT Jacobs? If LTJG Underwood had already discovered Danny's body during his inventory check, why did he hesitate to investigate or to tell someone immediately? Why did he later act as though he had just noticed something suspicious from a distance? Why did it take 15 minutes and a cold beer at the ballfield before LTJG Underwood thought to go back and check on a sailor he suspected might just be sleeping? These statements cast doubt on the sequence of events and warrant a closer examination of the timeline surrounding Danny's discovery.

LTJG Underwood stated that he vividly remembers Danny holding a large wooden cross in his right hand, a detail corroborated by multiple unit members who witnessed the scene. However, another troubling revelation is that LTJG Underwood does not recall being interviewed for any JAG investigation. While he acknowledged that the events took place a long time ago and that he is now in his 80s, his lack of recollection is not just a memory lapse; it was confirmed by LT Jacobs himself. LT Jacobs stated outright that he did not interview LTJG Underwood during his investigation. This raises a critical question: Why wasn't LTJG Underwood interviewed? He was the first person to find Danny's body, which should have made him a key witness and the primary lead in the investigation. Failing to interview the first person on the scene represents a significant oversight, one that could have compromised the integrity of the entire investigation. Although the above questions and the following may appear accusatory, it is entirely appropriate within the framework of a crime scene investigation, where all aspects of the scene and all possible explanations must be examined.

As the supply officer, LTJG Underwood was responsible for overseeing all incoming and outgoing supplies, managing the entire supply department, and supervising the petty officers assigned to his command. Given his role as a department head, it is highly unusual that he would personally conduct a routine or impromptu inventory check after normal working hours. Typically, such inventories would be performed by junior petty officers, not the head of the department. The fact that LTJG Underwood was the one to find Danny's body during an uncharacteristic after-hours inspection raises serious questions.

Why was he there at that specific time? Was this truly a routine check, or was there another reason he was in the supply laydown area? Under normal investigative procedures, the first person to find a body, especially under unusual circumstances, would be considered a person of interest until proven otherwise. I also find it deeply troubling that it seemingly never crossed LT Jacobs' mind to question or interview LTJG Underwood about the discovery of Danny's body.

In any thorough investigation, the person who finds the body is always questioned, often multiple times, to establish critical details about the discovery, timeline, and surrounding circumstances. I am not aware of any legitimate investigation where a key witness, the first person to find a body, was not questioned at all. This omission raises serious concerns about the thoroughness and credibility of the investigation into Danny's death. The information I have gathered regarding the circumstances surrounding Danny's death raises a series of critical investigative questions. 1) Did the search team conduct a thorough and complete search of the base to locate Danny's body? If they did, was the supply department's laydown area included in that search? According to LT Jacobs, the CBMU-302 compound was roughly the size of four football fields, and the entire CRB base was either equivalent to that size or slightly larger. Given this relatively small area, it should not have been difficult for a six-man team to complete an exterior search of the base in just a few hours. However, a search of the buildings and all interior spaces, where a body could potentially be hidden, requires significantly more time and careful attention. Did the search team account for every possible interior hiding spot? Were all buildings thoroughly searched, and were all potential locations for a person to hide (even those less obvious) examined in detail?

This leads me to my second question: 2) If the laydown area was thoroughly searched, how did the search team miss Danny's body in such an open area? The laydown area was not secluded or obscure, and it wasn't enclosed by any fences. Based on the size of the compound, it would take only a few minutes for one person to walk through and around the perimeter and rows of supplies in the laydown area. If the laydown area was indeed searched earlier in the day and Danny's body was not found, this opens up a troubling third and fourth question: 3) Was Danny's body placed there after the search had already been conducted? 4) Was it moved to that location after he was shot and killed elsewhere? The fact that the body was not discovered during an earlier search and was later found in the supply area raises serious questions about the integrity of the search process. The fifth question that arises is: 5) After Danny's body was found, did anyone ask whether the search party had checked the laydown area during their search? It seems almost impossible that a six-man team would miss such a small, open area. To clarify this, I posed the following question to LT Jacobs: "After Danny's body was found and during your JAG investigation, do you remember if you or anyone asked the six-man search team if they had searched the laydown area, or did they simply miss it?" In response, LT Jacobs stated: *"The search team probably did a quick pass on the laydown area because its regular configuration didn't offer 'hiding places.' Probably an error by whoever covered that area."* This response raises further concerns. Was the search of the laydown area really as thorough as it should have been? Did the team overlook potential hiding spots or fail to properly search the area? The fact that Danny's body was found there, despite being located in plain sight, suggests that the initial search may not have been as detailed as it should have been.

Gaining access to LT Jacobs' JAG investigation should provide a list of the CBMU-302 members who were assigned to the six-man search party. Each of these six individuals must be asked the critical question: Did you search the supply laydown area? The answer to this question is crucial in determining the manner of death. Based on the medical examiner's report, Danny's body exhibited post-mortem signs indicating that he had been dead for a significant period of time before being found. The official autopsy report should provide a clearer timeline of Danny's time of death. The examiner should be able to determine whether Danny died 12, 24, 48, or 72 hours prior to the supply officer discovering his body. Determining the time of death is crucial because it provides a time frame for when Danny was fatally shot and could help confirm if his body was placed in the laydown area after death. Additionally, the fact that no one reported seeing any blood at the scene could also support the theory of foul play or, at a minimum, require further investigation as to why no blood was found at the scene.

Because there is rarely more than one opportunity to obtain evidence from a crime scene, the investigation must be methodical and complete. In most of today's investigations, the CSIs and police use cameras as well as video recordings to capture every aspect of the entire crime scene to enhance the evidence collection. In particular, photographs of blood stain patterns obtained at the scene can provide critical details of what took place at the crime scene. The blood-pattern analysis, for example, can determine whether a victim was standing still, walking, or running at the time of death. Another great investigative tool is the use of toxicology. Toxicology evidence examines blood and tissues to ascertain the presence and quantity of drugs, alcohol, or poisons in a person's body. Toxicology reports can give the investigator insight as to the victim's condition at the time of death. A good example would be if a victim was found dead in a car crash, the toxicology report would let the investigators know if the victim was high on drugs, intoxicated from alcohol, or even poisoned before the crash. If the victim was poisoned, the car crash would have been an attempt to cover up the actual cause of death, which in this example would have been murder.

However, most crime scenes these days involve the use of a firearm. Crime scene investigators have been very successful when using the collection of firearm evidence to help solve a crime. During LT Jacobs' JAG investigation, a gun was found approximately 4-5 feet away from Danny's body (near his left hand). The gun was determined to be a non-military issued weapon, which LT Jacobs termed as an "orphan gun". According to LT Jacobs, the gun found near Danny's body had a single bullet casing jammed in the ejection port, and no other bullet casing was found at the scene. However, the initial medical examination of Danny's body revealed one entrance wound and one exit wound, which would suggest a single shot passed through his body. Additionally, the medical examiner also recovered a bullet from Danny's chest. The bullet extracted from Danny's chest, as described by LT Jacobs, was a completely intact full metal jacket (FMJ) bullet, with only a slight bend at the tip. If the bullet was fully intact, it raises an important question: What caused the 2 centimeter exit wound in the middle of Danny's back? Given that the FMJ bullet recovered from Danny's body was unfragmented, we can reasonably infer that this bullet did not cause the exit wound. This evidence strongly suggests that Danny was shot twice, with one bullet recovered from his chest and a second bullet (unlocated) that caused the exit wound in the middle of Danny's back. The fact that the bullet retrieved from Danny's body was intact, we can determine that it did not cause the exit wound. This fact raises significant questions about the circumstances surrounding Danny's death and whether the investigation accounted for all potential projectiles (i.e., bullets).

The fact that there was only one entrance wound and one exit wound suggests that the gun was pressed up against Danny's chest when the two shots were fired. From the medical report, it can be concluded that one bullet passed through Danny's body, leaving an exit wound in his back, while a second bullet remained lodged in his chest. It's also critical to note that the M1911 pistol is a single-action, recoil-operated, semi-automatic firearm. This means that the trigger must be pulled twice to fire two bullets, one shot for each pull of the trigger. This is in contrast to fully automatic weapons like the M16 rifle, where a constant pull of the trigger would fire multiple rounds. The fact that two separate bullets were fired with only one entrance wound suggests that the gun was held tightly against Danny's chest and that the trigger was pulled twice. This raises significant questions about the manner and intent of the shooting, suggesting a deliberate act rather than a single, accidental shot.

Every firearm will leave individual markings on the bullet and the bullet cartridge when it is fired. These markings can be examined to determine whether a specific bullet was fired from a specific weapon. Firearm evidence is critical in crimes against a person. While the comparisons of bullets and bullet cartridges are the most common examination requested by CSIs, there are other firearm-related examinations, such as distance determinations based on powder residue around the entrance wound. Additionally, evidence of who fired the weapon can be obtained from analysis of gunshot residue collected from a person's hands or other body surfaces. Other firearm-related evidence is the location and condition of the firearm in relation to the victim. The crime scene should be diagrammed with recorded distance measurements and photographed before recovering and securing the firearm. Only after all this is completed should the firearm be carefully unloaded and only handled by personnel with rubber gloves to preserve fingerprint evidence. All bullets, bullet cartridges, and magazines should be processed as evidence. Fingerprint evidence is also key to a crime (death) scene investigation. This evidence may include fingerprints on the firearm itself, the trigger, the magazine, the bullet, and the bullet casings.

From all the information I gathered surrounding Danny's death, it is evident that several critical pieces of evidence were not properly processed or archived. These pieces of evidence could have been crucial in helping investigators determine the manner of Danny's death, whether it was accidental, suicide, homicide, or murder. The key pieces of evidence that I believe were mishandled or not properly processed include: The gun(s), the bullet(s), the bullet casing(s), the fingerprints on the gun, gunpowder burns on the skin, gunpowder residue on the hands, the missing bullet (the one that exited Danny's body), the missing casing from the second bullet, blood splatter marks on surrounding supply crates, bullet holes in Danny's clothing (or the lack of bullet holes in Danny's clothing), bullet holes in nearby crates, the lack of blood observed at the scene, the location of the gun relative to the body, and the position of Danny's body relative to the blood splatter on the crates. Additionally, many other critical details were likely overlooked or not investigated in depth. The failure to process and preserve these pieces of evidence severely hampered the investigation and left many questions unanswered.

If this evidence had been collected, photographed, and properly processed, it would have been possible to reconstruct the crime scene and potentially provide valuable leads toward determining the actual manner of death (even years later). However, during the investigation, these critical items and other pieces of physical evidence were neither considered nor processed as evidence and were not retained for further examination. The lack of physical evidence collection for forensic examination hindered the ability of the investigator to accurately assess the circumstances surrounding Danny's death. Additionally, the gun and bullet casing were never fingerprinted, leaving unanswered questions about who handled or fired the weapon. This raises further critical questions: Whose fingerprints were on the gun? Were the fingerprints on the gun Danny's or someone else's? Why didn't the investigator collect fingerprints from the gun, bullet, and casing? Why wasn't a ballistics test conducted on the gun found near Danny's body, as well as on the gun recovered from the CM shop? Such testing could have determined which weapon was used to kill Danny, or whether neither weapon was involved. In the absence of a ballistics test, there is no way to determine whether the gun found at the scene or the one found in the mechanics workshop was the weapon that killed Danny. Since both guns were destroyed, we cannot rule out the possibility that the gun at the scene was deliberately placed to create the illusion of an accidental or self-inflicted death.

The failure to process the gun(s), casing, and bullets as evidence raises critical concerns. Why were the gun(s) destroyed by the base Master at Arms and not preserved for further examination? This oversight significantly undermines the investigation and leaves the truth in question. Additionally, did the forensic examination include gunpowder residue tests around the entrance wound, his clothing, and on his hands to help determine how close the weapon was to his chest and whether Danny had handled the gun before he was shot? A gunpowder residue test would have provided invaluable evidence as to who fired the weapon. Gunpowder residue on Danny's hands could suggest accidental discharge or suicide. However, if Danny had gunpowder residue on his hands, it might also indicate that he struggled with the assailant and attempted to grab the gun during the confrontation. Conversely, if Danny's hands did not show any gunpowder residue, it would likely suggest homicide or murder as the cause of death. Furthermore, we must not overlook the 2" x 1" burn mark on Danny's left elbow. Could this burn have been caused by the gun barrel after Danny was shot twice in the chest, or does it hold significance in the sequence of events?

It is important to note that if the barrel of the weapon was placed against Danny's body when the weapon was fired, then there would be a definite pattern around the entrance wound. If the weapon touched Danny's body, it would have left an 'abrasion ring' on his skin. The gun's barrel would have left an imprint of the weapon's barrel, which would be helpful evidence to identify the exact weapon used to kill Danny. If a weapon was in close proximity to Danny's body, then it would have left gunpowder residue on his clothes, skin, and possibly his hair. The distance of the gun from Danny's body can be estimated based on the size and pattern of the gunpowder residue deposited on Danny's skin. Why is the distance of the weapon to the victim important? The distance can either support or eliminate a manner of death. For example, if the gun powder residue is within 1-18 inches, it would support all manners of death and could support possible theories of accidental or suicidal death. However, if the gun powder residue is beyond this distance, it would limit the manner of death to homicide or murder. It's also important to understand that entrance wounds are normally smaller and symmetrical in comparison to exit wounds. Exit wounds are larger and often associated with hanging skin and tissue around the exit wound. Exit wounds will often bleed profusely as they are larger and often leave large amounts of blood at the crime scene.

Investigators need to look beyond the wounds on the body and search the crime scene, looking not only for the weapon involved but also for spent shell casings and loose rounds that were fired but did not hit their intended targets and embedded themselves in nearby objects like walls, doors, the ground, and, might I add, supply crates surrounding the body. Bullet holes in the victim's clothing are essential forensic evidence, used to reconstruct bullet trajectories, assess firing distances, and determine how many times the victim was shot. If an investigator finds a victim with multiple gunshot wounds and no shell casings in the area, this will indicate two plausible scenarios: 1) the body was not shot at that location, or 2) the perpetrator made sure minimal evidence was left behind at the scene.

From LT Baker's medical report and the fact that the bullet removed from Danny's body was fully intact (did not fragment), it is easy to conclude that Danny was shot twice. Of course, at the time they found Danny's body, LT Jacobs could not have made this determination until after he assembled all the crime scene information, including the medical death report. I believe that if LT Jacobs collected all the evidence, read the medical report, and consulted with LT Baker and other forensic pathologists, he would have come to the conclusion of a two-bullet theory during the early phases of his investigation. Therefore, it is imperative that the investigator complete a thorough search for evidence at the scene, to include all spent bullets and bullet casings, even bullets embedded in surrounding objects. If this was completed and no bullet was found lodged in the objects behind Danny and there was no spent shell casing at the scene, then this could indicate that Danny was NOT shot and killed in the laydown area. However, if the bullet that passed through his body was found along with the additional missing bullet casing, then this would help conclude that he was shot and killed where his body was found.

LT Baker reported one entrance wound and one exit wound on Danny's death certificate (medical death report). The exit wound was larger than the entrance wound. The investigative officer (LT Jacobs) stated that he did not see any blood at the scene and assumed that it was all internal bleeding. It was his belief at the time that there was only one bullet, and it was lodged in Danny's chest, and based on that theory, that is why he didn't remember seeing any blood around Danny's body. Also, LT Jacobs stated that he did not remain at the scene once the medical team arrived, so he did not see them move Danny's body or get the chance to see if there was blood or other bullet casings under his body or near the scene. From the medical report, we know there was an exit wound, and therefore, there should have been a significant amount of blood on Danny's clothes and on the ground under his body, again, only if he were shot and killed in that location.

Forensic pathologists conduct autopsies and can establish a cause, manner, and an estimated time of death, which often provide clues regarding the physical characteristics of the person or persons responsible for the crime [33]. An autopsy is critical in unexplained, unattended (no witnesses) deaths. The forensic pathologist will closely examine the wounds, blood, and tissue to ascertain how the person died. The documenting of post-mortem changes to the body assists the investigator in explaining the body's appearance following death. Inconsistencies between post-mortem changes and body location may indicate movement of the body and validate or invalidate witness statements. It's essential that forensic pathologists be given only facts pertaining to the case and access to any of the physical evidence (i.e., Gun, knife, etc.) directly related to the case (if needed). To ensure an unbiased determination of the cause and manner of death, they must not be influenced by circumstantial evidence, rumors, or hearsay, which can lead to false or misleading conclusions.

The first page of Danny's official CERTIFICATE OF DEATH, NAVMED M (REV. 4-58), reported that Danny's date and time of death on November 10th, 1969, was approximately 0500 (5:00 AM). The second page of this same document reported that Danny's date and time of death (as reported by the medical officer) was 1815 (6:15 PM) hours on November 10th, 1969. This was the first thing I noticed when conducting my research. The on-scene medical officer listed the time of death as 1815 (6:15 PM) hours because that was when he first examined Danny's body, and it was before a full autopsy was completed. Also, there is the fact that LT Baker was a doctor and not a forensic pathologist, specifically trained in these post-mortem forensics. I also reviewed a correction document that called for a change on the death certificate. This correction concerned the color of Danny's hair because the death certificate reported that he had blond hair, and the autopsy reported that he had brown hair. These two discrepancies may seem trivial, but they provide yet another indicator that the investigation was incomplete and that errors were made in the reporting.

The death certificate stated that the examination of Danny's body revealed acro-cyanosis (bluish or purple coloring of the hands and feet), dilated and fixed pupils, no respiratory or cardiac function, and segmented hemostasis of the retinal veins. It also reported that an anterior-posterior and a left lateral chest x-ray were performed and revealed that a metal projectile (FMJ Bullet) was still present in the right posterior chest cavity. During an autopsy, the Forensic Pathologist will conduct wound ballistics, which is the examination of a gunshot wound to determine and document the type, size, shape, site, and location of the gunshot wound. They are required to investigate and deliberate as to the cause and manner of the injury, the entry and exit characteristics, the distance and direction of fire, and the vitality of the wound.

The term 'vitality' refers to determining whether an injury has been caused ante-mortem (preceding death) or post-mortem (after death). Conducting the external body examination provides the investigator with objective data regarding the single most important piece of evidence at the scene, the body. This documentation provides detailed information regarding the decedent's physical attributes, his/her relationship to the scene, and possible cause, manner, and circumstances of death [34]. Evidence or clues may be discovered once the body is removed. Inconsistencies between post-mortem changes and body location may indicate movement of the body. In addition, post-mortem changes to the body, when correlated with other evidence or information, can assist the investigators in estimating the approximate time and location of death [34].

In many death investigations, the manner of death may not be determined. So, what does it mean when the manner of death is undetermined? If a death is classified as "undetermined," it means that the examiner cannot make a determination based on the physical body and the collected evidence. A medical examiner, at the conclusion of an autopsy, may classify the manner of death as undetermined if there is not enough evidence to make a determination. Despite substantial evidence, if a medical examiner cannot reconcile their findings with the intent surrounding the death, the examiner will often list the manner of death as undetermined in accordance with legal standards. In order to list a manner of death as homicide, the medical examiner must conclude from the autopsy that the death resulted from a violent act where a human being was responsible for the death.

As for the determination of suicide, the medical examiner must conclude beyond a reasonable doubt that the person intentionally meant to kill himself. Many suicides end up as an undetermined classification due to the lack of knowledge surrounding intent. Accidental cases present the exact same issues. Intent makes a big difference between homicide, suicide, and an accident. Since a medical examiner knows nothing of intent, if the autopsy does not conclusively tell him that a death was the result of an accident, he is still reluctant to call it homicide or suicide, resulting in an undetermined classification. [37]. As shown by the procedures required for a thorough crime scene investigation, completing such an inquiry demands extensive evidence collection and careful analysis. In contrast, the JAG investigation into Danny's death was incomplete and led by an investigator lacking the necessary qualifications. Numerous critical questions were left unanswered, significant mistakes were made, and evidence collection was either entirely neglected or mishandled. The limited evidence that was gathered appears to have been mismanaged, discarded, or destroyed. I have a great deal of sympathy for LT Jacobs, who was assigned an investigative task that far exceeded his training and experience in crime scene analysis. To make matters worse, there appears to have been a premature determination of the manner of death, either by the Commanding Officer or by LT Jacobs himself, before a proper investigation could be completed.

Will the truth behind the death of CMCN Daniel Joseph Davidenko ever be known? Based on the fact that no physical evidence was collected or preserved for future forensic analysis, the answer is no. We may never truly know how or why Danny died on the relative safety of the base at Cam Ranh Bay, Vietnam.

Chapter 11

Plausible Manner of Death Scenario's

The previous chapter (Chapter 10) outlined the fundamental procedures for processing a crime scene involving a death. It covered the requirements for collecting, recording, and documenting physical evidence and emphasized the importance of evidence collection in determining one of the five manners of death (MOD). In this chapter, the reader will explore four plausible MODs that may have contributed to Danny's death. Since medical reports confirm that Danny did not die of natural causes, this chapter will not include a scenario involving death by natural causes.

The following scenarios will focus on death by accident, death by suicide, death by homicide, and death by murder. These are the only 4 possibilities that could be considered for the events that took place in Vietnam on November 10th, 1969. I have yet to uncover any physical evidence suggesting how Danny was shot twice with a gun he allegedly found on base. Based on the medical report's findings, each of the scenarios that follow will begin with the fact that Danny was shot twice with a .45 caliber pistol. It is important to note that there were no witnesses to the events that led to his death. According to LT Jacobs, during his JAG investigation interviews, a few sailors claimed that Danny had found a pistol on the base. Post-death interviews are crucial to any investigation, but with the victim deceased, it is impossible to determine whether these witnesses were telling the truth or concealing it. Therefore, we must apply greater emphasis on the physical evidence collected and compare that evidence to the CBMU-302 personnel statements and their recollections of the events that may have led to Danny's tragic death.

In a death case when no witnesses are present, crime scene investigators must depend on the physical evidence collected at the scene to determine the manner of death (MOD). All my efforts to locate the official NIS death investigation report have yielded no results. My numerous requests for a copy of the JAG investigation report into Danny's death, submitted to the Office of the Judge Advocate General (OJAG), have also yielded no results. However, I still have one request for the JAG investigation that is still pending. Over the years, I have contacted numerous government agencies by letter, phone, and email, in an effort to secure Danny's official autopsy records. Despite these repeated inquiries, the autopsy and accompanying report have never been provided. However, I still have one last request for the official autopsy pending. To be fair, all the federal organizations I contacted made an attempt to help me find these documents. However, each responded with the same reply: *"We don't have these documents."* They did provide other organizations to contact, but those organizations also came up empty. Therefore, I will continue writing this book as I persist in my search for these documents, holding onto the hope of uncovering the truth. If and when I get these last few critical documents, I will use them to draft a final chapter that reconstructs the events surrounding that fateful day in November 1969, as investigated and recorded by LT Jacobs. Hopefully, there will be enough information and physical evidence to help determine Danny's actual manner of death (MOD).

In each of the scenarios below, I will ask you, the reader, to examine and debate all the collected research data, military records, as well as all the written and oral conversations I collected from my family, Danny's friends, and former members of CBMU-302. Most of the personal and firsthand information about Danny's tour of duty in Vietnam came from conversations through letters, emails, and phone calls with the following

CBMU-302 members, LT Jacobs (Executive Officer), LTJG Underwood (Supply Officer), LT Casey (Base Chaplain), ENS Kelly (Administrative Officer), and numerous other former enlisted members of CBMU-302.

In an effort to assist me with this book, LT Jacobs provided a hand-drawn map of the base, which included the locations of all facilities at Cam Ranh Bay. The firsthand information provided by LT Jacobs, the officer who conducted Danny's JAG investigation, will serve as the foundation for many of the questions both you, the reader, and I, the author, will need to consider as we explore the following MOD scenarios. Let's begin by looking at some probing questions that can help guide us in our investigation into Danny's death and the surrounding circumstances. Each one challenges the available information and assumptions, potentially leading to a deeper understanding of what happened.

Here's a short list (not inclusive) of questions an investigator might start with during an investigation:

1) How can we be certain that Danny discovered a non–military-issued pistol on base? A thorough investigation into the origins of the weapon found beside his body, along with the pistol recovered during the subsequent search, may confirm or deny how these firearms entered the CBMU-302 compound.

2) How did Danny obtain the ammo for this pistol he reportedly found? If the ammo came from an official or unofficial source, it raises questions about the chain of custody and security measures on the base.

3) What were the base policies and regulations regarding firearms and personal possession?

4) Why didn't Danny's shipmates report the alleged pistol to a higher authority? If the pistol was known among Danny's

peers, it could suggest either a breakdown in reporting or a deliberate concealment of information.

5) Why wasn't LTJG Underwood interviewed? According to LT Jacobs, he did not recall interviewing LTJG Underwood, the person who found Danny's body and the gun.

6) Did LT Jacobs have prior knowledge of the CO's letter sent to Danny's parents that predetermined the manner of Danny's death before he completed his investigation? This could reveal potential bias or premature conclusions that impacted the investigation's thoroughness.

7) When was Danny actually last seen?

8) Did Danny's roommate (CMCN Howard Decker) report him missing?

9) How did the six-man search party fail to locate Danny's body in the supply laydown area? This could point to issues with the search effort. Did the search team take the search seriously? Did they complete a thorough search of the base? If yes, then how did they miss seeing Danny's body in the supply laydown area?

10) Why wasn't the scene treated as a crime scene? Investigating the initial handling of the scene might reveal oversight or negligence in preserving crucial evidence.

11) Why weren't photos taken of the scene? Photos are vital for documenting the scene. This question raises concerns about whether critical evidence was lost or overlooked.

12) Why wasn't evidence collected, processed, and archived? This could suggest a failure to follow procedures or an intentional effort to minimize the scope of the investigation.

13) Why was evidence destroyed in the first place? The destruction of evidence raises serious concerns, suggesting the

possibility of foul play, negligence, or even an intentional cover-up.

14) Why didn't LT Jacobs notice or raise concerns about the exit wound on the medical report? From my conversations, LT Jacobs didn't recall the fact that Danny had an exit wound. Why were these key details overlooked?

15) Why didn't LT Jacobs understand that when the medical staff gave him the FMJ bullet retrieved from Danny's chest, it should have been processed as evidence, and not as a memento of this tragic event?

16) Who first reported the rumor that Danny received a 'Dear John' letter? Uncovering the source of this rumor could help determine whether emotional stress or interpersonal factors played a role in Danny's death. The source of this rumor and others could have been used to cover up a crime and to throw off investigators, making the scene look like suicide.

17) Why didn't anyone report actually seeing or reading the rumored "Dear John" letter? Examining the credibility of this supposed letter could shed light on its origins and whether it played any role in Danny's death. Yet no one interviewed by me, or by LT Jacobs, ever saw or read such a letter. Every person said the same thing: it was only something they had heard about.

18) Why would the supply department head (LTJG Underwood) conduct an after-hours inventory of the supply laydown area, instead of having one of his junior personnel perform these routine tasks? During my interview with LTJG Underwood, he stated that he found Danny's body while he was conducting an inventory of supplies.

19) Was NIS informed about Danny's death?

20) Who was ultimately responsible for informing NIS that a death occurred on base?

21) Did NIS conduct a formal death investigation? NIS involvement would have been crucial for a thorough investigation. The lack of an NIS investigation could suggest negligence or a failure to recognize the seriousness of the situation.

22) Why didn't LTJG Underwood immediately confront the suspected sleeping sailor he observed in the supply area?

23) Why didn't LTJG Underwood check to see if this sailor was hurt or incapacitated (simple well-being check)?

24) Why didn't LTJG Underwood immediately inform the chain of command that he saw a sailor in his supply area? Understanding what prompted this delay would help clarify whether the sighting was relevant, misinterpreted, or a possible cover-up.

25) Why did LTJG Underwood wait for ENS Kelly before going back to the supply laydown area to check on the sailor's condition or well-being? Why didn't he immediately approach the suspected 'sleeping sailor' and make him leave this area?

26) On his first sighting of a 'sleeping sailor', how close was LTJG Underwood to Danny's body when he said he could read the first two letters of Danny's last name and the word SEABEE on his uniform from the seat of his jeep? That detail matters because his ability to see that level of detail depends entirely on how far away Danny's body actually was from LTJG Underwood's seated position within the jeep and his vision. How good is his eyesight? Does he have 20/20 vision? Does he wear glasses to correct his vision, and was he wearing them at the time?

27) When did LTJG Underwood first notice the gun near Danny's body? What was the exact distance and position of the gun relative to his body, and more specifically, Danny's hands? Did he or anyone else touch or move the weapon?

28) Who was actually present at the scene? List of names.

29) Who actually moved Danny's body?

30) Where is the rumored 'Dear John' letter? Did they conduct a search for this rumored letter? If it existed, then who found it, and why wasn't it kept as evidence and submitted with the JAG investigation?

31) Why was there no blood at the scene?

32) Why were there no bullet holes in Danny's clothes?

33) Who searched Danny's body for potential evidence (search clothes pockets)?

34) Etc.

The answers to these questions listed above should raise additional questions about the events and the urgency and timing of the command's response to the situation. There are many more questions that need to be asked and answered, but the few listed above should help the reader to dissect and evaluate all available information. As you examine these findings, you will develop additional questions, both about what truly happened to Danny and whether the investigation was complete, accurate, or affected by a premature conclusion. Ultimately, you will have to decide for yourself what you believe happened on that fateful day and make your own determination about Danny's manner of death (MOD).

The base map below clearly illustrates that Danny's barracks and the CBMU-302 compound were on opposite sides of the Naval base at Cam Ranh Bay. Assuming that Danny's weekday routine was consistent; each morning, he would wake up in his barracks room, shower, and get dressed for work. After a quick breakfast at the chow hall, he would head to morning quarters before making his way to the auto body shop where he worked. The blue line on the first map represents what may have been

Danny's typical workday foot traffic, while the red line on the second map represents what may have been his typical security duty and liberty foot traffic while assigned to CBMU-302. On a normal day, Danny would follow this typical path during his daily routine. If he were running late, his normal routine might change, and his route to morning quarters could take him directly through the supply laydown area, traveling west. However, if Danny decided to leave early and go to the body shop before morning quarters, then his direct path to the body shop would take him along the main road. In addition to Danny's normal traffic pattern, the entire enlisted component of the command would have traveled a similar path, depending on their assigned company and workspace. The officer quarters were located just south of the main road, and due to their senior ranks, officers would have used vehicles (U.S. Military Jeeps) to drive to their respective offices within the CBMU-302 compound. According to LT Jacobs, he recalls that most of the troops would walk along the road after chow to reach the CBMU-302 compound.

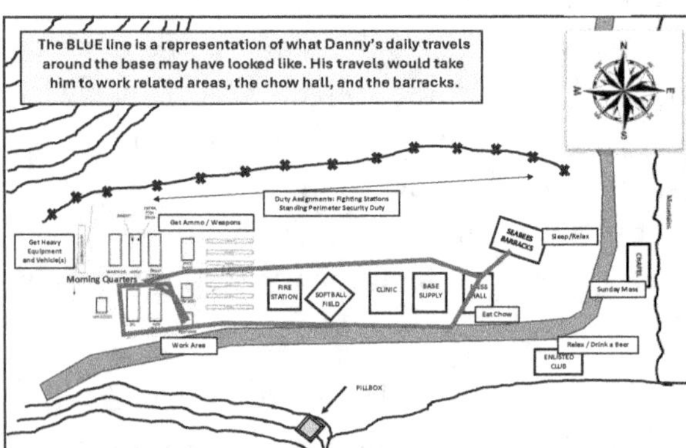

Representation of Danny's Probable Daily Work Travel Routine

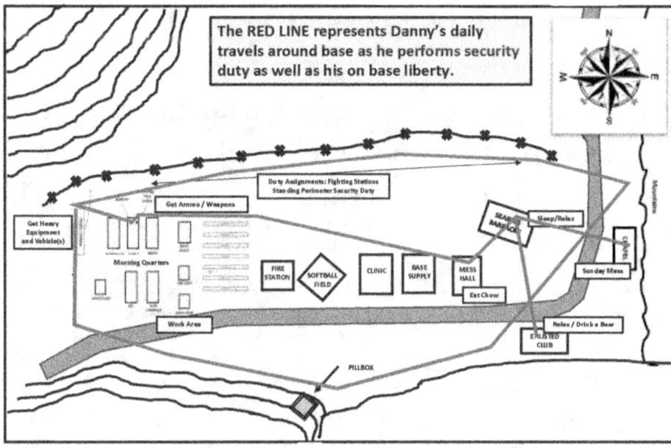

Representation of Danny's Probable Daily Duty and Liberty Travel Routine

From Danny's barracks room, it was a short walk to the base chow hall. From there, he had multiple routes to reach the CBMU-302 compound. One of those routes passed through the supply laydown area, a section filled with rows of supply crates and materials. The supply laydown area was situated between the main base facilities (chow hall, supply, clinic, enlisted club) and the CBMU-302 headquarters. It would be fair to assume that a majority of the enlisted CBMU-302 members (approximately 257 enlisted men), including Danny, would follow a similar path from the barracks to the chow hall, and from the chow hall to work, which would put them on a 'beeline' through or just south of the supply laydown area. The term 'beeline' is used in the military to describe walking in a straight line from one point to another. If a sailor was running late for morning quarters, taking a 'beeline' through the supply laydown area would be the fastest route. The laydown area had no fence around it, making it an easy and direct walking path to various company command buildings and other facilities adjacent to the supply laydown area. On weekends, only personnel on duty would be walking through or around the laydown area, as the rest of the command would likely be enjoying liberty on base and would not be near the laydown area or their work facilities.

Understanding the base layout and the locations Danny frequented during his work week is crucial. Given the base's relatively small size, the command's enlisted personnel may have opted for a direct path through the supply laydown area to reach their workshops from the barracks. A look at the base map reveals that this direct path was the shortest route from the east side to the west side of the base. On the morning of November 10th, 1969, it's reasonable to assume that some CBMU-302 personnel, who may have been running late, may have walked through the supply laydown area on their way to morning quarters outside the HQ building. On Monday morning, everyone was heading to the command's morning quarters for roll call, where they would hear the CO and XO outline the plan for the day and share any pertinent news. Given the usual traffic through the area, what are the odds that, on this particular day, no one walked through the laydown area on their way to morning quarters? Is it possible that a command this size (over 250 personnel) and not a single person walked through the supply laydown area that morning? Even if personnel had walked through the supply laydown area, they would have had to travel between the two specific rows of supplies to stumble upon Danny's body. However, what if they did travel between the two rows of supplies where Danny's body was found that evening, but nobody saw his body? This would suggest that his body may have been placed there after morning quarters. As an investigator, you must look at every possibility and search for evidence to confirm or deny its veracity.

During my conversations with LT Jacobs, he provided an answer regarding the command personnel walking through the supply laydown area. LT Jacobs explained, "*Everyone went to morning quarters on their own schedule. The main road was shorter than taking a 'beeline' through the supply area, with less dust and no maze of materials to navigate. As a result, most of the troops used the road to go back and forth, avoiding the supply laydown area because it was faster.*" LT

Jacobs' statement confirms that walking through the supply laydown area was not prohibited; however, it was not a common practice among the troops. This does not mean that no one passed through the area, only that LT Jacobs did not observe many troops using this direct 'beeline' path to work.

Within an hour or so after Monday morning quarters, and after an initial search by a single individual failed to locate Danny in his barracks room, the CO ordered a search team to be assembled to find the missing sailor. A six-man team searched the entire base but found no sign of Danny. As the lead and sole JAG investigator, LT Jacobs should have immediately questioned the search team to address a critical investigative question: Which person(s) from the search team physically searched the supply laydown area? The answer to this question could have significantly impacted the investigation. Did the team overlook the area, or did they search it and find nothing? Another essential piece of information was the exact time of the search, which could have been crucial in determining Danny's time of death and whether his body was placed there after the search. The answer to this question remains a mystery. If the search team searched the laydown area and Danny's body was not there, then when and how did his body get there? There is also the possibility that the search team simply forgot to search the supply laydown area. Was this a mistake or an oversight, or did team members assume that someone else had already checked that area, leaving it unchecked? If the area was not searched, Danny's body may have been there the entire time and was simply overlooked. However, if the area was searched and no body was found, this would suggest that Danny's body was placed there sometime after the search, pointing investigators toward foul play, and the investigation would trend towards homicide or murder.

The medical report indicates that Danny had been deceased for some time (approximate time of death was 0500 / 5:00AM), and no one reported hearing the two gunshots that morning. In the next few scenarios, we will examine four plausible MODs (accidental, suicide, homicide, and murder). Given the limited information and evidence at this point, it is ultimately up to the reader to determine the most probable conclusion regarding the manner of Danny's death (MOD).

Without witnesses or documented evidence, it is difficult to construct scenarios in which Danny may have committed suicide by shooting himself twice in the chest. Similarly, it is equally challenging to suggest that he accidentally shot himself twice while mishandling a gun. In both scenarios, the reader is asked to accept the theory that Danny, who was right-handed, either intentionally or accidentally pressed the barrel of a gun against his chest with his left hand and fired twice, all while holding a large wooden crucifix in his right hand. Likewise, without witnesses or documented evidence, it is difficult to construct a scenario in which Danny's death resulted from homicide, whether he was shot during an argument or disagreement, or to suggest that he was murdered by someone who intentionally used the gun found near his body. To compound the challenges surrounding Danny's manner of death (MOD), we do not have the official JAG investigation report, the official NIS investigation report, the official military police report, or the complete autopsy report. Accordingly, the following fictitious yet plausible scenarios were created based on the documentation collected and the interviews I conducted with members of CBMU-302. This data was then combined with examples from real-world crime investigative reports and statistics to develop plausible MODs. Please keep in mind that all of these scenarios are possibilities of what could have led to his manner of death.

Fictitious Scenario 1- Death by Accident

In this fictitious scenario, while spending time with friends on the beach, Danny wandered off to explore an old Vietnamese artillery bunker overlooking the shoreline. Inside, he discovered a .45 caliber pistol and some ammunition. Excited by his find, he brought the gun back to his barracks to clean and examine it more closely. After getting it into what seemed like working order, he decided to test it. Aware that he wasn't allowed to have the weapon, one early morning, he went to a secluded spot (the supply laydown area) where he decided to test-load the pistol. After loading the pistol, he accidentally pulled the trigger twice, shooting himself in the chest. In this accidental death scenario, we have to accept the premise that Danny found this gun and ammo. Furthermore, we must accept that he somehow, after being formally trained on this weapon in boot camp, mishandled the weapon, causing the semi-automatic pistol to accidentally discharge twice. During this accidental shooting, two bullets exited the gun and struck Danny in the exact same location on his chest, creating only one entrance wound and only one exit wound. After the two gunshots, Danny collapsed to the ground and dies from these two accidental self-inflicted wounds.

A well-trained investigator would find this scenario statistically difficult to believe and prove. A review of Danny's medical records, along with interviews with his friends and family, would confirm that Danny was right-handed (right-hand dominant). It's important to note that when Danny's body was found, he was clutching a large wooden crucifix in his right hand. With that in mind, one must ask: How did Danny accidentally shoot himself twice in the same location on his chest, using his non-dominant left hand with a semi-automatic pistol? Additionally, the position of Danny's body on the ground, in relation to the gun, located a few feet away from him on his left side and near his left hand (according to LT Jacobs memory), suggests that Danny used his non-dominant left hand to accidentally shoot himself twice.

In this scenario, Danny would have had to pull the trigger twice while holding the gun in the same position relative to his chest, with his non-dominant left hand. Consideration must also be given to the weapon's recoil, which would cause the gun to move away from the body. How could Danny's left hand maintain such precise control of the gun after the first shot? Given the intense pain, his body and left hand would have reacted violently, making a second shot in the same exact location nearly impossible. What is also perplexing is that after the two painful gunshots to the chest, Danny's right hand somehow never released the large wooden cross. The cross was found tightly clutched in his right hand, resting across the name tag on his uniform. Lastly, it's important to note that this scenario reflects exactly what LT Jacobs believed happened to Danny. However, his account assumed Danny was shot only once. When I mentioned the 2 centimeter exit wound in the middle of Danny's back, LT Jacobs became quiet and seemed confused. In that moment, I could sense that LT Jacobs realized his investigation was incomplete and hadn't been conducted in a proper CSI environment.

Fictitious Scenario 2 - Death by Suicide

In this fictitious scenario, the workload in the mechanics shop had risen at an alarming rate, becoming unbearable for the few mechanics in the unit. Danny struggled to keep vehicles repaired and returned to service in a timely manner. The oppressive heat and heavy rains only added to his frustration. Exhausted, hot, and overworked, Danny and his buddies often went to the enlisted club at night for a few drinks to ease the stress. One evening, as they sat drinking, Danny pulled out two letters he had just received, one from his brother and one from his girlfriend, Helen. At first, he was cheerful, but as he read, his mood shifted to quiet sadness. Moments later, he tells his buddy that he just received a 'Dear John" letter from his girlfriend Helen. After finishing his drink, he left the club, distraught, without saying goodbye to his friends. Danny returned to his barracks room to retrieve the .45 caliber pistol he had found.

Danny put the gun in his front pocket and his wooden crucifix in his back pocket, before he walks to the isolated supply laydown area. The workload, the weather, the stress of the war, and the letters were more than he could bear. Danny sat between a stack of pallets, clutching his wooden cross tightly to his chest with his right hand and gripping the .45 caliber pistol in his left. Moments later, he pressed the gun to his chest and fired two consecutive rounds. Within seconds, he was dead.

In this suicide scenario, we must accept the premise that Danny found both a gun and .45 caliber ammunition on base, and then willfully took his own life with them. If we are to presume that Danny committed suicide in this scenario, we must closely examine his physical environment and seek any evidence regarding his emotional state at the time of his death. If suicide is the conclusion, we need to explore the "why" behind the event. Was it due to depression caused by his physical environment, the war zone, the rain, the heat, etc.? Were there work-related issues, conflicts with his superiors, or the overwhelming stress of the Vietnam War unfolding around him? Could it have been loneliness and missing his family and friends, or possibly the result of a 'Dear John' letter from his girlfriend, Helen?

Sadly, it will be nearly impossible to determine whether any, or all, of these factors contributed to Danny's emotional state at the time of his death. Without firsthand knowledge of his emotions, it would be difficult to prove whether he was suffering from depression. During my research, I found no written documentation from Danny indicating that he was struggling with severe depression or contemplating suicide. His letters home reflected some highs and lows as he adjusted to his new job and life in Vietnam. However, I did hear from Danny's brother, Chuck, who mentioned receiving a letter from Danny that, in Chuck's mind, suggested Danny was in a poor emotional state. From that letter, Chuck felt that Danny might be suicidal.

During my interview with Chuck, he recalled that after reading the letter, he gave it to his mother, Mary. When I asked Chuck directly if the letter stated that Danny wanted to kill himself, Chuck responded, *"No, he just seemed distraught."*

After learning about this letter from my Uncle Chuck, I immediately asked a few of my other relatives about this letter that Chuck received. However, none of them had seen or heard of the letter; only Danny's brother, Chuck, had any knowledge of it. From numerous conversations with my family members, I learned that my grandmother, Mary, shared everything she knew about Danny's death with her daughters (Joan, Jean, and Jan). I also discovered that Joan, Jean, and Jan passed these stories down to their children, primarily the women in the family. This makes it all the more concerning that my grandmother, Mary, never shared this letter with her daughters.

From the numerous letters Danny wrote to his family, none suggested that he was severely depressed or suicidal. However, these letters did indicate that Danny experienced emotional highs and lows, primarily related to work issues, mostly concerning being late for work. Yet, these minor infractions do not indicate or suggest an emotional state severe enough to lead Danny (or anyone, for that matter) down a path toward suicide. In Danny's letters home to his parents, he mentioned that he was doing his best to fit in and acclimate to this new U.S. Navy way of life in a war zone. The bottom line is this: If I could find this letter from Danny to Chuck, it could have a significant impact on the possibility of suicide. However, for it to hold any weight, we would need to see the content of the letter, determine the date it was written relative to Danny's death, and then assess his mental state at the time he wrote it. With extremely limited information or facts to rely on, the suicide scenario hinges heavily on physical evidence. Unfortunately, no evidence was collected at the scene of Danny's death. Also, if we had the letter that Chuck received and gave to his mother, or if we had the rumored 'Dear John' letter from his girlfriend Helen, then we would have tangible 'physical evidence' to support the possibility of suicide.

I had been in contact with Danny's girlfriend, Helen, and during our many conversations, she never mentioned a breakup with Danny or sending him a 'Dear John' letter. In fact, she only expressed her love for Danny and the grief she suffered from his untimely death. The one question that weighed heavily on my mind, and on everyone else's, was: Did you break up with Danny and send him a 'Dear John' letter? Sadly, Helen passed away before I had the chance to ask her this final, crucial question. Throughout our conversations, Helen gave heartfelt and sincere answers to all my questions, and I never once sensed that she had broken up with Danny. I had no reason to believe that Helen was being anything less than honest with me, and based on my interviews with her, I truly do not believe that she ever wrote a 'Dear John' letter to Danny. Therefore, without any physical evidence to confirm Danny's emotional state, all this information is categorized as hearsay and circumstantial evidence. If a proper death investigation had been conducted to determine the manner of death (MOD), investigators would have searched for clues supporting the various possible MODs. Unfortunately, without such an investigation, we are left to piece together fragments of information and speculation. Lastly, any investigator assigned to this case would have conducted a thorough search of Danny's personal belongings. Given the rumors and hearsay surrounding a possible 'Dear John' letter, investigators would have been specifically looking for that letter. The actual letter would support the determination of suicide as the manner of death (MOD), and that physical letter would be preserved as evidence. Yet, to this day, no such 'Dear John' letter, or any physical evidence supporting the existence of one, has ever been discovered.

If we analyze suicide statistics and closely examine forensic data from suicide cases involving a single gunshot wound to the chest, some interesting findings emerge. My research uncovered forensic studies spanning a 20-year period, which revealed that suicides resulting from self-inflicted gunshot wounds to the chest are relatively uncommon. In fact, individuals who choose to end their lives with a firearm typically aim for the head, making chest wounds an unusual choice in suicide cases. Additionally, when comparing suicides to homicides, a notable pattern emerges: the majority of homicide victims suffer from two or more gunshot wounds to the body, whereas most suicide victims die from a single gunshot wound, typically to the head.

During my research, I uncovered intriguing statistics about suicide trends. For instance, the most common entrance sites for self-inflicted gunshot wounds in men are the temple, the mouth (or under the chin), and the forehead. Interestingly, self-inflicted gunshot wounds to the chest are far less common among men and are more frequently observed in female suicide cases. Statistically, men overwhelmingly prefer self-inflicted gunshot wounds to the head or mouth in the majority of suicide cases. When investigating multiple self-inflicted gunshot wounds, forensic experts face a significant challenge. Such cases are rare and present difficulties in determining the manner of death. The ability to assess how a person reacts following a gunshot wound is crucial in crime scene reconstruction. This can play a pivotal role in distinguishing between homicide and suicide, as the likelihood of a person being able to inflict multiple wounds on themselves, particularly in a precise manner, is extraordinarily low.

Questions surrounding the possibility of a person performing physical actions after sustaining a gunshot wound have been repeatedly raised in courts of law. Logically, committing suicide by attempting multiple self-inflicted gunshots to the chest seems improbable. Following the first gunshot, the body typically reacts immediately to the pain and trauma caused by the bullet. This response would likely prevent a steady hand from firing a second or third shot. Additionally, the body's typical reaction to a gunshot often triggers either a fight-or-flight response. The fight-or-flight response is your body's natural reaction to danger. In a fight response, the victim attempts to fight off the assailant. In a flight response, the victim attempts to flee in panic. However, in most cases, once a victim is shot in the chest, they often lose dexterity in their limbs, causing them to fall to the ground as their body goes into a state of shock.

Following a self-inflicted gunshot to the chest, the person would experience the immediate loss of dexterity in their limbs, and this would likely cause the person to drop the gun. Additionally, the recoil of the gun could also force the gun to fall from the hand. The intense pain and trauma from the initial injury would typically cause the body to go into shock, making it exceedingly difficult for the victim to regain enough coordination or dexterity to fire a second shot. While it is theoretically possible for a victim to attempt a second shot if the first was ineffective, the chances of them regaining composure and precision to shoot themselves a second time in the exact same location, creating the same entrance wound, are incredibly low. If Danny had shot himself twice in the chest in the exact same location (single entrance wound), this would suggest that he was able to press the barrel of the gun against his chest, fire once, and then, despite the pain, fire a second time in the same exact spot. For this to happen, Danny would have had to maintain control of the semi-automatic pistol, a weapon known for its substantial recoil and difficulty in managing after a shot.

The typical .45 caliber pistol has a tremendous kick when fired, and it usually requires two hands to control it properly, especially when fired repeatedly. According to Danny's medical records he was right handed (right hand dominant). Therefore, it would have been very difficult for him to control the weapon with his non-dominant left hand alone, particularly after the first shot, which would have caused a severe reaction of pain and shock to his body.

One piece of crucial information that was obtained from my interviews with LT Jacobs and other CBMU-302 members was that Danny was found clutching a 6-8" wooden crucifix in his right hand and that the right hand with the crucifix was held firmly against his right chest, covering his name tag on his shirt. LT Jacobs, LTJG Underwood, and ENS Kelly all reiterated this fact when they each stated that Danny's stiff right hand had to be moved to read his name tag on his uniform. This detail is incredibly significant because it raises serious doubts about the plausibility of Danny having fired a gun twice while holding onto a large wooden crucifix in his right hand. If his right hand was clutching the crucifix tightly at the time of death, so much so that it had to be moved to read his name tag, this strongly suggests that his right hand was not used to fire the weapon. Given that Danny was right-handed, this would mean he would have had to fire the gun with his non-dominant left hand, which is already a difficult scenario to accept. Furthermore, the natural reaction of the human body after sustaining a gunshot wound, especially to the chest, would likely result in a loss of grip strength. The fact that Danny was still clutching the crucifix tightly suggests that he may have already been holding it at the moment of his death, which complicates any theory of suicide or accidental discharge.

This raises critical investigative questions: 1) If Danny fired the gun himself, how did he maintain such a firm grip on the crucifix in his right hand after two gunshots to the chest? 2) If his right hand was occupied with the crucifix the entire time, how did he effectively control and fire a powerful .45 caliber pistol with his non-dominant left hand? How did he shoot himself twice in the exact same location with the extreme recoil of the pistol? 3) If this were a suicide or accidental discharge, how does the position of the crucifix factor into the mechanics of how he supposedly shot himself? Not to mention the severe pain inflicted from the first shot, which would have caused him to lose control of the hand holding the weapon.

When you closely examine these critical investigative questions, they tend to defy logic and basic forensic principles. If Danny had truly shot himself twice in the chest, his body would have reacted immediately to the immense pain, causing involuntary muscle contractions, loss of dexterity, and potentially even immediate incapacitation. There are several key inconsistencies that make the suicide theory implausible, and they include: 1) The Crucifix Grip, if Danny had fired the gun twice, his body's natural response to such trauma would likely have resulted in his hands either going limp or reflexively grasping at the wound. Instead, his right hand remained tightly clenched around the crucifix, an unusual reaction for someone in the midst of suffering fatal gunshot wounds. Also, no blood was observed on either of Danny's hands. 2) The gun's location was found near his left hand, a few feet away, indicating that he used his non-dominant left hand to shoot himself. This is inconsistent with typical suicide scenes. 3) The precision of the two shots; It is difficult to believe that Danny, using his non-dominant left hand, could have maintained enough control over a .45 caliber pistol to fire two bullets in the exact same entrance wound location. Given the weapon's strong recoil, it would be nearly impossible to keep the gun perfectly steady for a second shot. 4) The lack of evidence collection. If photos were taken and evidence was collected, we could determine if a struggle took place. If there was a struggle, there should have been additional signs, such as defensive wounds, bruising, or signs of a physical altercation. But due to the lack of proper crime scene investigation, we will never know if such evidence existed.

In most instances, death investigations allow for a straightforward differentiation between homicides and suicides.

Occasionally, however, such differentiation can prove to be quite difficult, with some cases being certified with an 'undetermined' manner of death (MOD). The investigation and autopsy findings help to differentiate a suicidal gunshot case from a homicidal gunshot case. The following are three case studies, where a homicide was initially believed to be suicide, but was reversed after a full investigation and assessment of all collected evidence.

Case Study 1: An 18-year-old man was found dead on the floor of his secured apartment, with blood on his head and face, and a small-caliber, semiautomatic handgun near his body. A blanket was partially covering his hands and face. A letter written to his girlfriend was found elsewhere in the apartment. In the letter, the decedent asked her for forgiveness. The initial police and death investigator impression was that the case represented a probable suicide. Subsequent body examination of the physical evidence revealed that the decedent's hands were clasped together, in a folded hands position. There was a penetrating contact gunshot wound (GSW) of the right temple. The cause of death (COD) was a contact GSW of the head. As there was no possible way for the decedent to have folded his hands after shooting himself through the brain, the manner of death (MOD) was changed to homicide. Police were immediately notified. Subsequent police investigation revealed that the decedent had been involved with dealing drugs. Another dealer confessed that he had shot him in the head while he was sleeping on the floor, leaving the gun adjacent to the body to make it appear that the decedent had shot himself. The letter to the girlfriend was considered coincidental [42].

Case Study 2: A 40-year-old female was found dead in bed by her husband after he heard a gunshot. Initial investigation, including interviews with the husband, suggested that the case represented suicide. A 9mm semi-automatic handgun was found adjacent to her body. Autopsy initially revealed two exit wounds on her left forehead region, as well as a single contact entrance wound of the right temple; however, during continued examination, a second contact entrance wound was found on her right temporal region. Both wound paths involved deep central structures of the brain. Additionally, blood spatter evidence on the husband's clothing suggested that the husband was in the vicinity of his wife at the time of the gunfire, a direct contradiction to his story. The cause of death (COD) was two GSWs to the head. Based on the fact that it was impossible for the "suicide" victim to have fired both of the shots in this case, the manner of death (MOD) was ruled homicide. The husband was subsequently charged and convicted of murder [42].

Case Study 3: A 35-year-old male was found dead in the basement room where he slept. He reportedly slept in the room because he and his wife were in the process of divorce. The wife reported that she and her husband had argued while in the same room. During the argument, she reportedly "thumped him on the head" because of something he said. He then produced a handgun and fired multiple shots, with one striking the door of the room and one striking her in the arm. She claimed that she then passed out and was eventually awakened when she heard another gunshot. Upon awakening, she found her husband's body lying on a mattress, which was on the floor, with a gun in his dominant right hand and blood on his head and face.

The COD was a gunshot wound to the head. Measurements of the decedent's arm length were taken at autopsy. Test-firing of the weapon revealed that the muzzle of the firearm was approximately 18 inches from the left side of the decedent's face when it was fired. Using a toy gun and using an assistant with similar body build, size, and arm length, the following conclusions were determined: In order for the decedent to have shot himself and produce the stippling pattern present evident at autopsy, he would have had to hold the weapon with his nondominant left hand (contrary to the wife's description), using an unconventional grip. In contrast, using the right hand, it was not possible to hold the gun far enough from the face to produce a muzzle-to-target distance of 18 inches.

Additionally, the scene and ballistics investigation showed that multiple rounds had been fired from multiple locations. Most importantly, a review of scene photographs revealed that fragments of the door were present on top of the victim's body and on top of pools of partially dried blood at the scene, indicating that the shot through the door occurred after the husband was dead, not before, as the wife had stated. Based on contradictory autopsy and scene findings, the MOD was ruled a homicide. The wife later confessed to having killed her husband and staging the scene to look like a suicide [42].

Discussion: In each of the cases listed above, each was initially considered to be a probable suicide. The first impression of suicide was related to various scene findings, as well as statements collected. The findings that suggested suicide in these cases included identifying a firearm adjacent to or within the hand of the decedent, an apparent contact or close-range gunshot wound, the presence of an apparent suicide note, false narratives made by witnesses and/or perpetrators, and indicators that the decedent had been depressed or otherwise mentally unstable or stressed prior to the death. Fortunately, important scene evidence and autopsy findings allowed forensic pathologists to determine that each case actually represented a homicide. Discovering a wound path (or paths) through the brain that would not allow for subsequent purposeful motor function was key in determining that cases 1 and 2 were not suicides. Proving the physical impossibility of the findings was key in all 3 cases, where the victim in case 1 would not have been able to fold his hands after sustaining the shot, the victim in case 2 would not have been able to fire two shots because each would have been immediately incapacitating, and the victim in case 3 could not have physically produced the GSW at the measured range-of-fire using the hand in which the gun was found. The recognition of scene findings that did not correlate with a suggested story was another factor for recognizing case 3 as a homicide, as the debris from the door shot obviously occurred after the fatal shot, not before.

When investigating a death, it is important to take into consideration the scene findings, the case history, forensic evidence, and the findings from the forensic autopsy. Certain cases initially thought to represent one MOD can eventually be determined to be incorrect after collaboration between death investigators, police, and the forensic pathologist. Whether a case is classified as an accident, suicide, homicide, or murder is obviously of vital importance, with wide-ranging potential ramifications.

Gunshot wound characteristics can provide valuable information to aid in correctly identifying the MOD. Identifying and knowing the difference between homicidal and suicidal injuries produced by firearms allows the forensic pathologist to make the best-informed decision regarding a suspicious death. Suicidal wounds are more likely to be at contact range to the head, whereas homicides can be from contact, near contact, or from short-range distances and often produce wounds at multiple locations on the victim. Intraoral (within the mouth) handgun wounds are more likely to be suicidal, while multiple GSWs from a handgun are more likely to indicate homicide. Multiple GSW suicides are not especially rare; however, whenever they do occur, they deserve additional scrutiny. Obtaining and documenting circumstance and scene information, including witness statements, can be extremely important. This can be especially useful when attempting to correlate autopsy findings and/or scene findings with subsequent reconstruction of events [42].

From Danny's medical death report, we learned that he suffered from what appears to be two gunshots to the chest. The report identifies a single 1.5 centimeter entrance wound to the mid-sternum and a single 2 centimeter exit wound mid-back. However, an anterior-posterior and a left lateral chest x-ray performed on Danny's body revealed a metal projectile still present in the right posterior chest cavity. If the bullet fragmented (separated) then the fragment could have caused the exit wound. However, according to LT Jacobs and ENS Kelly, the FMJ bullet removed from Danny's chest was completely intact with a slight bend at the tip. LT Jacobs remembers this because the bullet was given directly to him by the medical examiner. At that time, LT Jacobs did not understand that the FMJ bullet was to be saved as evidence for the investigation. Due to his lack of investigative knowledge and evidence collection, LT Jacobs simply placed the bullet in his desk drawer, not knowing that it should have been preserved as evidence. Knowing that the medical examination and X rays showed a single FMJ round lodged in Danny's chest and despite the presence of both an entrance and an exit wound, this should have alerted LT Jacobs to multiple possible manners of death scenarios. With a single bullet lodged in Danny's chest, one has to ask the obvious question: What caused the 2 centimeter exit wound in his back? This information about the fully intact bullet would lead even the most inexperienced investigator to conclude that Danny was shot twice.

In an effort to confirm my theory, I made a few calls to some forensic pathologists who, according to multiple website listings, had experience with my two-bullet theory surrounding Danny's death. I was fortunate enough to talk with Doctor Shiping Boa, MD, forensic pathologist, on February 28th, 2022. Over the phone, I reviewed the medical report with him and asked him about the 1.5 centimeter entry wound, the 2 centimeter exit wound, and the single bullet still logged in Danny's chest. He stated that, *"it is very rare to see this type of scenario, but it can happen if the bullet fragments"* (i.e., the bullet breaks up into two pieces). Before reaching out to Dr. Boa, I spoke with LT Jacobs again about the bullet removed from Danny's chest and asked if he had a photo of it. He replied, *"Dan, no photo of the bullet."* He explained that the medical examiner had given it to him, unrequested. Jacobs described the bullet as a full metal jacket (FMJ), completely intact but slightly bent at the tip, with no fragmentation. He believed the impact on Danny's sternum caused the bend before it lodged in his chest. Since LT Jacobs didn't consider the bullet to be evidence, he kept it in his desk drawer. On his last night in-country, he threw the bullet into the South China Sea.

According to Doctor Shiping Boa (forensic pathologist), he stated that, *"without actually seeing the complete autopsy findings, there is a very high probability that Danny received two gunshots to the chest."* His unofficial assessment supports my two-bullet theory as to the cause of Danny's death. Under the two-bullet theory, regardless of interpretation, the facts suggest an improbable outcome: either Danny possessed the ability to fire two consecutive shots through the same entrance wound using his non-dominant left hand, or an unknown assailant pressed the firearm against his chest and fired twice, with both rounds entering the exact same location. Given the factual evidence collected, how can we reasonably conclude that Danny managed to shoot himself twice in the exact same location, with his non-dominant left hand, while holding a large wooden crucifix in his right? Case studies and other research presented above indicate that the human body succumbs to involuntary muscle responses upon sustaining such a wound. This would cause the body to collapse and the hands to reflexively flex and release any objects they were holding, such as a gun or crucifix.

Fictitious Scenario 3 - Death by Homicide

In this fictitious homicide scenario, we can assume that Danny had a conflict or disagreement with someone on base over an issue such as a bet, a card game, or perhaps a loan repayment. Regardless of the reason, an argument broke out, and in the heat of the moment, someone drew a gun in anger. At some point, Danny and the assailant got into a disagreement that escalated into a fist fight. During this struggle, a gun falls to the ground. Immediately, both Danny and the unknown assailant, most likely another service member stationed on base, began fighting to gain control of the weapon. In the heat of the struggle, the assailant took control of the gun and either accidentally or intentionally pressed the gun against Danny's chest and fired two consecutive bullets into Danny's chest. The pain from the two gunshots triggered an immediate involuntary response, causing Danny's body to collapse to the ground. The assailant, now in a desperate state of panic, began to realize the gravity of what had just occurred and the potential consequences of his actions. Instead of providing first aid or seeking help, he panicked and fled the supply laydown area, leaving Danny behind to bleed to death. In an attempt to cover up the homicide and stage the scene as a suicide, the assailant returned to the scene and placed the gun near Danny's left hand, then removed the wooden crucifix from his back pocket and positioned it in his right. The assailant then arranged Danny's arm and hand across his chest to conceal his name tag. After staging the scene, the assailant fled the area, waiting for someone else to discover the body.

In this scenario, we assume an unintentional discharge of a firearm by someone who knew Danny and had access to the base and the CBUM-302 compound. Additionally, we also have to assume that Danny was shot by either the gun he reportedly found or by some other .45 caliber pistol. As is common in many homicide cases, the shooter typically does not have premeditated intentions to kill. Instead, the death results from an argument or disagreement that escalates, and in the heat of the moment, one person shoots the other. In cases like these, if there are multiple witnesses to the argument or disagreement, investigators will conduct interviews and collect witness statements. These statements are critical in determining whether a homicide or murder was committed. Without witnesses, however, it becomes much more difficult to establish the manner of death (MOD). It is important to note that we do not know which individual (among those LT Jacobs interviewed) made this statement about Danny finding the gun. Could it have been his roommate, Howard Decker? Given that they lived together and worked in the same shop, Howard likely spent most of his time with Danny and would be the most logical person to have this information.

Regardless of whether the shooting was accidental or purposeful, the manner of death would still be classified as a homicide because it was not a premeditated act. In this homicide scenario, Danny and the unknown assailant struggle over the gun, resulting in two point-blank shots being fired into Danny's chest. It does not matter who initially had possession of the weapon, whether Danny produced the gun he reportedly found on base, or if the assailant arrived with the firearm later found at the scene. Once the struggle began, the gun was fired twice in rapid succession, with both rounds striking Danny's chest. During the altercation, Danny also sustained a 2"x1" burn on the inside of his left arm near the elbow, possibly caused by contact with the hot barrel of the .45 caliber pistol. This burn was documented in the initial death report completed by the medical officer. One bullet entered and exited Danny's body, while the other remained lodged in his chest.

In the assailant's haste, he made critical mistakes: he placed the crucifix in Danny's dominant right hand and set the gun several feet away from Danny's non-dominant left hand. With the scene staged, it now appeared, at least superficially, that

Danny had taken his own life. The only piece of evidence that could have 'sealed the deal' in this homicide, staged to look like a suicide, would have been a handwritten 'Dear John' letter from Danny's girlfriend, Helen. The presence of such a letter would have steered investigators closer to ruling the manner of death (MOD) as suicide. However, no such letter was ever found, only rumored.

In reality, only a thorough crime scene investigation, combined with meticulous evidence collection and a detailed forensic analysis of both the scene and the autopsy report, could have revealed the true manner of death. According to crime statistics, homicide cover-ups are often set up to look like suicides. These attempted cover-ups have been repeated in a majority of cases, which have later been determined to be actual homicides. In fact, in the three case studies outlined above, we see this exact scenario play out. It is also important to point out that I only gave three examples of similar case studies; however, there are thousands of similar case studies that reveal a homicide took place and that the assailants made overt efforts to cover their involvement and to make the scene look like an accident or suicide.

Fictitious Scenario 4 - Death by Murder

In this fictitious scenario, the intentional discharge of a firearm by an assailant was a deliberate act intended to kill Danny, presumably by a fellow service member or possibly a local Vietnamese civilian working on base. As in all murder cases, the shooter acted with premeditated intent and likely took steps to plan when and where the murder would occur. In this murder scenario, Danny may or may not have known his killer. It's possible he angered someone or got into a heated argument with a co-worker, without realizing the gravity of the situation. Danny, still very young at the time, may not have fully understood how quickly conflicts can escalate with unstable individuals, people who can turn a minor disagreement into a deadly confrontation in the blink of an eye. Some people are able to control their anger and frustration, channeling these feelings into non-destructive outlets. Others, however, exhibit a terrifying lack of self-control, turning their emotions into violent actions, even murder.

In a murder investigation, the first step is typically to identify a motive. Understanding the killer's motive helps guide the investigator's search for evidence and clues. Purposefully taking someone's life is often the result of intense emotions tied to a specific event or close relationship. Jealousy, anger, revenge, love, hate, and other powerful feelings can drive people to commit unspeakable acts. Excessive alcohol consumption, which clouds judgment and amplifies anger, has led to many regrettable decisions, sometimes resulting in the tragic loss of another life.

According to Bryn Donovan and his paper on *"Why Do People Kill? 15 Motives for Murder,"* he outlines 15 different motives people use to drive them to murder. Below are the 15 motives he discusses in his paper: 1) A physical fight turns deadly. 2) Someone murders their spouse, partner, or ex out of jealousy. 3) The murderer hates the victim's politics, religion, or sexual orientation. 4) Someone murders an abusive partner or family member because they're sick of the abuse. 5) The murderer kills the victim in order to steal from them. 6) The murderer wants to collect life insurance money or an inheritance from the victim. 7) A spouse wants out of a marriage but doesn't want to pay alimony. 8) The killer wants to eliminate a rival. 9) The killer is taking revenge for being wronged in their career. 10) The murderer is taking revenge for a loved one who was wronged. 11) Someone takes revenge on a person for rejecting their romantic overtures. 12) Someone mistakes the victim for their actual intended victim. 13) Someone murders a family member who needs an extraordinary amount of caretaking. 14) Someone murders their pregnant partner due to extra stress and/or fear of responsibility. 15) The victim was a witness to a crime [47].

Based on the collected information and the fact that this event took place on a military base overseas, I will narrow down the list to five possible motives behind Danny's death. We will examine motives 1, 3, 5, 11, and 15 as possible scenarios of murder, as they would pertain to Danny's death. In each of these motives for murder, the assailant must have thought out or planned beforehand the death of Danny. Because, as we learned earlier, if it was not premeditated, then that death would fall under the homicide category for the manner of death. Below, we examine five of the fifteen plausible motives that could support a final determination of murder as the manner of death in Danny's case.

Motive Number 1: A physical fight turns deadly.

To apply this motive, we must assume that Danny had an argument with a fellow sailor, one that escalated into a physical altercation. The fight was brief, but Danny got the upper hand, and the sailor fled, ashamed and embarrassed by his loss. Over the next few days, the sailor became fixated on revenge and devised a plan to kill Danny. He waited for Danny to pass through the supply laydown area on base. As Danny approached, the sailor confronted him with a gun, pressed the barrel to his chest, and shot him twice. Afterward, the shooter staged the scene to look like an accident or suicide before fleeing the area. An alternate version of this fictitious scenario would have the murder taking place elsewhere on the base, with the assailant waiting until the base was dark and quiet before moving Danny's body to the supply laydown area and staging the scene to resemble a suicide.

Motive Number 3: The murderer hates the victim's politics, religion, or sexual orientation.

In this motive scenario, we consider the possibility that Danny's religious beliefs (his Catholic/Christian faith) were the driving factor behind his murder. To understand this motive, we first need to explore what drives religious violence and hate crimes. Some individuals become deeply obsessed with their own religious beliefs, often refusing to consider or respect other viewpoints. These intensely committed followers will go to great lengths to prove that their religion is the only true and valid path. Those motivated by religious hate are typically driven by a deep opposition to the beliefs of others, especially when those beliefs contradict their own. Different religious ideologies have historically been sources of conflict, and religious hate crimes have unfortunately increased in recent years. At the heart of the issue are the various religious institutions, each offering scriptures that are often vague and open to interpretation. These differing interpretations can lead to conflicts over which is true and which is false. To complicate matters, some religious texts contain verses that can be misinterpreted as justifications for violence in the name of a higher power. Furthermore, numerous studies have explored how extreme religious ideologies within various groups promote violence. Some of these groups use their religious writings and beliefs as a form of sanctioned permission to commit acts of violence and even murder in support of their cause.

In this scenario, we know that Danny was a devout Catholic who regularly attended mass on base. One of his most cherished possessions was a 6-8 inch wooden crucifix, which he carried with him and proudly displayed in his barracks room, often seen in photos. Let's assume that one day, while at the chow hall or the enlisted club, Danny was confronted by someone who questioned why he carried such a large wooden crucifix. What began as an innocent inquiry quickly escalated into a heated religious debate, with Danny defending his Catholic beliefs against those of the other person.

As we continue with this scenario, we'll assume that the discussion became highly contentious, with Danny and the other individual holding diametrically opposed views. They debated the topic at length, but their opinions remained irreconcilable. Eventually, the conversation ended, and they both went their separate ways. However, before leaving, the other individual became visibly agitated and angry with Danny. In a fit of rage, he threatened to kill Danny and stormed off, yelling a few expletives. In the following days, consumed by anger and resentment, this individual began to contemplate and plan how he would murder Danny. With his plan in place, the assailant waited for Danny to pass through a secluded area in an effort to catch him completely off guard. Within seconds, the barrel of a gun was pressed against Danny's chest, and two consecutive shots were fired. From the immediate pain, Danny's body went into shock, and he collapsed to the ground. Motivated by contempt for Danny's religious beliefs, the assailant then moved his body to the supply laydown area and staged the scene to resemble an accident or suicide. In a final act of disdain, he placed the crucifix in Danny's right hand and arranged his arm across his chest, as if to mock the very faith Danny had held dear.

Motive Number 5: The murderer kills the victim in order to steal from them.

In this motive scenario, we would have to assume that Danny possessed something of value that someone else desired enough to kill for. Based on my research, the only expensive items Danny owned were a tape recorder, a reel-to-reel player, and some assorted electrical equipment, like fans and a radio. Could these items have been what the assailant wanted, and would he go so far as to kill to get them? Or was there something else of value in Danny's possession that we're not aware of? In this scenario, let's assume that Danny did, in fact, possess something of greater value. The assailant, aware of the item, lured Danny to the supply laydown area under the pretense of buying the item, discussing fair payment, and completing the transaction. But when the moment came, the assailant brought no money, only a .45 caliber pistol. He shot Danny twice in the chest, sending him collapsing to the ground. As he lay dying, the assailant removed the wooden cross from Danny's back pocket, placed it in his right hand, and positioned it perfectly across his chest to block his name tag. The assailant then staged the scene by placing the pistol near Danny's left hand to resemble a suicide.

Motive Number 11: Someone takes revenge on a person for rejecting their romantic overtures.

The motive of revenge for the rejection of romantic or sexual advances has led to a multitude of murders throughout history. In this scenario, we would assume that a homosexual or bisexual male on base became sexually attracted to Danny and made unwanted advances toward him, attempting to persuade him into a sexual relationship. When Danny rejected these overtures, the individual's feelings of anger and humiliation could have driven him to seek revenge. We know for a fact that Danny was a practicing Catholic, heterosexual, and in a relationship with his girlfriend, Helen. Given this, any homosexual advances would have been rejected by Danny. Such a rejection would have created an uncomfortable situation for both Danny and the individual making the advances.

If we consider the possibility of a homosexual relationship scenario, we must assume that Danny was approached by a gay or bi-sexual male service member who made unwelcome sexual advances. If these advances came from a junior enlisted man, the threat of Danny reporting the incident up the chain of command would likely result in the service member's separation from the military and a return to civilian life, a consequence not likely severe enough to lead to murder. However, if the advances came from a senior enlisted or senior officer (a "lifer" as Danny would call them) who had aspirations of retiring from the military, the situation becomes more complicated. A reported assault in this case could destroy the individual's career and end their retirement plans, possibly leading them to take drastic measures to prevent exposure.

In this scenario, we find that Danny was repeatedly exposed to overt sexual advances from a senior member of the military. Each time Danny was offended, he did his best to avoid the advances. Eventually, the situation became unbearable, and Danny threatened to report the individual to his commanding officer. During this era in the military (1969), homosexuals were not allowed to serve, so the service member would have known that his career would be destroyed, and he would be discharged from the Service. The fear of losing his military career or being exposed as a homosexual could have been the driving motive behind the murder.

Days after Danny rejected the assailant's final sexual advances, he was confronted again. This time, unbeknownst to Danny, the assailant lay in wait late one evening near the enlisted club, watching for him to leave. The assailant approached Danny and appeared calm and acted as though he was seeking forgiveness. Danny, wanting to avoid any trouble, reluctantly agrees and attempts to walk away. However, the assailant had already planned what would happen next, and Danny would not be given the chance to report his sexual advances. In an instant, the assailant knocked Danny to the ground, and a struggle ensued. Danny was pulled behind the back of an adjacent building and forced to the ground. The assailant pulls out a gun and forces Danny to remove his shirt and to drop his pants. Danny is then forced to the ground and sexually assaulted. Afterward, the assailant orders Danny to pull his pants back up and to stand up. Within seconds of standing his attacker, presses a gun against his chest, and fires off two bullets that rip through Danny's chest. He collapses to the ground and dies almost instantly. The assailant knew he had to kill Danny to silence him and prevent him from reporting the assault. Determined to conceal both the assault and the murder, the assailant places Danny's body into the back of a jeep and drives off to a secluded spot near the supply laydown area. The assailant places Danny's body between some supply crates and leaves the scene. The assailant then returns to the site of the attack, where the ground was covered in blood. He covers the bloodstained ground with dirt and sand, attempting to erase any sign of what occurred. He then gathers Danny's olive drab shirt bearing his name tag, and the wooden cross that had fallen to the ground. The assailant returns to the supply laydown area, where Danny's body was hidden. He then puts Danny's shirt back on his body, places the cross in his right hand, and positions the hand against Danny's chest to obscure the name tag. He carefully wipes his fingerprints from the gun and places it next to Danny's left hand. Every detail was staged to create the illusion of suicide.

Motive Number 15: The victim was a witness to a crime.

The final plausible motive for murder is that Danny was a witness to a crime (illegal activity). In this scenario, we must assume that Danny inadvertently witnessed illegal activity and was confronted by those committing the crime. For this scenario, we will use illegal drug smuggling as the crime being committed. During the Vietnam War, there was extensive reporting of criminal activity surrounding drug smuggling, particularly involving U.S. military personnel. Some service members found ways to ship illegal drugs, such as marijuana and cocaine, back to the United States. They concealed the drugs inside military containers, returning coffins, and even their own personal luggage. Many of these containers, coffins, and bags were loaded onto military planes and ships bound for U.S. military ports and bases. During this time period, it was common for service members returning from the war zone to avoid thorough U.S. Customs inspections, making smuggling operations both feasible and lucrative.

In this context, if Danny witnessed or discovered evidence of such a smuggling operation, he would have posed a significant threat to those involved. Fearful that Danny might report them, the perpetrators may have decided that silencing him permanently was their only option. In this scenario, Danny unknowingly encounters several local Vietnamese laborers assigned to the CRB workforce supporting the U.S. government. They were in the supply laydown area, loading what appeared to be illegal drugs into U.S. shipping containers. As he continued to watch, one of the drug-smuggling men noticed Danny's presence and realized their operation had been compromised. The smugglers immediately confronted Danny, and an argument ensued. Sensing the danger, Danny attempted to get away to report what he had witnessed to his superiors. However, Danny was quickly subdued and held captive while they decided what to do with him. As the minutes passed, panic set in among the smugglers. They knew that being caught smuggling drugs by the U.S. Government could lead to lengthy prison sentences, or even worst, death from their drug bosses. Driven by fear and desperation, the smugglers concluded that the only way to protect themselves and their lucrative operation was to kill Danny.

Under threat of death and extreme duress, Danny was coerced into writing a suicide note suggesting he was depressed and intended to end his life. Danny was terrified, his hands shaking uncontrollably. Barely able to hold the pen, he hastily scrawled a few words on a scrap of paper, explaining how he had acquired the weapon and ammunition. He wrote nothing that suggested suicide, hoping that these men couldn't read what he wrote in English. His intuition paid off, one of his assailants grabbed the note from Danny and gave it a quick look before shoving the note into one of his pockets. With the supply crates stacked high enough to shield their actions from anyone passing by, the smugglers forced Danny against the crates. One of them pulled Danny's shirt over his face, drew a .45 caliber pistol fitted with a suppressor, and fired two rapid shots into his chest at point-blank range, killing him instantly. To conceal the crime, they moved his body between rows of crates on the opposite side of the supply laydown area. After positioning his body, they placed an unregistered .45 caliber pistol near his left hand and removed the wooden cross from his back pocket and placed it in his right hand, carefully arranging the scene to resemble a suicide. The scene was staged to make it appear as if Danny's death was a suicide, hoping to deflect suspicion and erase any trace of their involvement.

Discussion: In each of the five scenarios outlined above, it becomes clear that without an abundance of properly collected and processed physical evidence, it would be extremely difficult for any crime scene investigator or forensic pathologist to determine with absolute certainty the manner of death in this case. To further complicate matters, there were no witnesses to the actual events that took Danny's life. I remain deeply perplexed and frankly, very upset, by the complete lack of evidence collection surrounding Danny's death. At a minimum, the firearm, the bullet casings, and the bullet lodged in Danny's chest should have been preserved and processed as critical pieces of evidence. Instead, they were either discarded or destroyed. When you stop and consider the significance of these three key items, you have to ask yourself: why would someone discard or destroy them? You do not have to be a trained crime scene investigator or forensic criminologist to recognize that the discarding or destroying these three key pieces of evidence could easily be viewed as an attempt to cover up a crime. I am not suggesting that LT Jacobs intentionally sought to conceal a crime committed on base; rather, I am pointing out that numerous mistakes were made during this investigation, and the destruction of critical evidence is deeply alarming. What I am stating is that LT Jacobs was not a trained or qualified crime scene investigator. Had he been properly trained, he would have understood that the gun, the bullet casing, and the bullet removed from Danny's chest were all vital pieces of evidence that needed to be collected, processed, and preserved.

In defense of LT Jacobs, I would also submit that commissioned officers are not trained or qualified to conduct criminal death investigations, unless they are special agents within the Naval Criminal Investigative Service (NCIS). These special agents are armed federal law enforcement investigators with the authority and the forensic expertise required to investigate criminal acts. Danny's death investigation should have been initiated by the base Military Police (MPs) and then thoroughly investigated by the Naval Investigative Service (NIS) before Danny's body was moved from the scene. Proper protocol would have ensured that the crime scene was preserved, critical evidence was collected, and a professional investigation was conducted from the outset.

Actual CBMU-302 Event: A True Story from a Former Unit Members Interview

The following incident occurred in early February 1969 in Cam Ranh Bay, Vietnam. This is a firsthand account from a Petty Officer (PO) assigned to CBMU-302, who was serving as the dispatcher on duty the day this event took place. This CBMU-302 unit member wishes to remain anonymous, so I will not reveal his real name. Instead, I will refer to him using the pseudonym "PO Smith" to protect his privacy as he requested. By chance, I discovered PO Smith while reviewing CBMU-302's master muster list on the unit's website, in search of former members who were stationed at Cam Ranh Bay (CRB), Vietnam, around the time Danny was stationed there. Though this unit member wasn't at CRB at the time of Danny's death, their paths may have crossed on the day Danny arrived in country, because he was scheduled to depart the country on the same plane that Danny arrived on.

During my interview with PO Smith, he recounted an incident that took place at Cam Ranh Bay while he was on duty. He began by describing several semi-truck trailers located within the CBMU-302 compound, each filled with various types of small arms ammunition and weapons intended for emergency resupply of nearby combat units. These trailers were parked in the CBMU-302 motor pool, fully stocked and ready to be deployed at a moment's notice. According to PO Smith, one evening, a call came in from a unit that had been hit and was running dangerously low on ammunition, and they needed an immediate resupply of ammunition.

As the duty dispatcher that evening, PO Smith was responsible for assigning a duty driver from the duty section to deliver the emergency resupply. The duty driver was tasked with hooking up a semi-truck to one of the loaded trailers and delivering it to the unit that had called in the request. PO Smith could not recall the name of the requesting unit or its exact location, as so many years have passed since the incident. Prior to the duty driver's departure, PO Smith outlined the significance of the resupply mission and stressed the urgency of its completion. The unit under attack was in desperate need of ammunition, and there was no time to lose. The driver was ordered to proceed directly to the unit without making any stops. Additionally, he was instructed to bring along another member from the duty section, someone who would be armed with a rifle and ammunition to provide security during the mission. However, the duty driver ignored his orders and, on his own accord, picked up four armed sailors (fellow CBMU-302 unit members) before leaving the base through the main gate.

About a mile down the road, the driver veered off the main route and drove down toward the beach on the South China Sea. The truck, heavily loaded with ammunition, became stuck in the sand. Rather than attempting to get help or return to the base, the driver and his companions built a bonfire on the beach. In a scene reminiscent of old cavalry rituals, they stacked their M-16s together near the fire. The five men listened to music and danced around the fire while smoking marijuana. According to PO Smith, he only learned about the situation when he was informed by another duty driver, who was returning from the airport, after dropping off a service member heading back to the States. On his return trip back to the CBMU-302 compound, the duty driver spotted the semi-truck stuck in the sand near the beach and observed five men smoking marijuana and dancing around a bonfire. He immediately reported the incident to PO Smith and described everything he had witnessed.

The unit in need of the ammunition resupply repeatedly called CBMU-302, urgently asking, "*Where is the resupply truck?*" Without hesitation, PO Smith ordered an alternate duty driver to hook up the other semi-truck loaded with ammunition and deliver it at once to the unit running low on supplies. Shortly after the truck departed, he learned that the second truck had successfully reached the unit in need of resupply. Meanwhile, back on base, PO Smith reported the incident to LCDR Filbry, the Commanding Officer of CBMU-302 at that time and briefed him on the situation. In response, LCDR Filbry instructed PO Smith to drive the base Commanding Officer's jeep over to the officers' quarters. When PO Smith arrived with the jeep, a U.S. Navy Captain, whom he had never met before, was already waiting outside (PO Smith could not recall the Captain's name). He assumed that LCDR Filbry had briefed the U.S. Navy Captain on the situation and that the Captain had decided to handle the matter personally. The Captain climbed into the jeep and ordered PO Smith to drive him straight to the scene, where the ammo truck was stuck on the beach, and the sailors had been reported smoking marijuana.

The road running parallel to the beach sat approximately 8 to 10 feet above the sandy shoreline, providing PO Smith and the U.S. Navy Captain with a clear view of the semi-truck and the men below. When PO Smith brought the jeep to a stop, both he and the Captain stepped out to assess the situation. PO Smith recalled chambering a round in his M-16 and raising the rifle, training it on the five sailors gathered around the bonfire. He was prepared to open fire if necessary. But before he could act, the U.S. Navy Captain firmly grabbed the barrel of the rifle, pointed it downward, and said, *"Son, don't do that. You'll never forget it for the rest of your life, and we don't need to do that."* PO Smith replied, *"But Captain, this is desertion in the face of the enemy, Sir."* The Captain responded calmly, *"I'll take care of this."*

The Captain and PO Smith went to confront the men. The Captain instructed PO Smith to collect the men's weapons and ordered the sailors to return to base on foot immediately. PO Smith complied, gathered the M-16s, placed the rifles in the jeep, and then drove the Captain back to base without further incident. Upon returning to the base, PO Smith dropped the U.S. Navy Captain off at the officers' barracks. That was the last time he saw or spoke to the Captain, although he later heard that the Captain held an immediate Captain's Mast that same evening.

Base security personnel at the main gate had been instructed to take the five men into custody as soon as they returned. The sailors were found guilty of multiple offenses and processed out of the U.S. Navy, each receiving a dishonorable discharge. The following morning, they were ordered to pack their belongings and to report to the front gate. From there, they were directed to walk, unaided, to the airport, which was approximately 14 miles from the base. To further emphasize the severity of their punishment, the base commander addressed the entire command that morning, explicitly ordering all personnel not to assist the men in any way. They were to make the journey to the airport alone, carrying their own gear. To make matters worse, they were issued only space-available travel orders, placing them at the lowest priority for outbound flights. This meant they could be stranded for days or even weeks before securing a seat on a plane out of Vietnam. Later that same evening, personnel were dispatched to recover the first truck that was stuck on the sandy beach. The command sent heavy equipment capable of pulling it back onto the road. Once the truck was freed from the sand, it was driven back to the base without further incident.

After capturing PO Smith's initial story, I asked him about drug use within the command and whether he had witnessed anyone else using drugs during his time there. He said he was unaware of any drug activity among personnel prior to the incident involving the five men and the ammunition truck. However, he added, *"There is something that may help you with your story."* PO Smith went on to recount an alarming event that occurred just a few days after the five men were dishonorably discharged. One evening, he was suddenly awakened in the middle of the night by the sound of a single M-16 round being fired. Startled, he looked toward the doorway of his barracks room and saw several men struggling with someone who had a rifle. During the scuffle, one round was discharged, shooting up through the roof of the barracks. When PO Smith got a clearer look at the man holding the weapon, he immediately recognized him as someone who had been a close friend of one of the discharged sailors. According to Smith, the man was enraged over his friend's discharge and had come to the barracks to shoot and kill the person he believed was responsible.

PO Smith continued his story, saying that two days later, while lying in his top bunk in the barracks, he suddenly felt his head being pushed back. When he opened his eyes, he saw the same man who had tried to kill him just days before standing over him with a bayonet pressed against his throat. According to PO Smith, he always slept with his .45-caliber pistol hidden under his pillow. In that moment, he quickly pulled out the gun, aimed it at the attacker's face, and warned, *"You can't cut faster than I can pull this trigger."* After this second attempt on his life, the command decided to transfer PO Smith from Cam Ranh Bay to Saigon for his own safety, while they worked to remove the assailant and any accomplices from the unit. Approximately five months later, PO Smith returned to Cam Ranh Bay.

After hearing PO Smith's story, I asked him whether he thought the men who received dishonorable discharges, and the individual who attempted to kill him, might have been involved in the lucrative drug trade, where active-duty servicemen were allegedly shipping drugs back to the United States. PO Smith paused, then replied, *"That's a very real possibility, but I did not know for sure."* He also mentioned that the Naval Investigative Service (NIS) was actively investigating the group of men caught smoking marijuana, possibly looking into drug dealing or distribution on base. After this brief conversation, I couldn't help but find it ironic that while NIS focused on sailors smoking marijuana, they seemingly overlooked the unexplained death of a sailor, shot dead within what should have been the relatively safe confines of a military base.

During my interview, PO Smith brought up another interesting topic regarding his issued weapon. He explained that he had attended a military school in Virginia and was an avid hunter, very familiar with rifles. He never liked the M-16, disparagingly calling it a *"Mattel Special,"* implying it was a cheap, poorly designed weapon. Upon arriving in Vietnam and being issued an M-16, he told the armory clerk that he didn't want it and asked if any alternatives were available. The clerk informed him of a locker next to the armory containing captured VC weapons and said he could choose any weapon from there. In that locker, PO Smith found an M2 Carbine rifle with a selector switch for semi-automatic and fully automatic fire, bearing the name Winchester. A fan of Winchester firearms, he happily chose the M2 Carbine as his weapon of choice, along with his issued .45 caliber pistol. I found this part of PO Smith's story particularly intriguing. The main reason why I added PO Smith's story in this book was because it illustrated three key issues that might be pertinent to the events surrounding Danny's death. 1) How prevalent was the use of marijuana or other drugs among CBMU-302 members, and were any personnel involved in smuggling narcotics back to the United States?2) The fact that NIS conducted a full investigation into this marijuana event but failed to investigate Danny's death. 3) It was the first time I had heard that the base kept a locker full of captured VC weapons, located right next to the armory. This VC weapons locker made me pause and gather my thoughts, and I wondered: could this have been the source of the .45 caliber pistol used to kill Danny?

Why didn't NIS Investigate Danny's Death?

As my research into Danny's death deepened, and through my many conversations with LT Jacobs, I learned that no physical evidence had been collected. That realization led me to conclude that Danny's manner of death cannot be determined solely from limited circumstantial evidence and the unit member statements recorded during the JAG investigation. The lack of crime scene photographs, the failure to collect critical evidence, and the absence of any eyewitnesses should be enough for even an untrained investigator to conclude that, at a minimum, Danny's death is highly suspicious. It appears that CBMU-302's higher headquarters left LT Jacobs to fend for himself, assigning him the sole responsibility of investigating the death of CMCN Daniel Joseph Davidenko without the proper resources, training, or support. It could have also been just numerous investigative errors and a lack of crime scene investigative experience that led to the wrong conclusion. Or the worst-case scenario is that the facts surrounding Danny's death were never properly investigated, and his death was a complete cover-up from the beginning.

In my attempt to dig deeper into the reasons why NIS did not conduct an investigation, I came across an article from the NIS office stationed in Saigon, Vietnam. Based on a Naval Investigative Service Office Vietnam (NISOV) article by Tom Brooks and Bill Manthorpe (former Commanding Officers of NISOV), it leads me to believe that the Naval Investigative Service Offices in Saigon and Da Nang were too busy with other investigations and counterintelligence operations to formally investigate Danny's death. The NIS team in Saigon was approximately 225 miles (approximately 5 hours away by jeep) from Cam Ranh Bay, and the NIS team in Da Nang was approximately 355 miles away (approximately 9.5 hours away by jeep). Both NIS teams were just hours away by airlift. At a minimum, this NIS article suggests that NISOV was too busy to investigate Danny's death, which ultimately ended in a travesty of justice.

The following information was taken from the Naval Investigation Service Office Vietnam (NISOV) article by Tom Brooks and Bill Manthorpe. Then, LCDR William H. J. Manthorpe Jr. was assigned as NISO Vietnam's first CO, and Special Agent (SA) Ken Nickel was its first Supervising Agent. According to Tom and Bill's article (they were both 1630 Intelligence Officers), beginning in 1968, the Naval Investigation

Service Office Vietnam (NISOV) expanded beyond Saigon and Da Nang. NISOV established Field Offices in areas with sizeable U.S. Navy and U.S. Marine Corps presence. These one-man satellite units called Resident Agents (RA) were located at Quang Tri, Vung Tau, **Cam Ranh Bay**, and Can Tho. The main members of NISOV consisted of six officers, four in Saigon and two in Da Nang. NISOV also had a dozen or so enlisted U.S. Navy Yeomen and approximately twenty-one Special Agents (SA). When I learned that CRB had at least one NIS Resident Agent (RA) assigned to the base, I was stunned. Who was this agent, and why wasn't he tasked with conducting an independent NIS investigation? Was the RA ever contacted by CBMU-302? Did he even know that a death had occurred on base? This is yet another issue that leads me to believe that the investigation into Danny's death was flawed from the beginning.

These units were commanded by U.S. Naval Intelligence Officers and during the time of Danny's death, LCDR Bill Armbruster (1968-1969) and LCDR Tom Brooks (1969-1970) were in command of NISOV. Their mission was concentrated around investigations; however, they also supported both intelligence and counterintelligence operations. [61] Another responsibility of NISOV was the large AWOL population that lived in the Saigon suburbs. NIS didn't know how many AWOL personnel were out there, but there were many, from all military services stationed in Vietnam. These AWOL service members made their living via the black market (i.e., smuggling drugs).

NISOV was involved in cases ranging from murder, theft, smuggling, money manipulation, and other abuses, as well as standard narcotics and petty crimes. **According to former commanding officers of NISOV, they turned down many cases that could have been 'front-burner' cases in any other NISO.** They simply couldn't take all the cases they were asked to take.[61] Was the death of CMCN Daniel Joseph Davidenko one of those cases that NISOV's main office or the Cam Ranh

Bay NIS field office could not handle? To hamper NISOV investigative procedures even further, the Special Agent's work was limited due to travel and transportation delays. Once the SAs got to their destination, the crime scenes were seldom undisturbed (not preserved), and witnesses were either out on an operation, transferred, or simply unavailable or unwilling to talk to investigators.

USNAVINVSERVO VIETNAM

Within the content of this chapter, I provided you with all the information surrounding the death of CMCN Daniel Joseph Davidenko. We learned that there was no real effort put into the collection and processing of evidence or the preservation of the crime scene for further and/or future analysis. From conversations with LT Jacobs, he stated that *"he did not take crime scene photos as a part of his JAG investigation."* At this point in my research, all I have to rely on is LT Jacobs' recollection of events and the memories of numerous CBMU-302 unit members from that fateful day. None of the interviews or statements collected during LT Jacobs' JAG investigation came from individuals who actually witnessed the events that took Danny's life. A majority of the information gained from these statements consisted of rumors and hearsay regarding what the individuals believed happened that fateful day. Most of the statements are purely circumstantial.

At the time of writing this chapter, it remains unknown how many personnel were interviewed or whether anyone mentioned Danny finding a gun on base. It is also unknown how many actually recalled not seeing Danny over a 72-hour period. Once I have the actual JAG investigation conducted by LT Jacobs, it should provide me with the names of these individuals. I hope that when I do get the JAG investigation, many of these men are still alive and that I can interview them to get their stories firsthand. If I cannot locate these individuals for an interview, I will just have to review and analyze what they provided in their interviews and/or statements during the actual JAG investigation. Lastly, I have provided you, the reader, with all the available information I uncovered. I also provided you with the standard investigative processes as they relate to the crime scene and evidence collection. Then finally, I gave four manners of death scenarios as to what may have happened to Danny on that fatefully day in Vietnam on November 10th, 1969. I also presented my professional assessment of the information collected, and I hope it is clear to the reader that the evidence available in this case is insufficient. However, you, the reader, must evaluate all the information presented and decide for yourself whether there is sufficient evidence to determine the manner of death in this case. Based on this lack of supporting information and evidence, we must reflect back on the key missing elements surrounding this tragic event. First, the CO's letter to my grandparents specified the MOD as an accidental self-inflicted gunshot wound to the chest. This was reported within 24 hours of Danny's death and 11 days before the official JAG investigation was completed. Then there is the lack of physical evidence. The gun, the casing, and the bullet were all discarded or destroyed. No blood at the scene. No bullet holes in Danny's clothes. No 'Dear John' letter, and more importantly, there were no witnesses to the events that took Danny's life.

Despite years of research and analysis, the scarcity of reliable data leads me to conclude that Danny's manner of death may never be determined with certainty. Until additional facts or physical evidence emerge to support a clearer determination. I don't know if I can fully express the level of disappointment I feel. How did this terribly flawed death investigation into Danny's death go so horribly wrong? After painstakingly sifting through all available data, I cannot definitively determine MOD.

I provided you, the reader, with all the available data I collected on this tragic event, and from what you have read so far, is there enough information or evidence for you to determine what manner of death (MOD) should be assigned to the death investigation of CMCN Daniel Joseph Davidenko? I ask you, the reader, to apply critical thinking as you review the data presented in this book and to use it in your effort to determine Danny's manner of death. Remember to ask yourself the CSI questions we discussed. Keep in mind the forensic pathology principles outlined in the previous chapters as you evaluate the evidence required to reach a conclusion. Will it be enough?

Chapter 12

Danny's Long Journey Home

This chapter was developed from my many interviews with Danny's family, friends, and former shipmates. In these conversations, people recalled the events they had either witnessed themselves or remembered hearing about through stories passed down by others. Many of these accounts are traced back to Danny's mother, Mary, who often shared these stories with her sons and daughters, stories they in turn passed on to their own children.

On the morning of November 11th, 1969, the sun rose at 6:35 AM as Elmer got out of bed to brew a pot of coffee and retrieve the daily newspaper from the front porch. The day began with mild temperatures in the low 50s, mostly cloudy skies, and no sign of snow or rain in the forecast. A few minutes later, Mary woke up and headed into the kitchen. She grabbed two cups from the cupboard, poured coffee for both her and Elmer, and joined him at the kitchen table. As they sipped their coffee, their conversation drifted to the usual topics: the grocery list and household chores. It was an ordinary morning, apart from the unseasonably mild weather. By 7:10 AM, Elmer had finished his second cup of coffee and was dressed for work. Then, at approximately 7:30 AM, there was an unexpected knock at the front door.

Mary and Elmer Davidenko (Danny's Parents)

Danny's Home in Brookfield, Illinois

Elmer was already behind schedule and had little time for unexpected visitors. But since Mary wasn't dressed, he reluctantly went to answer the door while she hurried to the bedroom to put on a robe and brush her hair, wanting to look presentable, for their unexpected visitor. As Elmer opened the door, his face went pale. He knew immediately why two uniformed U.S. Navy men had come unannounced on this quiet Tuesday morning. There was no need for words; he understood at once that they were there to deliver heartbreaking news about their son Danny. Within a minute, Mary emerged from the bedroom, curious about their early morning visitor.

As Elmer stepped aside to invite the two U.S. Navy men into the living room, Mary entered just in time to meet their unexpected guests. At first, she was confused by the sight of the uniformed men standing in her home. But then she saw the look on Elmer's face, followed by the realization that the men were wearing U.S. Navy uniforms, and in that instant, she knew. Something terrible had happened to Danny. The two U.S. Navy men wore full dress uniforms; one was a Commissioned Line Officer, the other a U.S. Navy Chaplain (Catholic Priest). The U.S. Navy and all branches of the military, take the duty of notifying a deceased service member's next of kin with the utmost seriousness and respect. This solemn responsibility is carried out with formality and dignity, typically within 24 hours of confirming a casualty.

Once inside, the Chaplain asked them to be seated before delivering the news. Mary felt her knees go weak, and Elmer quickly guided her to the sofa, where they both sat down. Before the two men could say a word, tears welled in Mary's eyes and spilled down her cheeks. A moment later, Elmer's own tears followed as he wrapped his arm around her, bracing himself for the news about Danny. As part of the official protocol, he first asked Mary and Elmer to confirm their identities by stating their full names. Then, he introduced himself and the U.S. Navy

Chaplain. Taking a steady breath, the officer delivered the words no parent ever wants to hear:

"The Commander in Chief of the U.S. Navy has entrusted me to express his deepest sympathies on the passing of your son, Daniel Joseph Davidenko, who died in Cam Ranh Bay, Vietnam, on November 10th, 1969."

Mary let out a loud scream as Elmer sat quietly, stunned in silence. Then, through her tears, Mary softly stated something like; What happened to my son? How did he die? Was his base under attack? The Naval officer replied to Mary with the standard preformatted response: At this time, Ma'am, we do not have details regarding the circumstances of his death. However, I can assure you that an investigation is underway, and that the U.S. Navy will be providing you with the results as soon as they are available. Following the official notification, Mary, Elmer, and the notification officer reviewed and verified Danny's personal information. They then discussed his death gratuity payment (Danny's life insurance policy), the transportation of his remains, and the role of the military's mortuary affairs department. Before concluding, the officer explained that they would soon be contacted regarding Danny's funeral honors.

A few days later, Mary and Elmer received a letter from Danny's commanding officer, dated November 11th, 1969. In it, he expressed his deepest sympathies for their loss and stated that Danny had died from an accidental, self-inflicted gunshot wound to the chest. This was the first official account of how Danny had died in Vietnam. At first, Elmer and Mary were confused, their minds racing with questions. Mary, her voice barely above a whisper, murmured… *an accident*? He accidentally shot himself. How could this have happened?" She shook her head in disbelief. This doesn't make sense. I just talked to him on the phone a few weeks ago… I just got a letter from him the other day… Why? How? Why, my baby boy? Despite all my research, I have yet to find a single U.S. government document addressed to Mary, Elmer, or any of Danny's siblings confirming that an official NIS death investigation took place that formally determined Danny's manner of death (MOD).

Memorial Service CRB, Vietnam

As soon as Danny's body was discovered and the first responders arrived on the scene, it was clear that he had been dead for some time, with no signs of life. Almost immediately, word swept across the base like wildfire: a body had been found. Within minutes, dozens of CBMU-302 personnel converged on the scene, joining the medical team already attending to Danny's lifeless body. According to LT Jacobs, as soon as LT Denis Casey (Father Casey was the U.S. Navy Catholic Priest assigned to Cam Ranh Bay) arrived at the scene, he began to pray and administered Danny's last rites. Last Rites are the sacraments administered to a person nearing death. When someone is in danger of dying, a priest is called to offer these final sacraments, which include Confession, Anointing of the Sick, and Holy Communion. Confession allows the dying person to repent and seek forgiveness for their sins. The Anointing of the Sick involves anointing with holy oil, symbolizing healing and spiritual strength. Holy Communion, known as Viaticum, when given to the dying, provides spiritual nourishment and prepares the soul for its journey beyond this life.

According to the Catholic Church, Last Rites (also known as the Commendation of the Dying) are the final prayers and sacraments given to a Christian shortly before death. These rites cannot be performed on someone who has already passed away. However, if a practicing Catholic dies before a priest can administer the Last Rites, the priest may instead offer a selection of Catholic prayers on their behalf. However, in the letter Father Casey sent to Mary and Elmer, he states that, *"your son received the last Sacraments when his body was found."* After a long search, I found Father Casey and contacted him via email. Now in his 90s,

Father Casey was alive and well and living in Ireland. However, he could not recall any details about Danny's death, the Last Rites, or the memorial service he conducted before Danny was sent home to rest. Therefore, according to the Catholic Church, I believe that Father Casey would have been compelled to provide alternative prayers that the Catholic church uses if the practicing Catholic dies before receiving last rites. This may seem like a minor detail, but it's important to note that not long ago, the Catholic church would deny last rites, funeral services, and even a burial in a Catholic Church cemetery to anyone who committed suicide.

Through my research and interviews, I learned that Danny was a devout Catholic. His mother had sent him a Saint Christopher pendant to wear, and he also kept a large wooden crucifix in his barracks room. I understand that Father Casey is very old, and with age, our memories, especially of events from more than 50 years ago, can be difficult to recall. But what lingers in my mind is this: Why didn't the sight of that large wooden crucifix in Danny's right hand, held across his chest, spark any memory? One would think that as Father Casey bent over Danny's body, offering prayers, such a striking image would be unforgettable, especially to a Catholic priest. I would even go so far as to say that this would be an unforgettable moment, one that a Catholic priest might recall and share in sermons and stories throughout his life as a man of God.

For Danny to have received Last Rites, his death could not have been ruled a suicide. This leads me to presume that Father Casey must have been told that Danny's death was an accident. But who informed Father Casey of this determination? Catholic Priests are not qualified to make such a determination. Did LCDR Harrell or LT Jacobs ultimately conclude that Danny's death was accidental? Regardless of who made this determination, the critical question remains: What evidence or information was relied upon to reach this conclusion regarding

Danny's manner of death? The bottom line is that no determination of the manner of death should have been made until a full investigation and autopsy were completed.

According to LT Jacobs recollection of his JAG investigation (which took him 10 days to complete), he told me that he determined that Danny's death was accidental. So how did LCDR Harrell come to the conclusion that Danny's death was the result of a *"self-inflicted"* gunshot wound to the chest nine days earlier? None of these men had experience in death investigations, and according to all available records, there is no documentation suggesting that a formal NIS investigation was ever conducted. Was classifying Danny's death as accidental simply the most convenient way to close the case and move on? Doing so would have allowed the command to resume normal operations with minimal disruption. Or was there a deliberate effort to conceal more troubling findings? With the absence of critical evidence and official investigation reports, we may never truly know the full truth about how Danny's life ended that fateful day at Cam Ranh Bay, Vietnam.

On November 11th, 1969, one day after Danny's body was discovered and taken to the base medical hospital, Father Casey insisted on holding a memorial service and Catholic Mass in his honor. He also wanted to ensure that Danny received full military honors at the conclusion of the service. So, on November 13th, 1969, three days after Danny's body was discovered, Father Casey had Danny's body delivered to the church for a formal Catholic Mass in his honor. The service began with Father Casey's opening prayers and readings from the Bible. Another U.S. Navy Chaplain, LT Habiby, delivered a gospel reading, followed by a brief sermon from Father Casey. After the sermon, LCDR Harrell, the commanding officer of CBMU-302, offered a few remarks before Father Casey concluded the service with a final prayer. At the conclusion of LT Casey's Mass, everyone gathered outside the church to observe Danny's military honors. As the command stood at attention, a six-man rifle detail fired an eighteen-gun salute in his honor. As a lone trumpet played TAPS, the American flag covering Danny's casket was carefully folded. LT Jacobs recalled that every seat in the church was occupied, with the entire CBMU-302 unit in attendance to honor Danny's memory.

Preparations to Return Danny Home

On the very morning that Danny's parents were being informed of his passing, his body remained at the base hospital morgue, awaiting final processing and the official autopsy. The autopsy was scheduled to take place at the Armed Forces Medical Examiner's office at Dover Air Force Base in Delaware. Due to the high casualty rate during the Vietnam War, in-country hospitals prioritized treating the wounded, and autopsies were typically conducted stateside. Danny's body was placed in a metal military transportation casket, draped with the American flag, in preparation for his journey home to Brookfield, Illinois, for interment. Military caskets are made of aluminum for strength. They are designed for durability and to provide protection for a fallen service member's body. These specialized caskets often feature patriotic designs and symbols, honoring the deceased's service. To preserve the body during transport, military caskets are refrigerated and maintained at a temperature of approximately 36–40 degrees Fahrenheit.

The Armed Forces Medical Examiner System (AFMES) provides the Department of Defense (DOD) and other federal agencies with comprehensive forensic investigative services. The AFMES was established following the assassination of President John F. Kennedy in 1963. Its Forensic Pathology Investigations Division is responsible for determining the cause of death (COD) and manner of death (MOD) for all active-duty service members who die within federal jurisdiction, such as on a naval base. Once the official autopsy report is released to me through my FOIA request, I will dedicate a separate chapter in this book to outlining the full report, analyzing its contents, and assessing whether the forensic pathologist was able to determine Danny's actual manner of death.

The Long Journey Home

Determined to uncover the truth and ensure accurate reporting of Danny's death, I committed to leaving no stone unturned. It required extensive time and effort to research and gather essential documents, as well as to conduct numerous interviews with family, friends, and former members of Danny's unit in Vietnam (CBMU-302). Through this process, I worked to piece together the events leading up to and following this tragic incident. I was fortunate to have Danny's two surviving brothers, Raymond (Ray) and Charles (Chuck), share their firsthand accounts of the day they were notified by their U.S. Army commands that their brother Danny had been killed in Vietnam. In the following paragraphs, you, the reader, will hear directly from these interviews, gaining insight into their experiences and the remarkable stoicism they maintained as they processed the devastating loss of their youngest brother.

SP4 Raymond Garold Nichols, USA, The Day he Learned About Danny's Death

Ray learned of his brother Danny's death while stationed at Camp Greaves in South Korea. On November 11th, 1969, one of his superiors instructed him to report to the Commanding Officer's (CO) office. Assuming he was in trouble for accidentally breaking a window at work earlier that day, Ray walked in unaware of the devastating news awaiting him. Upon entering the CO's office, the officer asked, *"Do you know why you're here?"* Still thinking it was about the broken window, Raymond answered, *"Yes."* But it wasn't about the window at all. Instead, he was met with the heartbreaking news that his youngest brother, Danny, had been killed in Vietnam. Ray was stunned, the weight of the news settling heavily on his shoulders. Yet, he showed no emotion. Was it shock that numbed his reaction, or was it the discipline ingrained in him through military training? Perhaps it was a little of both. The CO arranged transportation for Ray to Seoul, South Korea, and handed him official orders for his immediate return home. Ray was to board a military transport plane and reunite with his family to attend Danny's funeral.

During my phone conversations with Ray, he did not reflect on his emotions from that day. Perhaps the passage of 50-plus years had dulled the immediacy of those feelings, or maybe the pain was still too deep to revisit. Either way, he chose not to discuss his emotions. Instead, he focused on recounting what he remembered about this tragic event. With orders in hand, Ray was transported to Seoul, South Korea, to await a flight home. Upon arrival, he was surprised to encounter his brother Charles (Chuck), who was also serving in the U.S. Army. Chuck was stationed at Camp Peleliu, a smaller installation within Camp Red Cloud, located in the city of Uijeongbu.

Ray and Chuck were relieved to see each other, but under the weight of their shared grief, their reunion was filled with tears and sorrow. Within a few hours, both brothers would be boarding a military airlift flight back to the United States, carrying the heavy burden of loss as they prepared to reunite with their family under heartbreaking circumstances. Ray recalled Chuck mentioning that their mother, Mary, had specifically asked him (Chuck) to escort Danny's body home for the funeral. It wasn't a difficult task; in fact, Chuck accepted the responsibility without hesitation. It's unclear whether the U.S. Army intentionally arranged for Ray and Chuck to meet in Seoul, Korea, or if it was merely a coincidence. Regardless, Ray and Chuck were happy to see each other and proud to be on that plane, accompanying Danny's body back to the States.

After several long-haul flights from Korea, Ray and Chuck finally landed at Dover Air Force Base in Delaware, only to discover that Danny's body was not on their plane. Ray and Chuck had no idea what had gone wrong or where the mix-up had occurred. More troublingly, they didn't know where Danny's body was. Concerned, Chuck immediately questioned the staff at Dover Air Force Base about Danny's whereabouts. They were informed that his body was still in Vietnam, awaiting transport to the United States. As the designated escort, Chuck's orders required him to remain at Dover AFB until Danny's body arrived. Since Ray's orders did not include escorting Danny's body, he was instructed to take the next commercial flight to Chicago O'Hare Airport and return home. Meanwhile, Chuck remained at Dover AFB, waiting for a plane from Vietnam to arrive with Danny's body on board.

SGT Charles Fredrick Davidenko, USA, The Day he Learned About Danny's Death

After numerous emails and phone calls with my Uncle Chuck, he finally agreed to share his story about his time in the U.S. Army and what happened after he learned of his brother Danny's death. Initially, Chuck was reluctant to discuss the issue, as it stirred up painful memories and caused him significant personal stress. Even after my long initial interview with Chuck, recalling his memories of that tragic event, I still had many unanswered questions and gaps in his story regarding his enlistment in the U.S. Army. I tried numerous times to get him to expand upon his military service, but it was just a very sad time in his life, and he did not wish to reopen the old wounds and the pain he suffered from the loss of his baby brother.

Due to this stress, Chuck discontinued our conversation about Danny. Chuck stated, "*It was becoming too difficult to relive this pain all over again.*" At that moment, my investigative instincts kicked in, and I recalled an important detail: Chuck was the one who had introduced his brother Danny to his girlfriend, Helen, all those years ago. I remember learning this during my interviews with Helen, when she described being introduced to Danny by Chuck at the Woolworth's diner. Could the weight of the rumored "Dear John" letter, and the knowledge that he had introduced Danny and Helen, have been the reason why Chuck found it too painful to relive this tragedy? Sadly, we will never truly know, because I did not get the chance to ask him about this issue before he passed away. During our last conversation, I told him that I appreciated all the information he shared with me and fully understood his decision to stop contributing to this book. When I last spoke with my Uncle Chuck on the phone, I said to him, *"If you change your mind, I'm here, and I'm willing to listen. I have so much to ask you, and only you can provide me with the answers to the information I seek."* I ended the phone call with the words, *"I love you."* Chuck replied, *"Love you too, kid."* To continue Chuck's story, I will rely on the details from the letter he sent me and our multiple conversations where he provided the stories about his military service and the death of his youngest brother, Danny.

On November 11th, 1969, Chuck's daily routine began like any other day, until he was ordered to report to his company commander's (CC) personal quarters. Like his brother Ray, Chuck must have wondered, 'What on earth did I do' to warrant a meeting with my CC, and why is this meeting in his personal quarters instead of his HQ office? When Chuck arrived, the CC welcomed him in and offered him a drink from a small selection of whiskey he had stashed in his quarters. At this point, Chuck was completely confused, as it was not your typical military protocol to call upon one of your enlisted troops for a casual drink. Chuck politely declined the drink and stood tall before his commander. He knew something was up, something important, especially if it couldn't wait until his CC was in his office. Under the circumstances, and in a somewhat formal military manner, Chuck's commanding officer solemnly stated:

"As your commanding officer, it is my duty to inform you that your brother, CMCN Daniel Joseph Davidenko, was killed in Vietnam on November 10th, 1969."

In that moment, I can only imagine what raced through Chuck's mind: his emotions raging as he silently asked himself how and why this could have happened. Chuck remained silent as the loss of his baby brother consumed his thoughts. Chuck told me that, immediately after hearing the bad news, he thought to himself, *"Wow, I shouldn't have turned down that drink,"* but he knew better than to ask for one. Chuck's mind was moving a million miles a minute. How could this be? Danny wasn't on the front lines; he was in the U.S. Navy for God's sake! In their letters to each other, Danny was safe on a U.S. Navy base doing auto repairs within the relative safety of the CBMU-302 compound. Chuck did not share with me all of the thoughts that went through his mind that day. All that he expressed was the fact that he felt completely numb and helpless. He had no idea how long he stood there, motionless and silent, grappling with the loss of his little brother. As a tear ran down his face, his commanding officer informed him that orders had been prepared for his return home for the funeral. Chuck remained motionless, attempting to take everything in, when the CC handed him an envelope with travel orders and told him that there were two platoon squad leaders standing by to drive him to Kimpo Air Base in Seoul, Korea. Once Chuck gathered his thoughts, he asked to be dismissed. He rendered a firm hand salute, did an about-face, and began to exit the CC's personal quarters. Before Chuck made his exit out the door, the commanding officer said, *"Wait, there is one more thing I have to tell you, your mother requested that you be the one to escort your brother's body back home to rest."*

Chuck was given an hour to pack his bags for the trip home before he was taken to Seoul, Korea. The U.S. Air Force at Kimpo Air Base (designated K-14) had C-141 Starlifter aircraft, which were flown by the 57th Military Airlift Squadron. From this Air Force base in Korea, Chuck would start the long journey home.

Kimpo Air Base, Seoul, Korea, C-141 Starlifter, & Tachikawa Air Base, Tokyo, Japan

Later that day, when Chuck arrived at Kimpo Air Base, he was surprised to find his brother Raymond waiting for him. Raymond had been transported from Camp Greaves, near the DMZ, where he was stationed. It wasn't until they arrived at the Kimpo Air Terminal that Chuck and Ray learned they would be traveling under high-priority orders and granted VIP status on their journey back to the States. Though there was no time to rejoice at the opportunity of travelling in VIP status, it was nice that the U.S. Army made it available under the circumstances. After checking in for their flights, Chuck and Ray were brought into the VIP waiting room to await departure.

Waiting in the VIP lounge felt strange to Chuck and Ray, as the room was typically reserved for senior officers from all branches of the military. Chuck could sense that he and his brother, two lower-ranked enlisted men, were not particularly welcome in the VIP area. What Chuck didn't realize was that every officer in the room knew they were there only because of an extreme event, one the U.S. Army deemed significant enough to warrant VIP status. Every senior officer in that waiting room knew that these two enlisted men were in the VIP area because of an emergency back home. For junior enlisted personnel, the most common emergency warranting VIP treatment was a death in the family, usually a parent or sibling. Without seeing their orders, the senior officers in the VIP area couldn't have known that these two soldiers were going home to bury their baby brother, a U.S. Navy serviceman, who had been killed in Vietnam.

While waiting to depart, Chuck felt that everything was going smoothly until it was time to board the plane. Out front, a bus was designated for most passengers, while a staff car was reserved for VIPs. Among those waiting in the VIP room was a U.S. Army Colonel and his wife. When boarding began, the Colonel and his wife made their way toward the VIP car, preparing to depart. However, a U.S. Army Staff Sergeant redirected the Colonel and his wife to the general passenger bus operated by the terminal staff, while another senior enlisted member personally escorted Chuck and Ray to the VIP staff car. The thought of bumping a senior officer onto the bus must have made the hairs on the back of Chuck and Ray's necks stand up. As they were escorted to the VIP car, Chuck says to Ray, *"Holy shit, did we just bump a full-bird colonel to the back of the bus?"* While inside the VIP car, both Chuck and Ray had a quick laugh, knowing that they bumped a colonel from the VIP car onto the bus. Both were momentarily amazed by the VIP treatment, thinking how surreal and special it felt, but the feeling quickly faded as they were reminded of the reason behind it. They were being honored that day not for celebration, but because of the death of their baby brother Danny.

From the VIP staff car, both Chuck and Ray boarded the C-141 and flew to Tachikawa Air Base in Japan, located in the western part of Tokyo. On or about November 12th, they both boarded a second Air Force C-141 in Tachikawa and flew to Elmendorf AFB in Anchorage, Alaska. Once they landed in Alaska, they waited to board yet another flight that would take them directly to Dover AFB in Delaware.

Throughout their journey from Korea to Japan, then Alaska, and finally Dover, Delaware, Chuck and Ray believed they were traveling with Danny's body and casket. However, upon landing in Dover on November 13th, they learned that Danny's body was not on the plane. Confused and concerned, they wondered, where was Danny? At first, the situation felt alarming, not knowing his whereabouts. But Chuck soon discovered that Danny's body had never been missing or misplaced. Danny's body was awaiting processing and was to be sent directly from Vietnam to the Department of Defense (DOD) mortuary in Dover, Delaware, for a forensic autopsy. Chuck also understood that with the ongoing war in Vietnam, troops and supplies took priority on all air travel, including the transport of casualties returning to the States. He knew Danny wasn't the only fallen soldier coming home; the death toll was staggering.

The mortuary affairs department in Dover, Delaware, was likely overwhelmed, and processing the remains of all deceased servicemen would take time. As it turned out, Chuck would have to wait for Danny's body, and Ray would catch the next commercial flight home to Chicago. From my conversations with Danny's former executive officer, LT Jacobs, and the memorial service pamphlet produced by the base chapel, Danny's body was present at his military memorial service and catholic mass held by Father Casey (Catholic Priest) on November 13th, 1969, at the Cam Ranh Bay, Vietnam base Chapel. Based on these facts, Danny's body would not have been scheduled to be shipped back to Dover until the following day (November 14th).

From Dover AFB, Chuck was escorted to the Marine Barracks at the Philadelphia U.S. Navy Yard, where he was assigned a room. The barracks could accommodate approximately 100 men, yet Chuck found himself the sole occupant. After the past 24 to 48 hours, Chuck desperately needed rest. However, the vast, empty barracks only amplified his loneliness and misery. He soon learned he would have to stay there until Danny's body arrived for processing. Chuck was treated very well during his stay at the Marine Barracks; however, Chuck stated that *"it took about four days for Danny to be processed, and those four days felt like an eternity."* During those four days, Chuck had multiple phone conversations with his parents about the status of Danny's body and when he would be bringing Danny's body home. Alone in his room, his mind raced as he imagined the worry, stress, and fear his parents must have been feeling. He, too, felt the weight of his family's suffering, and it pressed heavily on his heart. Adding to Chuck's stress, he was given the solemn duty of escorting Danny home. The only thought consuming him was that his mother had personally chosen him to bring her baby boy home to his final resting place. Chuck's stress grew with each passing day until finally, he was told that Danny's body was ready, and it was time for Chuck to bring Danny home.

Danny's remains returned to the United States around November 14th. The autopsy was performed at Dover AFB, carried out sometime between the 15th and 16th of November 1969. Monday morning, November 17th, was the day Chuck had been waiting for; he was finally bringing Danny home to his family and his final resting place in Illinois. A huge sigh of relief washed over Chuck when he finally received the word; it was time to take Danny home. At Dover Air Force Base, in his full dress uniform, he stood at attention, holding a sharp military salute as he watched Danny's flag-draped coffin being carefully loaded onto the aircraft. Once the cargo door closed, Chuck sharply dropped his salute, did an about-face, and boarded the plane. It must have been difficult for Chuck to watch them load his brothers' casket onto the plane. However, Chuck remained stoic; he did not shed a tear, remained professional, and provided Danny with the military honors he deserved. Shortly after they loaded the casket, Chuck and Danny took flight on the last leg of their travels home to Brookfield, Illinois.

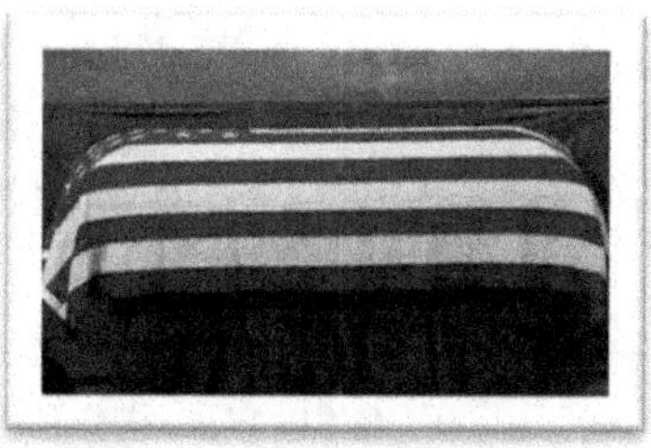

U.S. Flag Draped Over a Fallen Service Members Casket

When the Continental Airlines Douglas DC-9 touched down in Chicago, Chuck remained seated, waiting for all passengers to disembark. Through the window, he spotted a hearse pulling onto the tarmac, stopping near the rear of the plane. After he exited the plane, Chuck stood at attention, holding a crisp hand salute, watching solemnly as Danny's flag-draped casket was carefully unloaded and placed into the hearse. Sadly, there was no six-man military detail to formally carry the casket into the hearse. Once it was loaded by airline personnel, Chuck climbed into the hearse for the journey to Paloucek Funeral Home in Brookfield. He watched as Danny's casket was carried from the hearse into the funeral home. Once Danny's body was respectfully delivered, Chuck knew he had to get back home. His parents and family were all waiting for his return. Chuck was also anxious to tell his mother that he did what she asked and that he had brought Danny back home.

After such a long and emotional ordeal, Chuck couldn't remember who dropped him off at the house. Overwhelmed with grief, he believed the funeral home director had driven him home, but the details remained a blur. The mental stress was nearly unbearable, and Chuck felt the urge to break down and cry, but now was not the time. He knew he had to stay strong and fulfill the request his mother had requested of him. He could still hear his mother say: *"Chuck, I want you to bring my Danny home"*.

Outside his parents' house in Brookfield, Chuck took a moment to straighten his uniform and compose himself before stepping inside. As he entered, he saw his mother, Mary, seated at the dining room table, surrounded by his father, Elmer, his brothers and sisters, and other family members. The moment his eyes met theirs, he could see the overwhelming grief etched on his parents' faces, as they were devastated and heartbroken. Chuck was immediately embraced by his mother, and the tears began to fall. The entire family was overcome with emotion, but at last, they were all together. Mary had her whole family around her, especially her boys, Jimmy, Ray, Chuck, and Alby. On the kitchen table lay a Western Union telegram from the Commander of Dover Air Force Base. Addressed to Mary and Elmer, it confirmed that Danny's remains had arrived in Chicago, Illinois, at 8:00 PM on November 17th, 1969, aboard CA Flight 495. That evening, the entire family finalized the funeral arrangements, and Danny's memorial service was scheduled for the morning of November 20th, 1969.

On the morning of November 18th, 1969, Chuck and his family accompanied their parents to the funeral home to see Danny's body. During the car ride, it was clear that Mary remained distraught, overcome with grief and emotion. She repeatedly insisted on having a closed-casket service, which led to a family discussion. Everyone was confused by her decision, but in Mary's mind, all she could imagine was Danny's body being in poor condition after ten days. She couldn't bear the thought of anyone seeing him that way. Finally, the family convinced Mary to see Danny before making a final decision. As they entered the funeral home, everyone braced themselves for the moment. But when Mary saw Danny, her fears eased, and she changed her mind, allowing for an open casket viewing. With the decision made, the final funeral arrangements were made. Danny's funeral service and Mass would take place on November 20th, 1969.

Funeral Service, Mass, and Burial

The morning of November 20th, 1969, began like any other crisp fall Thursday in Brookfield, Illinois. The skies were mostly clear, but gusty winds of 10-15 knots cut through the air, making the already frigid 36 degree temperature feel closer to 26 degrees. The biting wind slipped through even the finest Sunday church clothes, adding to the overall discomfort and stress of the day. Family and friends gathered at Saint Barbara's Church in Brookfield, Illinois, for Danny's funeral Mass, scheduled for 10:00 AM. The church was adorned with candles and flowers, creating a solemn yet reverent atmosphere. The Catholic priest led a beautiful service filled with scripture readings, hymns, and a heartfelt eulogy. Several loved ones stepped up to the podium, sharing cherished memories and expressing their love for Danny. As each speaker's words echoed through the silent church, all eyes were drawn to a young blonde-haired girl sitting near the front, Helen, Danny's girlfriend. Just seventeen years old, she was too overcome with grief to speak.

Following the Catholic Mass, the entire family escorted Danny's body to its final resting place at Saint Mary's Cemetery for interment. Surrounded by loved ones, he was laid to rest with a traditional Catholic burial, accompanied by a military honors detachment paying tribute to his service. The solemn notes of the bugle and the precision of the flag-folding ceremony ensured that Danny received the honors he deserved. After the funeral, in true Irish tradition, the family gathered to share food, drinks, and stories, celebrating Danny's life, however brief, with love and remembrance.

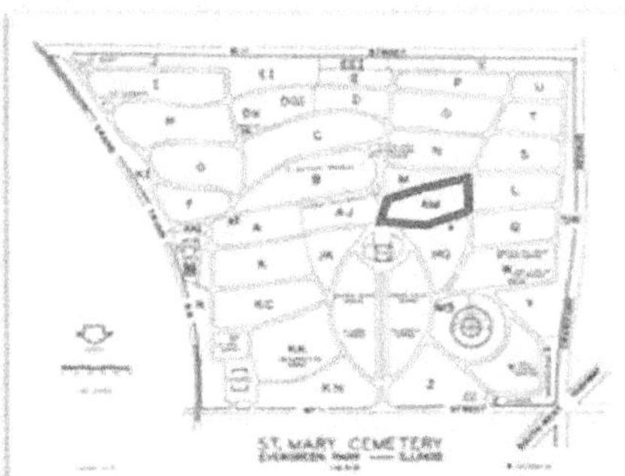

Danny's Final Resting Place - Saint Mary Catholic Cemetery

PLOT: Lot 1135 / Section AM / MEMORIAL ID 126725663

The Memorial in Chicago Ridge

In 1966, the American Legion Hall (Glenn Maker Post 1160) located at 10739 Ridgeland, Chicago Ridge, Illinois, purchased the land adjacent to their building for use as a parking lot. In 1968, the parking area was paved, and a Memorial was erected and dedicated to all the fallen service members from the local area. This American Legion in Chicago Ridge was a favorite drinking spot for Danny's father, Elmer, and his brothers-in-laws, Bill Stockdale and Herbert Roberts, all of which were members of this American Legion.

During my research for this book, I made an effort to visit Danny's grave site in Evergreen Park, IL, his memorial in Chicago Ridge, IL, his memorial in Brookfield IL., the Vietnam Memorial in Washington, DC, and the SEABEE Memorial in Washington, DC. On my visit to Danny's memorial outside the American Legion, a few family members and I gathered near this stone with Danny's name, and with a moment of silence, we all closed our eyes and said a private prayer. Immediately after viewing the memorial monument, we went inside the American Legion and had a few beers in Danny's honor.

While enjoying a few beers at the American Legion, a few old military veterans were sitting around the bar and wondering what this group of people (Danny's family) was doing at their American Legion. I spoke with one member (a former U.S. Navy veteran) and told him we were there honoring our fallen family member. The Legion members were a bit awestruck and very pleased to see our family showing our love and respect for Danny and the ultimate sacrifice he made while serving his country. Upon leaving the American Legion, the gentleman I was chatting with about Danny stopped me at the door, and then all the men around the bar lifted their glasses of beer and said, *"In memory of Daniel Davidenko,"* and each one took a sip of beer in Danny's honor. It was a special tribute to Danny, and I'm sure the look on all of our faces was quite incredible. I just wish someone had taken a picture of us at that moment. I was honored to share that moment with the following family members: Chuck Davidenko, Ray Nichols, Richard Roberts, Bill Stockdale Jr., Jerry Stockdale, Rita Barnes, and Cassie Barnes. Shown below are several photos from outside and inside of the

American Legion in Chicago Ridge, where Danny's name was added to the city's veteran's memorial.

The Memorial in Chicago Ridge (American Legion - Glenn Maker Post 1160)

Chuck Davidenko & Dan & Dan & Ray Nichols

Bill, Ray, Rich, and Jerry

The Vietnam Wall

The Vietnam Veterans Memorial, commonly called the Vietnam Memorial, is a U.S. national memorial in Washington, D.C., honoring service members of the U.S. armed forces who served in the Vietnam War. The two-acre site is dominated by two black granite walls engraved with the names of those service members who died or remain missing as a result of their service in Vietnam and Southeast Asia during the war. The names on the Wall, originally numbering 57,939 when it was dedicated in 1982, are listed in chronological order of the dates of casualty. Additional names have been added throughout the years, and there are currently 58,318 names (as of January 2026). Inscribed on the Memorial are the names of service members who were classified as dead, missing, or prisoners.

Over the years, a majority of our family and I have visited the Vietnam Memorial and touched or traced Danny's name. When you read about the Vietnam War and about the 58,318 American service members who lost their lives, it's hard to put the death toll into perspective. However, once you are at the Vietnam Memorial Wall, you gain a new perspective when you can see, feel, and touch the names that appear on that wall. When you walk from end to end of this memorial, you understand the magnitude of the lives lost to this senseless war. Reading a number from a history book does not have the same effect; however, when you actually see and touch the names of the fallen on the Vietnam Memorial Wall, you leave this hallowed ground with a real understanding of the magnitude of what was truly lost.

The portrait (below) was painted by Lee Teter, and the painting is called Reflections. Lee painted this shortly after experiencing loss in his own family during the Vietnam War. The painting depicts a man leaning forward with his hand on the Vietnam Memorial Wall, and in his reflection, you see a soldier's hand reach out and touch his from the other side of the Vietnam wall.

Vietnam Veterans Memorial "Reflections", Lee Teter, 1988.

The American Gold Star Family

The American Gold Star Family title is reserved for families of military members who have died in the line of duty. It is meant to honor the service member's ultimate sacrifice while acknowledging their family's loss, grief, and continued healing. The phrase 'Gold Star Family' dates back to World War I, when military families displayed service flags featuring a blue star for every immediate family member serving in the Armed Forces. Over time, Mary had five blue star flags, one for Jimmy, Raymond, Albert, Charles, and Daniel. The star's color would be changed to gold if the family lost a loved one in the war, hence the term, 'Gold Star Family'. In November of 1969, Mary received a separate, second flag, her first gold star flag, for the loss of her son Danny.

The last Sunday of September has been designated as National Gold Star Mothers' Day. Most Americans do not realize this, but 12% of the U.S. population served during WWII, 9.7% of the U.S. population served during the Vietnam War, and today, only 1% of the U.S. population serves in the military, based on U.S. population statistics. So, to put the American Gold Star Mothers from the Vietnam War into perspective, approximately 2,709,000 Americans served during the Vietnam War, and 53,318 died, and each of their deaths made one more gold star mother and family.

This chapter was originally intended to be the last chapter of this book. However, shortly after completing it, while in the final stages of editing, I received a package in the mail. Initially, I didn't pay close attention to the address on the package. I assumed it was from the National Archives, containing documents related to other family members' military service. But upon closer inspection, I noticed the package was from the Department of the Navy, Office of the Judge Advocate General. My eyes widened. Could this be the JAG investigation report? Not wanting to get my hopes up, I took a deep breath and carefully opened the package. To my surprise, the package contained the official JAG investigation into the death of CMCN Daniel Joseph Davidenko. Included among the documents was a copy of a report titled 'Autopsy Protocol'. Although it was not the full official autopsy report I had requested, I was excited and nervous at the same time. Although the documents were in poor condition, most were still legible. I will transcribe them so the reader can review precisely what was recorded. In the next chapter, I will analyze these documents and compare the official reports with LT Jacobs' recollection of the events in November 1969. I will also offer investigative commentary on each documented finding.

Chapter 13

The Official JAG Investigation

After years of research and multiple FOIA requests, I finally obtained the official JAG investigation completed by LT Jacobs on November 22nd, 1969. However, the package did not include the full autopsy report or photographs, only a summarized version titled the Autopsy Protocol Report. This final chapter will take an in-depth look at the official JAG investigation conducted by LT Jacobs, along with the limited autopsy protocol report that was submitted as part of the JAG investigation. The facts of the case, as gathered by LT Jacobs, will be presented exactly as reported. In addition, I will include information I collected independently through letters, emails, chats, personal interviews, and phone conversations with CBMU-302 personnel. When and if necessary, I will also reference comments from Danny's family. As I review LT Jacobs' JAG investigation report, I will also provide investigative commentary, highlighting the procedures that were carried out correctly, as well as those that were not. A few days after receiving the JAG investigation documents, I received a letter from the National Archives responding to my final FOIA request. Unfortunately, my last and final request for the official Naval Investigative Service (NIS) report was met with the following response:

"After reviewing your requests and consulting with the Naval Criminal Investigative Service, we have determined that we do not have any responsive records. However, it is possible that the National Archives and Records Administration (NARA)/National Personnel Records Center (NPRC) may still maintain the records you are seeking. Accordingly, we have referred your requests to NARA/NPRC for processing and direct response to you."

My initial FOIA requests to NARA and NPRC produced no results. However, since the U.S. Navy Fleet Judge Advocate General's office forwarded my request, there is still a chance I might eventually see these documents, if they exist. As for the official autopsy report (with photos), it remains a complete mystery as to why it cannot be located. What I did receive was a five-page document within the official JAG investigation titled 'Autopsy Protocol', which provides a summary of the autopsy findings. The autopsy protocol report is a document that outlines the findings of an autopsy, including the cause and manner of death. It is structured to ensure that all important findings are documented accurately and concisely. The protocol report is distinct from the autopsy report, which provides a broader overview of the postmortem examination results. The protocol report is used to produce a shorter more concise report, while the official autopsy findings serves as a comprehensive medical document summarizing the postmortem examination details.

One of the biggest challenges now is working with these poorly preserved documents from over 55 years ago. Much of the text has faded, making it extremely difficult to read. As a result, I must recreate the text into a clear, readable format. This process will take some time, but it is necessary to accurately reconstruct the events of November 10th, 1969. Additionally, the documents provided in the FOIA response were not organized in any particular order, so I arranged them chronologically based on the dates they were drafted and signed. My biggest concern was that these documents would ultimately prove to me that a premature determination of Danny's manner of death (MOD) was hastily determined. Early in my research, it became clear that there had been a rush to judgment followed by an incomplete, somewhat informal criminal investigation surrounding Danny's death. Any lingering doubts I had about the premature manner-of-death determination and the quality of the investigation were confirmed when I read the fourth line of the official personnel casualty report, which read, "APPARENT SUICIDE." Despite years of interviews with LT Jacobs and other CBMU-302 personnel, no one had ever indicated to me that either the casualty report or the JAG investigation classified Danny's death as a suicide. LT Jacobs told me several times (verbally) and once in a letter, that Danny's manner of death was determined to be "*ACCIDENTAL*." Therefore, you can only imagine my disbelief when I read his JAG investigation report. What follows are transcriptions of the messages contained in the JAG investigation report. For clarity and transparency, some of the transcriptions will be accompanied by a full copy of the original document.

The First Document - PERSONNEL CASUALTY REPORT

The Personnel Casualty Report – Paragraph 1 Transcribed

"IMMEDIATE 101220Z NOV 69

1) LAW REF A FOL INFO SUMITTED:

ALFA: CMCN DANIEL JOSEPH DAVIDENKO, USN

BRAVO: ACTIVE DUTY ATTACHED TO CBMU-302, CRB, RVN

CHARLIE: DECEASED ***(APPARENT SUICIDE)***

DELTA: 10 NOV 1969, APPROXIMATELY 0500H, NAVSUPPPFAC, CRB, RVN, CAUSE OF DEATH, GUNSHOT WOUND ***APPARENTLY INTENTIONALLY SELF-INFLICTED.*** *DAVIDENKO FOUND NEAR NORTHWEST CORNER OF NSF, CRB SUPPLY COMPOUND BY LTJG UNDERWOOD, AT 1755H, .45 CALIBER PISTOL LOCATED NEAR BODY."*

IMMEDIATE
O 101220Z NOV 69
FM CBMU THREE ZERO TWO

TO SECNAV

INFO CHNAVPERS
CHBUMEDSURG
NAVFINCEN CLEVELAND
CONNAVFORV
COMNINE
CONNAVSUPPACT SGN
CONCBPAC
NAVFACENGCOM
CINCPACFLT
COMSERVPAC
CDR 31ST NCR
FLEET HOMETOWN NEWS CENTER
CNO
COMUSMACV
BUDOPAC
NAVDISP 12ND
US ARMY MORTUARY, SIAGON

UNCLAS E F T O

PERSONNEL CASUALTY REPORT
A. BUPERS MAN ART C-9801

1. IAW REF A FOL INFO SUBMITTED:
ALFA) CMCN DANIEL JOSEPH DAVIDENKO, USN, [redacted]
BRAVO) ACTIVE DUTY, ATTACHED TO CBMU-302, CRB, RVN
CHARLIE) DECEASED (APPARENT SUICIDE)
DELTA) 10 NOVEMBER 1969, APPROXIMATELY 1800H, NAVSUPPFAC, CRB, RVN. CAUSE OF DEATH, GUNSHOT WOUND APPARENTLY INTENTIONALLY SELF-INFLICTED. DAVIDENKO FOUND NEAR NORTHWEST CORNER OF NSF, CRB SUPPLY COMPOUND BY LTJG [redacted], AT 1755H. 45 CALIBER PISTOL LOCATED NEAR BODY.
ECHO) GRAVE REGISTRATION SECTION, [redacted], APO [redacted]
FOXTROT) [redacted] DAVIDENKO, (FATHER), [redacted] DAVIDENKO

NEU:2)/PERS(1) ...ACT
JAG(1) 030

C13389/1/9WF | 1 | 10/1317Z | 101220Z NOV 69

Personnel Casualty Report Dated November 10th, 1969

The personnel casualty report was the first of many documents drafted and sent to higher headquarters. This official message was sent out by the commanding officer of CBMU-302, and it was addressed to the Secretary of the Navy (SECNAV). This message was released on November 10th, 1969, at 1220Z. In military communications, ZULU time (coordinated universal time) is used to prevent confusion and maintain consistency across different time zones, especially when coordinating with allied forces around the world. In Vietnam, 1220Z translates to 1920 (7:20 PM) local time in Vietnam. This initial message was drafted, signed, and released by the commanding officer of CBMU-302 approximately one hour and twenty-five minutes after the discovery of Danny's body. How could they make this determination before the completion of the autopsy and the JAG investigation? Another document, NOT included within the JAG investigation documents, was the November 11th, 1969, letter from the commanding officer to Danny's parents stating that he died from a self-inflicted gunshot wound. Both of these documents were released within 24 hours of Danny's death and both stated that Danny's MOD was due to a self-inflicted gunshot wound to the chest.

As we will see from the following review of the JAG investigation documents, a letter Danny wrote but never sent home as well as a handwritten note found in Danny's pocket, directly contributed to the immediate MOD decision equating to Danny's death to suicide. In my opinion, the manner of death (MOD) determination appears to have been based solely on these two handwritten pieces of evidence and a rumored 'Dear John' letter. No additional physical evidence or analysis appears to support this conclusion. It is important to note and to refresh in the readers mind, that the JAG investigation was not completed until 12 days after Danny's death and yet the MOD determination was made in less than 90 minutes, based on two pieces of evidence and a rumor. Furthermore, the manner of death (MOD) determination was rendered before the official autopsy was completed, again raising significant concerns about the reliability of that conclusion.

The Second Document - CERTIFICATE OF DEATH

This document, dated November 10th, 1969, was completed by LT Baker, the medical officer, who was one of the first responders to the scene. The report states that the cause of death (COD) was an *"injury of sternal region, penetrating gunshot wound (missile) 036-4141."* This was the first document that led me to believe the manner of death (MOD) was inaccurately classified and that the command rushed to judgment in handling this event. An official copy of Danny's certificate of death, was transcribed and analyzed in chapter 9. From all my conversations with CBMU-302 members, as well as Danny's friends and family, everyone was under the assumption that Danny had been shot once. In fact, many CBMU-302 members interviewed stated they did not see any blood at the scene. LT Jacobs later explained that he believed all bleeding was internal, noting that only one shell casing and a single bullet lodged in Danny's chest were recovered. However, when I revealed to him that LT Baker's death report documented a 2 centimeter exit wound in the middle of Danny's back, it indicated to me that Danny had been shot twice. LT Jacobs appeared confused by this revelation, as the recovery of only one bullet and one shell casing had led him to conclude that Danny had only been shot once.

The Third Document - LT Jacobs Official JAG Appointment Letter

In this letter, LCDR Harrell officially appointed LT Jacobs as the sole JAG investigator into Danny's death. In paragraph two, LT Jacobs was directed to *"Conduct a thorough investigation into all the circumstances connected with the deceased."* The letter also referred to the requirement to follow procedures outlined in the JAG Manual during the course of his investigation.

The Fourth Document - AUTOPSY PROTOCOL

This autopsy protocol document was completed by an Air Force Captain (0-3) in the Air Force medical corps assigned to the hospital at CRB, Vietnam. It remains unclear whether this officer was a standard medical practitioner or a medical examiner with specialized training in criminal autopsies and forensic analysis. The autopsy was performed on the "trunk only" (meaning only the upper torso of the body) and did not include a full body (including the lower extremities) autopsy. Based on this fact, it is my belief that this was the preliminary (initial) autopsy report and not an official autopsy that should have been performed at Dover AFB. A formal autopsy report would contain a thorough head-to-toe forensic examination, including detailed anatomical diagrams and photographic evidence documenting the injuries that resulted in Danny's death. On page one, the first portion of this autopsy form was titled the CLINICAL DIAGNOSIS (Clinical Summary), and reads as follows:

"19 year old male, with U.S. Navy Seabees, found on 10 November 1969 about 1815, lying on left side between 2 piles of lumber. ***A .45 cal pistol was found about 15 feet from the body,*** *and a note on the body stated that the gun and shells had been obtained secretly by the deceased. The victim had apparently been sitting upright with his back against a side of box;* ***a shirt was blood-stained, but without holes in either front or back. Information in relation to circumstance is <u>scanty and somewhat contradictory</u>;*** *a guard listening post is situated 200-300 yards from the scene, and* ***a gunshot was heard early Sunday morning*** *and investigated without findings. A friend of the deceased states they were talking on Sunday night.* ***The deceased is said to have received a Dear John recently.*** *He was found late Monday afternoon; obvious rigor mortis is stated to have been present at that time".*

When I first read the opening entry in the Autopsy Protocol report, it became immediately apparent to me that the medical examiner had been provided with information that appeared to include hearsay, assumptions, opinions, and rumors, all factors that may have contributed to a biased assessment and a premature manner-of-death (MOD) determination. This medical examiner should never have been told any details (whether opinions, assumptions, or hearsay) about the deceased body he was assigned to perform an autopsy on.

Having the preconceived notion about suicide may have influenced his findings and led him to the premature conclusion as to the manner of death. The information about how he might have obtained the gun and the hearsay from a friend about a 'Dear John' letter are irrelevant to the autopsy and would only further compound the MOD determination. However, there are a few interesting pieces of information in this clinical summary that are quite alarming. The examiner's understanding that the gun was found 15 feet from the body does not align with other CBMU-302 testimonies, which indicate multiple and conflicting accounts regarding the weapon's location and distance from Danny's body. In a majority of the crime statistics surrounding suicide; most often, the gun is only a few feet away or in the deceased person's hand. Was this location of the gun just another rumor, or was it factual?

From my interview with ENS Kelly, he stated that he never saw a gun near the body. LT Jacobs stated that the pistol was found 4 feet away from his left hand. Additionally, the medical examiner stated in his report that ***"a gunshot was heard early Sunday morning."*** Based on his own time of death estimation of 0500 (5:00 AM) on Monday morning, he should have concluded that the shot heard on Sunday evening was not the shot that killed Danny. The most troubling statement in this clinical summary is the fact that his ***"shirt was blood-stained, but without bullet holes in either front or back."*** Then the examiner states that the ***"information in relation to circumstances is scanty and somewhat contradictory."*** This medical examiner was conflicted about what he was told and from what he observed during his autopsy. The location of the gun in relation to Danny's body alone should have raised serious concern. When coupled with the presence of both an entrance and exit wound, no bullet holes in his shirt, and an absence of blood on his clothing, the scene simply does not make sense.

The Autopsy Pathological Diagnoses - Transcribed

"1. GSW, anterior chest, perforation of aorta.

2. Right hemothorax; hemopericardium.

3. Early decomposition.

Cause of Death: Gun shot wound anterior chest, with perforation of aorta.

Conclusions:

1. Immediate death from contact pistol ***gunshot wound, anterior chest, with perforation of ascending aorta*** *and resultant hemopericardium and right hemothorax.*

2. Probable time of death was in the early hours of Monday, 10 NOV 69.

3. ***Probable manner of death was suicide.****"*

The term Hemothorax is a medical condition characterized by the accumulation of blood in the pleural cavity, the space between the lungs and the chest wall. The term Hemopericardium refers to the presence of blood in the pericardial sac, the thin, two-layered membrane surrounding the heart. The term perforation of the aorta is when the aorta is punctured, torn, or bruised, and is called traumatic aortic injury. This type of injury most often occurs as a result of a gunshot or knife wound or as a result of a motor vehicle accident.

Autopsy Excerpts - Transcribed

"GROSS DESCRIPTION / External Description: Caucasian male 69 inches in length, estimate weight 150 pounds. Dependent livor mortis, dorsal; slight peripheral rigor mortis. ***Brown hair,*** *hazel eyes, mustache, dentition unremarkable. Some sand about face and ears."* (Note: Danny had Blonde hair)

"ANTERIOR CHEST: 1 centimeter missile entrance wound, to the right lateral side of the sternum over the junction of the right 3rd rib and

manubrium. 1mm rim of abrasion, contusion and possible powder. Small abrasions are noted 1 and 2 centimeters to the left of this wound. The exit wound is present on the back, at the T6 level, 2 inches from the midline; 18mm diameter and appears as entrance wound, with 2mm rim of abrasion and contusion. The slug is found just beneath the skin."

"<u>INTERNAL DESCRIPTION:</u> *The missile tract is as follows: 1. Through articulation of right third rib and* (unreadable word) *with fracture. 2. Anterior mediastinum, into pericardial sac, into aorta about 1centimeter above aortic valves; exited from aorta posteriorly, through pericardium, into posterior mediastinum and into body of T6, to right of midline. 3. Through body of T6, with fracture, out back as described. Massive hemopericardium (200cc) and right hemothorax (1.51) resulted from injuries."*

"<u>MICROSCOPIC DESCRIPTION:</u> *Sections from the wound of entrance show particles of gunpowder along the missile tract and extending deeply into the superficial soft tissues, with adjacent tissue destruction and necrosis, no inflammatory response present. Tissue sections are essentially unremarkable, showing various degrees of congestion; autolysis is evident and widespread."*

"<u>SPECIAL STUDIES:</u> *Photographs taken, external. Toxicology studies are pending; preliminary reports include blood alcohol of 6mgm%. X-Ray report – See enclosed sheet."*

"<u>X-RAY FINDINGS:</u> *AP and Lateral Chest:* **No broken bones demonstrated. A non-deformed bullet far posterior right chest medially.** *Cannot determine if it is just inside or just outside thoracic cage. Measures 13mm in diameter with a little magnification. Left chest negative. Some free air in the neck to the right of the traches. So much fluid in right chest, cannot see lung parenchyma as such nor right heart boarder of right mediastinal boarder."*

"<u>SUMMARY:</u> **Autopsy finding support a conclusion of immediate death by suicide**, *accomplished by a contact pistol gunshot wound of anterior chest;* **the gun was <u>likely</u> held in the right hand.**

The ascending ***aorta was perforated resulting in rapid exsanguination and cardiac tamponade.*** *The time of death was likely in or about the early morning hours of Monday 10 November 1969."*

Let's break down the key findings within the autopsy report. First, let's examine the exit wound present on Danny's back at the T6 level. Nowhere in the autopsy does the examiner identify or even suggest a second bullet theory. He describes the entrance wound and notes the bullet lodged in Danny's chest, but he never addresses what caused the 2 centimeter exit wound in the middle of Danny's back. The examiner seems somewhat perplexed by the fact that there was both an entrance wound and an exit wound, yet no corresponding holes in Danny's shirt. How could that happen? A deeper analysis should have led the examiner to question the cause of the 2 centimeter exit wound. However, he did not pursue this line of inquiry. It appears he may have formed a preconceived conclusion about the manner of death (MOD) based on the story he was told prior to conducting the autopsy. Had the examiner not been influenced by rumor or hearsay, he might have been able to approach the autopsy objectively and without prejudice.

The examiner noted the presence of gunpowder residue extending deeply into Danny's tissue but failed to address a critical detail: the size and pattern of the residue. This information would have been vital to a CSI in determining the distance between the firearm and Danny's body at the time of discharge. In forensic investigations, the analysis of gunpowder residue, specifically its size and distribution, can often distinguish between suicide and homicide. A tight, concentrated residue pattern typically suggests a close-contact shot, supporting a suicide theory, whereas a larger, more dispersed pattern might suggest the possibility of homicide. Unfortunately, the examiner made no mention of the size or pattern of the residue found on Danny's chest. Even if the residue pattern had suggested suicide, a serious, unexplained discrepancy remained: the presence of a single entrance wound with a bullet lodged in the body, alongside a significant exit wound. A seasoned forensic pathologist would have flagged this inconsistency, or at the very least, acknowledged the possibility that Danny had been shot twice.

With or without access to the original X-rays, determining which bullet caused the fatal damage to Danny's aorta would be difficult. However, it should have been immediately clear to the examiner that the bullet lodged in his chest could not have produced the exit wound. The medical examiner noted that no broken bones were observed during the autopsy, and that the bullet recovered from Danny's chest was not fragmented, meaning it was completely intact. This intact condition indicated that the bullet did not fragment or cause the exit wound. This was further confirmed during my interview with LT Jacobs, who stated that the bullet removed from Danny's chest was a full metal jacket (FMJ) round, fully intact except for a slight bend at the tip. The absence of broken bones eliminates the possibility that a bone fragment created the exit wound, further deepening the mystery surrounding Danny's injuries. The most troubling statement by the medical examiner came in his summary and determination of Danny's manner of death (MOD), where he concluded:

"Autopsy findings support a conclusion of immediate ***death by suicide****, accomplished by a contact pistol gunshot wound of the anterior chest;* ***the gun was likely held in the right hand.****"*

How could this examiner come to this conclusion? It appears the examiner was influenced by external narratives, including the story of the "Dear John" letter. It is also clear that the examiner had been informed about the note found in Danny's pocket, or he might have been the person who actually found the note, explaining how the deceased had obtained the gun and ammunition. This mixture of factual details and hearsay likely shaped the examiner's conclusions and led him down a path toward a pre-determined manner of death ruling Danny's death as a suicide. However, there are two critical facts the medical examiner missed. First, in his summary, the examiner suggested that the *"gun was likely held in the right hand."* What he did not know was that Danny was found deceased with a crucifix tightly clenched in his right hand.

This raises serious doubt and investigative questions on how Danny could have shot himself twice while holding a rather large crucifix in his right hand. If the examiner had crime scene photos of Danny's body with the crucifix in his right hand, it would have led him to think that Danny used his left hand to shoot himself twice. Second, the examiner had been informed that the gun was found approximately 15 feet away from Danny's body. Whether that detail was accurate or not, a thorough and unbiased examiner should have questioned how Danny could have fired two shots into the same entrance wound and then somehow propelled the weapon such a distance from his body. Had the examiner been aware of the crucifix or approached the autopsy without being influenced by hearsay, rumors, and selectively presented facts, he likely would not have reached such a definitive conclusion of suicide. In fact, the appropriate determination, based solely on the autopsy findings, should have led the medical examiner to record Danny's manner of death as UNDETERMINED.

Lastly, this autopsy protocol report is dated November 11th, 1969, meaning that this was a preliminary autopsy report conducted by the hospital located in Vietnam. This examiner had less than 24 hours to reach a manner of death determination. Given this compressed timeline, it raises serious questions about whether sufficient time was available to conduct a thorough and independent MOD evaluation. Based on these circumstances, it appears possible that the medical examiner's conclusion was influenced by the commanding officer's premature opinion that Danny's death was a self-inflicted suicide. An autopsy can be complicated and take a long time, depending on the condition of the body. A standard autopsy generally requires two to four hours to complete. However, in cases involving extensive trauma, such as multiple gunshot or stab wounds, the examination may take considerably longer. The medical examiner must also allocate additional time to review findings, analyze photographs, and prepare the written report. Based on the date of this autopsy protocol report, it is abundantly clear that this autopsy was not the official autopsy conducted by the medical examiner (forensic pathologist) a Dover Air Force Base.

Imagine for a moment that this examiner had no previous knowledge of the scene, hearsay, rumors, or facts in the case. He would have quickly determined the cause of death (COD) as a gunshot wound to the chest. He would have immediately questioned how one entrance wound and one exit wound could coexist with a full metal jacket (FMJ) bullet still lodged in the victim's chest. Under those circumstances, reaching a definitive manner of death (MOD) determination of accidental or suicide would have been exceedingly difficult to justify. Regardless of the hearsay, rumors, and facts in this case, there were no witnesses to this tragic event, and therefore, the facts to determine MOD should have come solely from the body itself.

The fifth document – THE OFFICIAL JAG INVESTIGATION MESSAGE

This report included numerous enclosures, several of which we have already discussed and analyzed. Additional enclosures included eleven statements from CBMU-302 personnel, the handwritten note found in Danny's pocket, a photograph of Danny's lifeless body with a gun nearby, and a portion of a letter (first page only- found among his personal items within his barrack room), Danny began on August 27th, 1969, but never mailed home. We will examine the findings of the JAG investigation by reviewing each entry line by line and applying basic crime scene analysis techniques to assess the validity of the conclusions. Below is the JAG investigation documents. Each section of the JAG investigation will be transcribed in italics (each paragraph/ line by line), followed by investigative commentary.

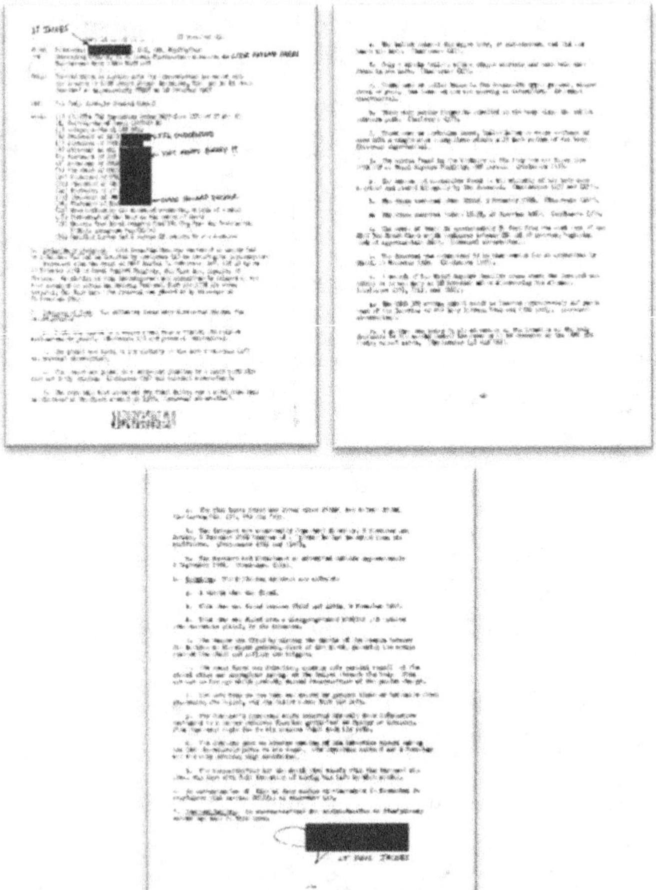

Official JAG Investigation Report Completed on November 22nd, 1969.

JAG Investigation Report - Transcribed

DTG of Message 22 November 1969
From: Lieutenant Paul Jacobs, CEC, USN
To: Commanding Officer, U.S. Naval Construction Battalion Maintenance Unit THREE ZERO TWO
Subj: Investigation to inquire into the circumstances connect with the decease of CMCN Daniel Joseph Davidenko, USN which occurred at approximately 0500H on 10 November 1969.
Ref: (a) Judge Advocate General Manual
Enclosures:
(1) CO, CBMU 302 Appointing Order 5830 Ser: 1170 of 10 Nov 69
(2) Certificate of Death (NAVMED N)
(3) Autopsy Protocol (SF 503)
(4) Statement of LTJG (Name Redacted - LTJG William Underwood)
(5) Statement of BUCN (Name Redacted)
(6) Statement of EO2 (Name Redacted)
(7) Statement of YNC (Name Redacted-Possibly YNC Monty Burkey)
(8) Statement of CMCN (Name Redacted)
(9) Statement of CMCN (Name Redacted)
(10) Statement of MR1 (Name Redacted)
(11) Statement of CE1 (Name Redacted)
(12) Statement of LT (Name Redacted - LT Paul Jacobs)
(13) Statement of CMCN (Name Redacted - CMCN Howard Decker)
(14) Statement of CMCN (Name Redacted)
(15) Note written by the deceased concerning origin of weapon
(16) Photograph of the body at the scene of death
(17) Excerpt from Naval Support Facility Cam Ranh Bay Instruction
(18) Unmailed letter dated August 27 written by the deceased

1. Preliminary Statement. This investigation was conducted as prescribed by reference (a) and as directed by enclosure (1) to investigate circumstances connected with the death of CMCN Daniel J. Davidenko, USN on 10 November 1969 at Naval Support Facility, Cam Ranh Bay, Republic of

Vietnam. Completion of this investigation was significantly delayed by the time required to obtain the Autopsy Protocol from the 12th Air Force Hospital, Cam Ranh Bay. The Protocol was picked up by messenger on 21 November 1969.

Investigative Commentary: The Autopsy Protocol was completed by the 12th Air Force Hospital on November 11th, 1969. This autopsy was not completed by an experienced criminal forensic pathologist. This Autopsy Protocol was conducted by the Air Force hospital medical examiner (medical doctor). The main difference between a medical examiner and a criminal forensic pathologist lies in their scope of work and the level of expertise required. Medical examiners are typically physicians who perform autopsies and determine the cause of death (COD), while forensic pathologists are medical doctors who specialize in the examination of deaths that occur under unnatural or suspicious circumstances, to determine the manner of death (MOD).

Forensic pathologists undergo additional training and certifications to perform autopsies and analyze evidence, whereas medical examiners may not have the same level of expertise in forensic science. LT Jacobs mentions that the autopsy report was delayed, which caused a delay in completing his JAG investigation. It is not clear how this delay occurred, but it was probably a communication issue or misunderstanding between CBMU-302 and the hospital as to who was responsible for sending and/or requesting the Autopsy documentation. Regardless, LT Jacobs did not get this information until eleven days after Danny's death, and most likely rushed to complete his investigation the following day.

"2. Findings of fact. The following facts were discovered through the investigation:

a. Death was caused by a bullet fired from a M1911A1 .45 caliber semi-automatic pistol. (Enclosure (3) and personal observation)."

Investigative Commentary: This entry (2. Findings of fact/paragraph a) addresses the cause of death (COD). The COD was officially determined to be a **single** gunshot wound (GSW) to the chest. However, after reviewing the Medical Officer's Certificate of Death dated November 10th, 1969, and the Autopsy Protocol dated November 11th, 1969, which both acknowledged the existence of an exit wound and an intact FMJ bullet lodged in Danny's chest, it becomes clear that serious questions should have been raised surrounding the exit wound. Everyone involved, from CBMU-302 up through the chain of command, should have immediately inquired about what caused the exit wound in the middle of Danny's back. The presence of both a lodged bullet in the chest area and a distinct exit wound directly contradicts the single bullet theory. A proper investigation into this discrepancy would have halted any premature conclusions regarding the manner and cause of death.

"b. The pistol was found in the vicinity of the body (Enclosure (16) and personal observation)."

Investigative Commentary: This entry (paragraph b) lacks critical specificity regarding the relationship between the body and the weapon. It should have clearly addressed both the exact distance and the exact location of the gun in relation to Danny's body. A thorough review of all official documents, including JAG investigation statements and my interviews with numerous CBMU-302 members, revealed conflicting reports about the gun's location in relation to Danny's body. 1) LTJG Underwood stated the weapon was ***"next to the body,"*** but provided no specifics. 2) LT Jacobs reported the gun was ***"approximately four feet away from Danny's left hand".*** 3) ENS Kelly did not recall seeing the gun at all at the scene. 4) The medical examiner, in the autopsy report, claimed the ***"gun was found fifteen feet away from the body."*** 5) The **photographic evidence shows the weapon near Danny's body, resting neatly on a pallet near his feet,** suggesting it may have been placed there intentionally rather than dropped naturally after the gunshots. The discrepancies in these accounts should have raised red flags immediately, prompting a deeper investigation into the actual sequence of events at the scene.

"c. The weapon was jammed in the semi-open position by a spent cartridge case not fully ejected. (Enclosure (16) and personal observation)."

Investigative Commentary: This entry (paragraph c) discusses the only cartridge found at the scene that was jammed in the gun's ejection port. The finding of this single cartridge (physical evidence) combined with the medical report of the entrance wound, exit wound, and the FMJ bullet lodged in Danny's chest, should have led the investigation to research the supply laydown area for a second cartridge on the ground and the second bullet possibly lodged in one of the surrounding supply crates.

"d. The cartridge that contained the fatal bullet was a steel case type manufactured at Frankford Arsenal in 1945. (personal observation)."

Investigative Commentary: This entry (paragraph d) suggests that the cartridge found at the scene was from the fatal bullet found inside Danny's chest. The medical report and the autopsy protocol concluded that a bullet went through Danny's aorta (this is what actually killed him), a single bullet remained lodged in his chest, and that there was a 2 centimeter exit wound in the middle of his back. Therefore, it is not possible to conclude that this was the cartridge that caused the fatal wound. A complete forensic test on the gun, cartridge, and the FMJ bullet would be required to reach such a conclusion.

"e. The bullet entered the upper body, at mid-sternum, and did not leave the body. (Enclosure (2))."

Investigative Commentary: This entry (paragraph e) discusses the full metal jacket (FMJ) bullet observed on the X-ray of Danny's chest. This is the same bullet that was removed during the autopsy and later handed over to LT Jacobs. From an investigative perspective, this bullet should have been preserved as forensic evidence. However, LT Jacobs lacked sufficient experience in death investigations and did not realize that he was responsible for documenting, properly processing, and preserving the bullet, the bullet casing, and the gun. The decision not to preserve these items as evidence constitutes a serious breach of investigative protocol and undermined the integrity of the investigation. As a result, any opportunity for later forensic testing (ballistic testing) to determine whether this was the weapon used in Danny's death was lost. The bullet removed from Danny's chest, was kept in LT Jacobs desk drawer, until his last day in Vietnam. Before leaving the country, LT Jacobs tossed the bullet into the sea.

"f. Only a single bullet, with a single entrance and exit hole were found in the body (Enclosure (2))."

Investigative Commentary: This entry (paragraph f) highlights yet another critical error in the handling of this death investigation. If a single bullet remained lodged inside Danny's chest, then what caused the 2 centimeter exit wound in the middle of his back? According to established forensic principles, exit wounds are created by some form of projectile, typically a bullet, a fragmented bullet, shrapnel, or another solid object. In this case, the bullet recovered from Danny's chest was a full metal jacket (FMJ) round that remained completely intact, eliminating the possibility that fragmentation caused the exit wound. Additionally, the autopsy report confirmed there were no broken bones, ruling out bone fragments as a possible cause. Therefore, some separate object or projectile must have caused the 2 centimeter exit wound in Danny's back, a major inconsistency that was never properly addressed during the investigation.

"g. There were no bullet holes in the deceased's upper garment, either front or rear. The deceased was not wearing an undershirt. (personal observation)."

Investigative Commentary: This entry (paragraph g) should have been a major red flag to the investigator. The evidence shows that Danny had a 1 centimeter entrance wound, a 2 centimeter exit wound, and a bullet lodged in his chest, yet there were no bullet holes in Danny's shirt. According to LT Jacobs, Danny's shirt was unbuttoned and opened in the front, suggesting that the shirt may have been moved aside when the shot was fired, allowing direct contact between the gun and the skin. However, this explanation still does not account for how a projectile exited Danny's back without creating a corresponding hole in the back of his shirt. This inconsistency raises serious doubts about the accuracy and thoroughness of the original investigation. Even the medical examiner was concerned about the fact that no bullet holes were observed in Danny's shirt. The medical examiner stated, *"the shirt was blood-stained, but without holes in either front or back. Information in relation to circumstance is scanty and somewhat contradictory."* The medical examiner recognized that something seemed suspicious, because the exit wound should have had a corresponding bullet hole on the back of Danny's shirt.

"h. There were powder fragments imbedded in the body along the bullet entrance path. (Enclosure (3))."

Investigative Commentary: This entry (paragraph h) examines the handling of gunpowder residue, another critical piece of forensic evidence that was mismanaged during this investigation. Gunpowder residue on a victim's body can help determine the distance between the firearm and the body at the time of discharge. However, the medical examiner failed to document the size or density of the gunpowder pattern on Danny's skin and made no attempt to estimate the gun's distance from the body. This omission was a significant failure in forensic analysis. Furthermore, when considering the two-bullet theory with a single entrance wound, the presence of direct contact residue would suggest that two bullets entered the body through the same location in rapid succession. In Danny's case, one bullet remained lodged in his chest, while the other exited through the upper part of his back. Without proper residue analysis and documented measurements, investigators, both then and now, were deprived of critical data necessary to reconstruct the events that led to Danny's death.

"i. There were no cartridge cases, bullet holes or other evidence of more than a single shot being fired within a 15 foot radius of the body. (personal observation)."

Investigative Commentary: This entry (paragraph i) suggests that a search of the immediate area was conducted for physical evidence, such as bullet casings or FMJ bullets embedded in nearby objects. However, a quick review of the one and only crime scene photo showing where Danny's body was discovered reveals several critical oversights. Visible in the background are large pallets of supply materials, stacked in wooden crates with vertical gaps and knot holes, potential entry points where an expended round could have lodged or passed through. Additionally, these wooden structures appear to be elevated a few inches off the ground, providing yet another potential hiding place for either a bullet or a spent casing. To further complicate matters, the scene was not secured or treated as a crime scene. The area was compromised by the individuals who initially discovered Danny's body, numerous onlookers swarmed the scene, and by the time first responders approached the scene, crucial physical evidence was disturbed, contaminated, or destroyed. As a result, any physical evidence that could have shed light on the circumstances surrounding Danny's death was either overlooked or irretrievably lost.

"j. The weapon found in the vicinity of the body was not taken from CBMU-302 or Naval Support Facility, CRB assets. (Enclosure (4))."

Investigative Commentary: This entry (paragraph J) states that the gun found near Danny's body was not a weapon from the CBMU-302 inventory, which was determined by comparing serial numbers against the command's armory records. LT Jacobs also states that the gun did not belong to the Naval Support Facility or to any other units assigned to CRB. How was this confirmed? Did the investigation actually check with other commands to see if the weapon belonged to one of these units stationed at the CRB base? This entry should have listed by name the other commands that completed an inventory check to see who did or did not have this weapon in their inventory. Additionally, we learned from a former CBMU-302 sailor that the unit armory contained a locker full of captured Viet Cong (VC) weapons, which could potentially have been the source of the .45-caliber pistol found near Danny's body. Notably, the investigation contains no reference to the VC weapons locker as a possible source, nor is there any indication that an inventory of the captured VC weapons was ever considered or conducted.

"k. The weapon and ammunition found in the vicinity of the body were acquired and stored illegally by the deceased. (Enclosures (15) and (17))."

Investigative Commentary: This entry (paragraph k) states 'as fact' that Danny acquired the weapon and ammunition illegally. Yet this conclusion is based solely on the pocket note found on his body, where allegedly, he claimed that he had stolen the gun from the Army side of the base. What remains unknown is whether Danny truly wrote this note as a factual admission or if he was coerced into writing it as part of an attempt to stage his death as a suicide. As countless crime scene case studies have shown, so-called suicide notes can sometimes be fabricated or manipulated pieces of evidence, deliberately crafted to steer an investigation in a particular direction. From a crime scene investigation (CSI) perspective, the vast majority of suicide notes left behind by the deceased rarely, if ever, address how the weapon was obtained. Instead, suicide letters typically focus on the reasons behind the act, often describing the emotional pain or depression the individual was experiencing, and usually conclude with an expression of remorse for what they did, or failed to do.

"l. The death occurred after 2000H, 9 November 1969. (Enclosure (14))."

Investigative Commentary: This entry (paragraph l) provides only an estimated time in which LT Jacobs believed Danny's death occurred. This contradicts the time of death (TOD) determination provided by the medical examiner, which suggested Danny's TOD at approximately 0500 (5:00 AM) on November 10th, 1969. By stating that the death took place sometime after 2000H (8:00 PM) on November 9th, 1969, it appears that LT Jacobs was trying to associate the gunshot heard that evening around 2000 by the tower watch with Danny's death. It would appear that this was an attempt to correlate the shot heard that evening, as the fatal shot in this investigation.

"m. The death occurred before 1815H, 10 November 1969. (Enclosure (2))."

Investigative Commentary: This entry (paragraph m) is just alluding to the fact that Danny's lifeless body was found just before 1815 hours local time on the CRB base on November 10th, 1969.

"n. The scene of death is approximately 50 feet from the east tent of the CBMU 302 Brick Plant which employes between 20 and 30 persons, beginning work at approximately 0800. (personal observation)."

Investigative Commentary: This entry (paragraph n) initially appears irrelevant, unless there is a connection between the brick plant personnel or the facility's location and Danny's death. What prompted LT Jacobs to include such seemingly insignificant information in an official JAG investigation? Could there be more to the story surrounding the brick plant and the sailors who worked there? Did the brick plant personnel, or even the plant's proximity to where Danny's body was found, hold information that might have shed light on the circumstances of his death? Or is this yet another piece of information that would alert a seasoned criminal investigator to look further? It appears that LT Jacobs was trying to provide context as to the location of the supply laydown area relative to other facilities on base. This statement just creates confusion and leads to more questions from anyone reviewing the investigation.

"o. The deceased was determined to be unaccounted for at approximately 0900H, 10 November 1969. (Enclosure (10))."

Investigative Commentary: This entry (paragraph o) is pure speculation about what Danny was thinking or doing at that specific date and time. Once again, LT Jacobs inserts a seemingly insignificant statement into an official JAG investigation without factual support or witnesses to the event that actually ended Danny's life. Based solely on the autopsy protocol, Danny was most likely already deceased by 0500 (5:00 AM) and therefore unable to report to morning quarters. Additionally, how could anyone conclude that Danny was "determined" to be unaccounted for that day? A more appropriate statement would have been: The deceased was unaccounted for on the morning of November 10th, when he failed to report for morning quarters at 0730.

"p. A search of the Naval Support Facility areas where the <u>deceased was likely to be</u>, was made on 10 November 1969, after discovering his absence. (Enclosure (10))."

Investigative Commentary: This entry (paragraph p) makes it abundantly clear that the search team only focused on areas where they believed Danny might have been hiding, rather than conducting a thorough search of the entire base, its facilities, and the surrounding areas. This could explain why Danny's body went undiscovered for so long in the supply laydown area within the CBMU-302 compound. While it is understandable that the search began with Danny's normal work areas and his usual hangouts, a proper missing person search requires a wider deeper sweep of all areas, including less obvious locations like the supply laydown area. The search and the search team were both inadequate and flawed from the beginning, largely because the search team lacked proper training for such events. The search team was made up of U.S. Navy servicemen with backgrounds in construction and not in investigations or searches. This statement by LT Jacobs in his JAG investigation further confirms my initial belief that the supply laydown area was never included in the base-wide search. A seasoned crime scene investigator would have directed all search team members to reconvene at the HQ building by mid-afternoon to present their preliminary findings. The team would then assess which areas had been covered and determine any locations that remained unsearched, before being redeployed to complete a thorough sweep of those sections.

"q. The CBMU 302 roving patrol watch is located approximately 100 yards west of the location of the body between 1900 and 0700 daily. (personal observation)."

Investigative Commentary: This entry (paragraph q) initially appears irrelevant, unless there is a connection between the roving patrol, the supply laydown area where Danny's body was ultimately found, or the suspected gunshot reportedly heard by security personnel. What prompted LT Jacobs to include such a seemingly insignificant detail in his official JAG investigation? Were all of the roving patrols required to walk through the supply laydown area as part of their official duty? Did the roving patrol see or hear something that night, or did he pass through the supply laydown area during his rounds? Could he have had knowledge of the events leading to Danny's death or insight into why his body was found in that location? The inclusion of this entry raises more questions than answers, leaving its true significance to anyone reviewing the investigation unclear.

"r. A gunshot was heard in the direction of the location of the body (relative to the watch stander) the evening of 10 November by the CBMU 302 roving patrol watch. (Enclosures (5) and (6))."

Investigative Commentary: This entry (paragraph r) references Enclosures 5 and 6. Enclosure 5 is the statement from the roving patrol on duty from 1900 to 2300 (7:00 PM – 11:00 PM) on the evening of November 10th, 1969, and Enclosure 6 is the statement from the duty dispatcher on duty the same evening. Interestingly, nowhere in the roving patrol's statement does he indicate that the gunshot he heard came from the area where Danny's body was later found. In fact, the roving patrol explicitly stated that the gunshot he heard "*sounded far away*," directly contradicting the suggestion made in paragraph r. As for Enclosure 6, the duty dispatcher, a Second Class Officer, stated that *"I received a phone call from the tower watch sometime between 2100 and 2200 reporting a gunshot."* This raises important unanswered questions: Did the roving patrol and the tower watch hear the same gunshot? Or were there two separate gunshots that evening? During my interview with LT Jacobs, he indicated that at the time, there was little concern about reported gunshots, because it was common for the base armory to discharge jammed weapons. Nevertheless, the JAG investigation file contains no statement from the sailor who was standing tower watch that night, an omission that leaves a critical gap in the timeline and raises further doubts about the thoroughness of the original investigation.

"s. The shot heard fired was fired after 2100H, but before 2300H. (Enclosures (6), (7), (8), and (9))."

Investigative Commentary: This entry (paragraph s) references Enclosures 6, 7, 8, and 9, which are statements from CBMU-302 members. Enclosure 6 notes that a gunshot was heard at approximately 2130 (9:30 PM) on Sunday evening, November 9th, 1969. Enclosure 7, however, makes no mention of any gunshots heard that evening. Enclosures 8 and 9 contain statements from two roving patrols who were on duty between 2300 (11:00 PM) Sunday and 0700 (7:00 AM) Monday morning; both reported that they did NOT hear any gunshots during their security watches. What's notably absent is Enclosure 5, which includes a statement from a separate roving patrol reporting a gunshot heard at approximately 2000 hours, described as sounding *"far away"*. The discrepancy between this report and the 2130 (9:30 PM) gunshot mentioned in Enclosure 6 amounts to a significant time gap of about 90 minutes. This difference should have prompted LT Jacobs to conduct a more thorough investigation, specifically to question both individuals in greater detail about what they heard, when they heard it, and the direction from which the sound originated. Based on the information available in these statements, it is clear to me that there was no meaningful investigative probing into the details of these two separate events.

"t. The deceased was emotionally depressed Saturday, 8 November and Sunday, 9 November 1969 because of a letter he had received from his girlfriend. (Enclosures (13) and (14))."

Investigative Commentary: This entry (paragraph t) references Enclosures 13 and 14, which are statements made by two construction mechanics who worked with Danny in the mechanics' shop. These statements are compelling and support the narrative that Danny was upset and possibly depressed over a possible 'Dear John' letter he allegedly received from his girlfriend. In Enclosure 13, one of Danny's shipmates stated, *"I talked to DAVIDENKO Saturday night at the EM Club. He was depressed because of a letter he got from his girl Friday or Saturday. We* *talked over his troubles. At one point he* (Danny) *said, I could just kill myself."* This is a critical piece of information that supports the possibility of suicide as the manner of death (MOD). Enclosure 14 is the statement from Danny's cube mate, CMCN Howard Decker, likely Danny's closest friend during his time in Vietnam. According to Howard, on Saturday evening (November 8th), he, Danny, and several others were at the enlisted club. Howard states that Danny had received a letter from his girlfriend, but he did not describe the letter's content nor indicate whether he personally saw or read it. Howard further stated that Danny *"tore up a couple of her pictures,"* another significant indicator of emotional distress that could support the suicide theory. However, it is important to note that these statements are hearsay and not physical evidence. To fully confirm these critical claims, the actual letter and the torn photographs should have been collected and preserved as evidence to corroborate his story. Without them, these statements, while suggestive, remain hearsay and uncorroborated. It is important to note that the alleged letter from Danny's girlfriend was never recovered, nor did anyone formally report physically seeing or reading it. As the sole investigator, LT Jacobs should have assembled a search team to locate these letters and torn up photographs so they could be properly processed as evidence, even if it required searching through the trash at the barracks or the enlisted club. The absence of these critical pieces of evidence weakens the reliability of the statements in enclosures 13 and 14. Without the actual letter, any conclusion drawn from its alleged content remain speculative at best.

"u. The deceased had threatened or attempted suicide approximately 6 September 1969. (Enclosure (18))."

Investigative Commentary: This entry (paragraph u) refers to a letter that appears to have been written by Danny and submitted as evidence to support an evaluation of his emotional state. Though I am not a certified handwriting forensic expert, my quick review and analysis of the handwritten note suggests that the letter was written by Danny. The note supposedly found in Danny's pocket described how he obtained the weapon, and ammunition. The authenticity of both handwritten documents needs to be examined by a forensic handwriting specialist to confirm that they originated from Danny himself. The letter bears the date *"Aug 27"*, and its content appears to reflect Danny's emotional state approximately two months before his death. However, the opening line references September 7th, indicating that the letter may have been written on that date. From an investigative perspective, the presence of two different dates introduces uncertainty regarding the document's authenticity and accuracy. It is also possible that Danny, still affected by alcohol or drugs from the previous night, simply recorded the wrong date. A significant detail about this letter that immediately drew my attention was the date, August 27th, located in the top right corner. What's troubling is that Danny never dated any of his letter's home, whether to his parents or other family members. Why did he make an exception here? What changed, and why is this the only letter he ever dated? Why was the letter dated August 27th, but the content in the first paragraph describes events that occurred on September 7th, some eleven days later? Compounding these concerns is the fact that several pages of the letter are missing and the fact that Danny never mailed it home. This raises additional questions: Why was it never sent? And what information might have been contained in the missing pages? It is also unclear as to where and when this letter was found (was it in his desk, his locker, etc.). The notation at the bottom of the letter, submitted as part of the JAG investigation, identifies LT Jacobs as the individual who located this letter among Danny's personal effects. LT Jacobs should have identified the who, what, when, where, and why behind the finding of such important evidence. The letter strongly suggests that Danny was in a very emotional and vulnerable state at that time, indicating that some event prior to the writing of that letter had pushed him toward irrational and desperate thoughts. However, what is most concerning is that this emotional crisis occurred months before his death and months before he reportedly received the rumored 'Dear John' letter from his girlfriend. This raises a very important question: what happened in early September that caused Danny such profound pain and depression that he drank heavily and took a bunch of pills? The significant time gap between these two events (September 6th through November 10th) suggests that there may be deeper or additional factors at play beyond a simple breakup with a girlfriend. If the 'Dear John' letter is being presented as the primary catalyst for his death, then investigators must also ask: What event(s) triggered his anxiety and depression in September? Were there other stressors, external pressures, illegal activities, or unresolved issues contributing to Danny's emotional deterioration? Another critical piece of missing evidence is the fact that this particular letter contained multiple pages. From the portion that remains, it appears Danny may have been trying to explain what led him to this event. Do the missing pages reveal Danny's feelings in greater depth, or describe the events and issues that drove him to take those pills and drink heavily, as suggested in the surviving portion of the letter? Do these pages offer insight into the factors influencing his emotional state? If so, were the remaining pages of the letter inadvertently misplaced, discarded, lost, or were they intentionally removed and destroyed to obscure the reasons behind his distress? Without these pages, we are left with a major gap in understanding Danny's thoughts and emotions during this difficult period, and whether the events recorded in these missing pages, had any influence to the circumstances that led to his death months later. Another major piece of missing evidence in LT Jacobs' JAG investigation is the failure to identify, locate, and interview CBMU-302 personnel who were reportedly with Danny when he allegedly ingested numerous pills and alcohol. In his own words, Danny stated in the letter, *"Some of the guys tried to take me to the hospital."* Why didn't LT Jacobs seek out these individuals to learn what they witnessed and to determine why Danny was in such a distressed emotional state? Their statements, even if considered hearsay, could have provided valuable insight into what was troubling Danny at that time. The discovery of this letter, coupled with the fact that there were missing pages to the letter, would have immediately prompted a seasoned investigator to probe deeper into Danny's emotional history, as such context would have been critical in determining a proper MOD in this investigation.

Danny's Partial Letter Dated August 27

Below is the transcription of a partial letter recovered from Danny's personal effects, which is believed to have been authored by him. A photocopy of the partial letter follows the transcription.

"Dear, (Name Redacted) *Letter Dated: Aug 27th*

I feel really rotten today. Its Sunday, the lord's day, the day of rest, the 7th day of September. I got a letter today from the Ill. Vet. Comm., I have to fill out some papers for the $100. (Name Redacted) *and I got drunk last night. What's even worse is before I got drunk, I took a half bottle of aspirins, and a full bottle of these other pills that I found. I wanted to die so much last night, as you can tell I didn't succeed. Some of the guys tried to take me to the hospital but I would not let them. I was throwing up for hours it seems, I either fell asleep or passed out, out in the field. When I woke up, I was in...."* **The rest of this letter is missing. These additional pages were either lost, never found, or discarded by accident or on purpose.**

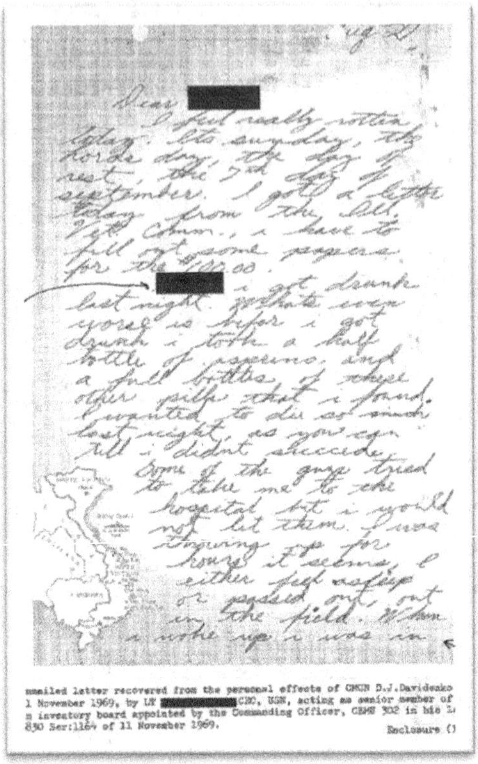

Aug 27

Dear [redacted]
I feel really rotten today. Its sunday, the lords day, the day of rest, the 7th day of september. I got a letter today from the Ass. Vet. Comm., i have to fill out some papers for the $100.00.
[redacted] i got drunk last night. whats even worse is befor i got drunk i took a half bottle of asperins, and a full bottles of these other pills that i found. I wanted to die so much last night, as you can tell i didnt succede. Some of the guys tried to take me to the hospital but i would not let them. I was throwing up for hours it seems, I either fell asleep or passed out, out in the field. When i woke up i was in

nmailed letter recovered from the personal effects of CMCN D.J.Davidenko 1 November 1969, by LT [redacted] CEC, USN, acting as senior member of n investatory board appointed by the Commanding Officer, CBMU 302 in his 1 830 Ser:1164 of 11 November 1969.

Enclosure (1

Danny's Partial Letter Dated August 27 (This letter was never mailed home)

In the section above, we reviewed the "Findings of Fact" outlined in paragraph 2, along with the associated subparagraphs (a) through (u) from LT Jacobs' JAG investigation. In the section below, we will examine the "Opinions" presented in paragraph 3 and the corresponding subparagraphs (a) through (i) from LT Jacobs' JAG investigation report.

"3. Opinions. The following opinions are offered:

a. A single shot was fired."

Investigative Commentary: This entry (paragraph a) represents the personal opinion of LT Jacobs rather than a fact-based conclusion. The absence of properly collected physical evidence, combined with the lack of forensic analysis of the medical report and autopsy records during the investigation, often results in poorly developed and unsupported opinions. Even with the autopsy, LT Jacobs does not reach the conclusion that Danny was shot twice. Instead, he stands firm with his opinion and suggests that the round fired was defective causing only partial recoil of the pistol slide and an incomplete passage of the bullet through the body. Once again, he glosses over the 2 centimeter exit wound located in the middle of Danny's back. He attempts to explain this wound as the result of a gaseous blast or hydraulic shock preceding the bullet. But from an investigative standpoint, this explanation is flawed. How could a defective round, which supposedly lacked sufficient force to allow the bullet to fully exit the body, generate enough energy to produce a 2 centimeter exit wound from gas or shock alone? The logic simply does not hold up under scrutiny. During my criminal forensic pathology research, I could not find one instance or example of an exit wound being caused by anything other than a projectile.

"b. This shot was fired between 2100H and 2300H, 9 November 1969."

Investigative Commentary: This entry (paragraph b) reflects the personal opinion of LT Jacobs rather than a fact-based conclusion. There was insufficient supporting evidence or convincing testimony from the roving watches to conclusively determine that the gun which killed Danny was fired between the hours of 2100 and 2300 (9:00 PM - 11:00 PM) on the evening of November 9th, 1969. LT Jacobs appears intent on linking the gunshot heard that evening (November 9th) to Danny's death, despite the medical examiner's official time of death determination of 0500 (5:00 AM, November 10th), which contradicts his timing correlation.

"c. This shot was fired from a misappropriated M1911A1 .45 caliber semi-automatic pistol, by the deceased."

"d. The weapon was fired by placing the muzzle of the weapon between the buttons of the upper garment, clear of the cloth, pressing the muzzle against the chest and pulling the trigger."

Investigative Commentary: These entries (paragraphs c and d) reflect the personal opinion of LT Jacobs rather than a fact-based conclusion. There were no witnesses, no cameras, and no video evidence of the actual shooting to confirm if Danny fired the gun or if someone else fired the gun that killed Danny. To suggest that the shooter, whether it was Danny or someone else, took the time to ensure the gun was clear of his shirt before firing is purely an assumption intended to support the investigator's theory of events. There is no statistical data or forensic evidence to support this opinion. Most shootings occur in the heat of the moment, with little to no regard for avoiding a victim's clothing during the act. According to LT Jacobs' opinion, the .45 pistol located near Danny's body was presumed to be the weapon used in his suicide. Yet the mere presence of a firearm near Danny's body does not definitively establish its involvement. A certified ballistic examination of the pistol, bullet, and casing would be required to verify whether it was the weapon responsible. In the case of a homicide disguised as a suicide, the pistol found near Danny's body could have served as a "throw-away gun" to direct investigators toward a conclusion of suicide. A second .45-caliber pistol was also discovered in the CM shop. It is possible that this was the weapon used to end Danny's life. We may never know for sure, because no ballistics tests were performed on either firearm, leaving the question unresolved. Additionally, all of this physical evidence (both guns, the bullet, and the casing) were destroyed and not preserved as evidence for future analysis.

"e. The round fired was defective, causing only partial recoil of the pistol slide and incomplete passage of the bullet through the body. This was due to its age which probably caused decomposition of the powder charge."

"f. The exit hole in the body was caused by gaseous blast or hydraulic shock preceding the bullet, and the bullet never left the body."

Investigative Commentary: These entries (paragraphs e and f) reflect the personal opinion of LT Jacobs rather than a fact-based conclusion. While there is some statistical data supporting the possibility that an older ammunition round could have been defective, potentially resulting in a partial discharge and the bullet remaining lodged in Danny's chest. However, the suggestion that the 2 centimeter exit wound was caused by a *"gaseous blast or hydraulic shock"* is unfounded. These terms are not used in professional forensic analysis. In forensic science, exit wounds are a well-documented occurrence in gunshot cases. Exit wounds are typically larger and more irregular in shape than entrance wounds. At close range, a .45 caliber bullet can easily penetrate and pass completely through the human body. According, forensic pathologists, a physical projectile is required to produce both entrance and exit wounds. Yet LT Jacobs claimed that a *"gaseous blast"* created the exit wound. To a trained CSI, this assertion reads less like a forensic conclusion and more like an unsupported guess offered to reinforce his investigative conclusions. It appears that, toward the end of his JAG investigation, LT Jacobs finally recognized that he had overlooked the 2 centimeter exit wound and felt compelled to develop a plausible explanation to support his MOD conclusion. Had he acknowledged that the exit wound was caused by a second bullet, his entire narrative of suicide would have been in jeopardy. To an experienced crime scene investigator (CSI), it would appear that LT Jacobs chose to offer an alternate explanation for the exit wound rather than restart the investigation, a course of action that might have led to a manner of death determination of homicide. What LT Jacobs didn't realize was that, had he acknowledged the exit wound and the second-bullet theory earlier, the entire case would have been immediately referred to the Naval Investigative Service (NIS) for a full criminal investigation.

"g. The deceased's depressed state resulted directly from information contained in a letter received from his girlfriend on Friday or Saturday. This depressed state led to his actions which took his life."

Investigative Commentary: This entry (paragraph g) reflects the personal opinion of LT Jacobs rather than a fact-based conclusion. If the two statements from Danny's friends regarding the letter from his girlfriend are accurate, and if this letter did, in fact, place Danny in a depressed state, then it is possible that he may have been suicidal. However, not a single unit member ever reported seeing or reading this rumored "Dear John" letter. If such a letter existed and was among Danny's personal effects, a copy should have been included in the JAG investigation. These statements centered around the rumored "Dear John" letter do not resolve the critical concerns in this case. There were no witnesses to the shooting; the gun's position in relation to Danny's body was inconsistent; there was no blood at the scene; his shirt did not have any bullet holes; and the exit wound itself supported the theory that Danny was shot twice. Furthermore, the only photographic evidence, a single photograph, shows the gun placed neatly on a pallet near Danny's feet. This detail was never mentioned by any CBMU-302 unit member during my interviews.

The photo clearly establishes the position of Danny's body in relation to the gun, once again raising serious questions about the manner of death and prompting even the most seasoned investigator to doubt the official conclusions in this case. Additionally, over the years and during numerous interviews, LT Jacobs consistently told me that his JAG investigation concluded that Danny's death was the result of an accidental discharge while he was handling the gun improperly. LT Jacobs never once mentioned the rumor about the supposed 'Dear John' letter, Danny's unsent letter home, or the note allegedly found in his uniform pocket. Was LT Jacobs trying to spare my feelings? Was he aware from the beginning of our conversations that his investigation determined that Danny's MOD was an apparent suicide? Or had his memory faded so much that he couldn't remember the contents of his own JAG investigation report? LT Jacobs relied on statements from two sailors to assert the existence of a "Dear John" letter, despite having no physical evidence to substantiate this claim.

"h. The deceased gave no advance warning of his intention toward taking his life immediately prior to his death. His depressed state 8 and 9 November was the only external sign manifested."

Investigative Commentary: This entry (paragraph h) reflects the personal opinion of LT Jacobs rather than a fact-based conclusion. According to enclosures 13 and 14 of the official JAG investigation, two of Danny's friends stated that he appeared emotionally depressed following a rumored letter from his girlfriend. One friend (enclosure 13) reported that, while drinking at the enlisted club on the evening of November 8th, 1969, Danny said, *"I could just kill myself."* The next day, Danny's cube mate, CMCN Howard Decker, stated in his official statement (enclosure 14) that on Sunday, November 9th, 1969, *"He* (Danny) *seemed pretty depressed,"* and that during chow, *"he didn't eat anything."* What LT Jacobs failed to recognize, however, is that Danny supposedly, gave advanced warning of his deteriorating mental state to at least two shipmates in the days immediately preceding his death and further warning approximately two months earlier as mentioned in his partial letter dated on the 27th. (see enclosure 18). These facts suggest that several members of CBMU-302 were fully aware of Danny's ongoing emotional distress, yet none took action to report it up the chain of command.

This oversight represents a significant failure in the command's response to a known mental health issue. The fact that the command (CBMU-302) was unaware of Danny's emotional state is not uncommon, and in fact, a majority of the military commands back then and still today lack the ability to identify, track, and monitor mental health issues among the ranks. This is not an accusation against the command. Rather, it highlights that Danny's behavior in the months before his death may have been missed warning signs of his mental state. And if he were suicidal, it would have taken only one person raising a concern for the chain of command to initiate some form of intervention. Currently veteran suicide rates have increased due to mental health issues. According to the U.S. Department of Veterans Affairs, in 2023, there were approximately 6,398 suicides among veterans, averaging about 17.5 suicides each day.

"i. The responsibility for the death lies solely with the deceased who fired the shot with full intention of taking his life by that action."

Investigative Commentary: This entry (paragraph i) reflects the personal opinion of LT Jacobs rather than a fact-based conclusion. Once again, there is no physical evidence to support LT Jacobs' opinion, only circumstantial evidence. Circumstantial evidence is indirect; it does not directly prove a fact but may give rise to a logical inference. However, in this case, without any witnesses to the event, it is impossible to definitively determine how Danny was fatally shot. To attribute sole responsibility for Danny's death upon him, without any supporting physical evidence, is merely an opinion and is not substantiated by forensic analysis or by any eyewitness accounts of the events that led to his death. Additionally, the majority of physical evidence was never collected or processed for forensic analysis. LT Jacobs relied solely on circumstantial evidence, without corroborating physical proof, which weakens the validity of his conclusions. A seasoned investigator reading LT Jacobs's opinion in paragraph (i) would immediately recognize it as an attempt to reinforce a suicide narrative without the support of physical evidence. This part of the report is built NOT on facts, but on circumstantial evidence, and LT Jacobs's own assumptions (opinions) about the events he believed led to Danny's death. After reviewing this assumption in paragraph (i), I questioned how LT Jacobs could conclude that Danny was responsible for his own death without any witnesses or supporting physical evidence.

"4. No determination of line of duty status or misconduct is furnished in compliance with section 0811(c) of reference (a)."

"5. Recommendations. No recommendations for administrative or disciplinary action are made in this case."

"Signature Block Signed by LT Paul Jacobs, Executive Officer CBMU 302"

Investigative Commentary: These entries (paragraphs 4 and 5) are required in the JAG investigation report and only apply in certain circumstances. No investigative commentary necessary.

JAG Investigation Witness Statements

In this next section, we will review and examine the eleven statements from CBMU-302 personnel that were recorded in the official JAG investigation. These eleven statements were provided by unit personnel who were either on duty during the period when Danny went missing or had interactions with him prior to his death. The investigation also includes a statement from LTJG Underwood, the officer who discovered Danny's body. Over the years, LT Jacobs consistently recalled that he did not formally interview LTJG Underwood. After reviewing the statements and interviewing LTJG Underwood myself, it appears that LT Jacobs requested only a written statement and did not conduct a formal interview or pursue investigative questioning.

On an important investigative note, what is clearly missing from the JAG investigation is a statement from ENS Kelly. He accompanied LTJG Underwood back to the scene, to check on the suspected sleeping sailor. But, once they arrived at the scene, they both discovered that there was no sign of life, and that Danny was dead. Also, absent is a statement from the Commanding Officer, LCDR Harrell, who was reportedly on scene providing guidance and direction. Below you will find the transcribed statements from each of the eleven individuals, followed by a copy of their original statements. After each statement I will provide my investigative commentary as to the content of each statement as it would be dissected and questioned in a CSI investigation. Not one of the statements below was dated as to when they were recorded for the investigation. So, I can only assume that they were collected sometime between the 10th and 22nd of November 1969.

The first official statement was provided by LTJG William Underwood (CBMU-302 Supply Officer), the person who found Danny's body.

Statement from LTJG William Underwood

Transcribed Statement: *"At approximately 1740, 10 November 1969, I left the CBMU 302 Supply Office, locked the warehouse, and proceeded alone in the Supply Jeep to the back of the brick plant where some recently delivered lumber had been off loaded. After inspecting the lumber and starting the jeep, I observed a man dressed in greens apparently sleeping behind some NSF Base Supply crates. I stopped the jeep and without getting out, observed from a distance the words Seabee and the first two letters of the name "DA" on his shirt. I then proceeded to the ballfield where I purchased a beer and joined the Commanding Officer of CBMU 302, in watching the ball game. I continued trying to reason logically why a Seabee would be hiding and sleeping at the base camp after working hours. Deciding to investigate I started to return to the sleeping man, but after recalling I had over-heard* *during the day that a man had been missing I asked the C.O. what the missing man's name was. The C.O. replied Davidenko and I answered that I might know where he was and would investigate and return to him with the results. Upon leaving the ballfield in my jeep, I gave a ride to Ensign Kelly the 302 personnel/admin. Officer who was returning to the base camp from supper. I asked him about the missing man and he verified Davidenko's name and I asked him to accompany me to the crates behind the NSF Supply Complex. We arrived approximately 1755 and I straightened Davidenko's shirt to read his name and we then realized this was the missing man. He was extremely stiff, and his hands and face were beginning to turn purple around the edges. At this time, we* (unreadable text) *out at the ballfield to inform the C.O. Upon telling HM2* (Name Redacted) *the duty corpsman, the details, he followed in an ambulance back to the scene. Immediately following the discovery of the body of CMCN Davidenko, the* *__weapon discovered beside his body__* *was ascertained to be a U.S. Government 45 caliber pistol with serial number 729391. Immediately the armory records of CBMU 302 and NSF Cam Ranh Bay were checked and this number proved negative in both activities custody files. The above statement is given freely and represents the facts to the best of my memory."*

Signed by: Name Redacted (LTJG William Underwood, SC, USNR)
Witnessed by: Name Redacted (LT Paul Jacobs, CEC, USN)"

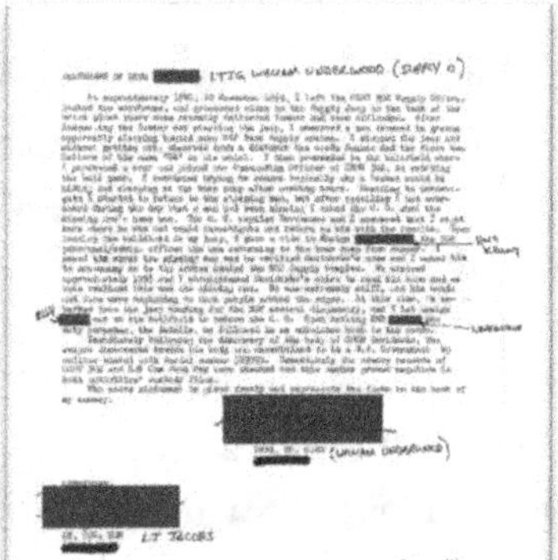

Enclosure 4 - Actual Copy of Statement from Name Redacted - LTJG William Underwood

Investigative Commentary: According to several CBMU-302 personnel, nobody was allowed in this supply laydown area unless you were assigned to the supply department. According to his statement to LT Jacobs, LTJG Underwood observed a man lying behind supply crates but took no immediate action to determine why the sailor was there or to have him removed from the supply laydown area. Additionally, from his position in the Jeep, he claims to have visually identified the word "SEABEE" on the uniform and the first two initials of the man's name, yet he chose to drive off without any apparent concern or effort to confirm the individual's welfare or identity. After obtaining and reviewing LTJG Underwood's statement, LT Jacobs should have conducted follow-up questioning to clarify the officer's actions, observations, and decision-making. For instance, questions he should have asked LTJG Underwood: 1) Why didn't you confront the sleeping SEABEE and instruct him to leave the restricted supply area? 2) How far away was your Jeep from the individual, and based on that distance, how clearly could you make out the uniform markings? 3) Do you have 20/20 vision, or do you wear prescription glasses? 4) Were sleeping sailors in the supply laydown area a common occurrence? 5) Were you at all concerned about the sailor's well-being or whether he might have been injured or unconscious?

Most individuals, when observing someone lying motionless in an isolated area, would feel compelled to approach and ensure the person was not in distress or in need of medical attention. This was a large supply laydown area, with rows upon rows of crates and materials, an environment where someone could have easily struck their head on an object (like a supply crate) and been rendered unconscious. From an outside perspective and especially through the lens of an investigative mindset, LTJG Underwood's actions appear questionable, raising the possibility that there may be more to the story regarding how he first noticed the so-called 'sleeping SEABEE'.

After leaving the so-called 'Sleeping SEABEE', LTJG Underwood nonchalantly and without concern for the sailor's welfare, departs the scene and heads to the ballfield to buy a beer and to sit with the CBMU-302 Commanding Officer (CO), who was watching a ball game. While sitting next to the CO, LTJG Underwood begins to think about the sleeping sailor in his supply area. Only after a short period of time (approximately 10-15 minutes) and a cold beer, does he decide to go back and investigate the 'Sleeping SEABEE'. What is especially concerning about LTJG Underwood's statement is his initial reaction to finding the 'Sleeping SEABEE', he casually dismisses the situation as routine and unworthy of follow-up. Later, while drinking a beer, he recalls overhearing someone mention that a sailor was missing. Only then does he ask the commanding officer for the name of the missing individual.

Upon learning that the missing man was named DAVIDENKO, LTJG Underwood tells the CO, *"I may know where this sailor is,"* and offers to investigate. This delayed reaction and passive attitude raise serious questions about his response and his overall awareness of the situation. Upon returning to the scene, LTJG Underwood picked up ENS Kelly and asked him to accompany him to the location where he had seen the 'Sleeping SEABEE', most likely CMCN Davidenko. They arrived at the scene approximately 15 minutes after the initial sighting. With ENS Kelly observing, LTJG Underwood straightened the sailor's shirt to reveal the name DAVIDENKO, confirming the identity. In my previous interview with ENS Kelly (referenced in an earlier chapter), he stated that LTJG Underwood attempted to wake the sailor by nudging him and snapping his fingers in front of Danny's face before making physical contact, an important detail that LTJG Underwood omitted from his own statement. According to Underwood, Danny's body was stiff, and his hands and face were beginning to turn purple, indicating early signs of death.

Both officers returned to notify the commanding officer, who then ordered medical personnel to respond to the scene. LTJG Underwood further stated that a weapon was recovered near Danny's body (with no precise location documented), and that upon examining the command's weapons inventory, the pistol's serial number was not listed among CBMU-302's registered firearms. Another critical piece of information missing from LTJG Underwood's statement, was the fact that Danny was clutching a rather large crucifix in his right hand, which covered his name tag. After reviewing LTJG Underwood's statement and analyzing its contents, LT Jacobs should have conducted a follow-up interview and posed several critical investigative questions to clarify the events and better understand the scene. These follow-up questions should have included: 1) Where exactly was the pistol located in relation to the body? 2) What was the approximate distance between the weapon and the body? 3) Did you touch or move the weapon at any point? 4) Did you touch or reposition the body in any way? These questions (and many other questions) would have provided essential context regarding the integrity of the scene, the handling of potential evidence, and whether the original positioning of the body and weapon was preserved or altered, all issues central to determining an accurate manner of death.

The second official statement was provided by BUCN (Name Redacted), the person on roving patrol Sunday evening on the 10th of November 1969.

Statement from BUCN (Name Redacted)

Transcribed Statement: *"I had the 1900 to 2300 roving patrol watch on Sunday the 10th. About an hour after assuming the watch I heard a gunshot from the direction of the CONSTA. It sounded far away. I reported it by phone to the dispatcher, I don't' know the exact time because I don't have a watch. The above statement is given freely and represents the facts to the best of my memory.*

Signed by: BUCN Name Redacted, USN
Witnessed by: Name Redacted (LT Paul Jacobs, CEC, USN)"

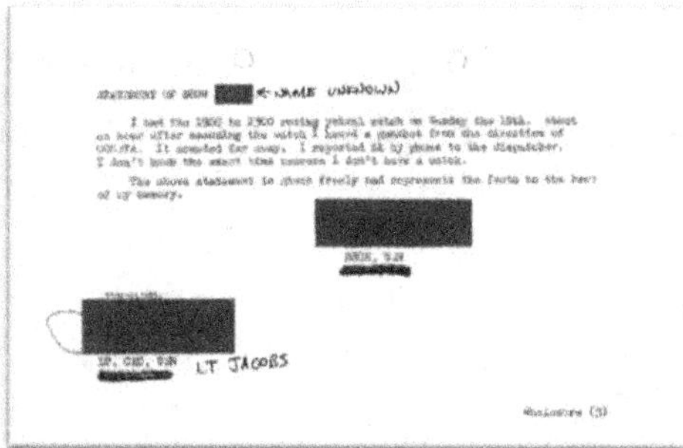

Enclosure 5 - Actual Copy of Statement from BUCN (Name Redacted)

Investigative Commentary: From the outset, this statement contains a clerical error. If the sailor stood the 1900–2300 (7:00 PM – 11:00 PM) roving patrol watch on Sunday, the correct date would have been November 9th, 1969, not November 10th as recorded. This discrepancy is significant because the watch stander reports hearing a gunshot around 2000 (8:00 PM) on the evening of November 9th, 1969. This reported gunshot was approximately 9 hours before the autopsy protocol determined Danny's time of death to be approximately 0500 (5:00 AM) on November 10th. We know from the autopsy that the bullet shot through Danny's chest went through his aorta and exited his back, which led to his immediate death. We also know that another bullet was lodged within Danny's chest, again confirming the two-bullet theory. According to this sailors' statement, he only heard a single gunshot, and it sounded far away. Additionally, the sailor's reference to "CONSTA" is unclear and undefined. It is uncertain whether he meant to refer to "NAVSTA" (Naval Station), "COMMSTA" (Communications Station), or another term entirely. While the exact meaning of the acronym may be irrelevant, its geographic location relative to the supply laydown area, where Danny's body was discovered, is critically important.

After reviewing BUCN's [Name Redacted] statement and analyzing its content, it was clear to me that LT Jacobs should have conducted a follow-up interview to ask the following questions: 1) Please confirm the exact date and time of your roving patrol watch. 2) Where were you physically located on base when you heard the single gunshot? 3) From your location, in which direction did the sound of the shot originate? 4) What did you mean when you referenced "CONSTA"? 5) Where is CONSTA located in relation to the supply laydown area? 6) Did your patrol route include the supply laydown area that evening? These questions are fundamental to determining whether the reported gunshot was in any way connected to Danny's death, or whether it was an unrelated event occurring in a different area or timeframe.

The third official statement was provided by EO2 (Name Redacted), the duty dispatcher on Sunday evening on the 9th of November 1969.

Statement from EO2 (Name Redacted)

Transcribed Statement: *"I was the Duty Dispatcher Sunday evening* (November 9th, 1969). *I received a phone call from the tower watch sometime between 2100 and 2200 reporting a gunshot. The call was after I had given him mid-rats about 2000. The above statement is given freely and represents the facts to the best of my memory.*

Signed by: EO2 Name Redacted, USN
Witnessed by: Name Redacted (LT Paul Jacobs, CEC, USN)"

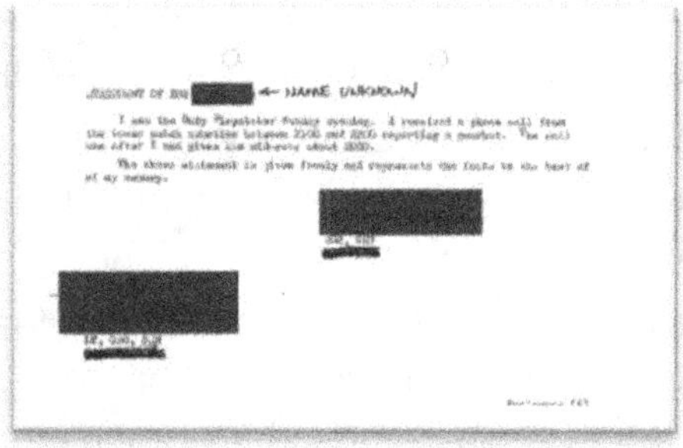
← NAME UNKNOWN

Enclosure 6 - Actual Copy of Statement from EO2 (Name Redacted)

Investigative Commentary: The statement from EO2 (Name Redacted) indicates that he received a phone call from the tower watch reporting a single gunshot sometime between 2100 and 2200 (9:00 PM – 10:00 PM) on Sunday, November 9th, 1969. However, within the JAG investigation documents, there is no accompanying statement from the tower watch stander. This raises a critical question as to who was on duty in the watch tower? Was this an oversight by LT Jacobs during his investigation, or merely a clerical error regarding who actually heard the gunshot that evening (was it the roving patrol, the tower watch, or both)? What was the name and rank of the tower watch stander? From an investigative standpoint, this was a significant mistake. Why did LT Jacobs get a statement from the roving patrol and not think it necessary to get a statement from the tower watch stander?

Both statements are critical to the investigation and could have provided a key piece of corroborative evidence. The next logical step in any thorough investigation would have been to identify and interview the sailor assigned to the tower watch that evening to confirm whether they heard the same gunshot and to clarify any details regarding its direction and timing. EO2 (Name Redacted) also mentioned that he gave the tower watch stander "MIDRATS" around 2000 (8:00 PM) on Sunday evening. In U.S. Navy terminology, "MIDRATS" refers to midnight rations, meals provided to night shift personnel or personnel on duty.

This detail introduces another important question: When the tower watch stander left for MIDRATS, was the watch post temporarily filled by someone else, or was it left unattended? If it were manned by another sailor, that individual should have been identified and interviewed. If the post was unmanned during that period, this would represent a serious lapse in security protocol and could mean that the gunshot heard by the roving patrol was not heard (missed) during a critical window of time when the tower was empty.

Clarifying who was actually standing the tower watch at the time the gunshot was heard, and whether the post was manned continuously, is essential to reconstructing the timeline and understanding the events surrounding Danny's death. After reviewing EO2's [Name Redacted] statement and analyzing its content, it is evident that LT Jacobs should have conducted a follow-up interview to clarify several key points. The following questions would have provided critical context and filled in gaps in the timeline: 1) Who was assigned to stand the tower watch on the evening in question? 2) When did the tower watch hear the gunshot, and when did he make the report? 3) Did the tower watch stander have a person relieve them of duty before he left to get MIDRATS? 4) At what time did the tower watch leave their post for MIDRATS? 5) If there was no personnel to relieve the tower watch stander, then how long was the tower watch absent from their post while eating MIDRATS? These clarifying questions are essential to determine whether the tower post was ever left unattended and to confirm who heard the gunshot that evening. If the post was unmanned, that lapse could explain the lack of a corroborating witness statement from the tower. If the tower was staffed and not left unattended, then the individual standing the watch and the person who relieved them should have provided a formal statement to the investigation. The failure to obtain these statements represents a significant oversight.

The fourth official statement was provided by YNC (Name Redacted), possibly YNC Burkey, the Officer of the Deck (OOD) on Sunday, the 9th of November 1969. I made several attempts to locate and contact YNC Burkey to get is story, however, none of the emails, addresses, or phone numbers were still in service.

Statement from YNC (Name Redacted)

Transcribed Statement: *"On Sunday, 9 November, I was assigned duty as CBMU 302, OOD. During the period approximately 2030-2100 I made a walking tour of the CBMU 302 shop and storage area. I went over and past the CM shop, between the lumber storage yard and the brick plant, behind the SW shop, Supply and Warehouse and up to the tower where I was challenged by a tower guard, admitted to proceed and then spent approximately ten to fifteen minutes in conversation regarding security lights, etc. I then went down and around the side of the Supply Warehouse and returned to the Admin Office. I have a log entry indicating that this action was done at 2100. The 2100 hour entry would be the time of completion rather than commencement. I was unaware of anything out of the ordinary during this tour. The above statement is given freely and represents the facts to the best of my memory.*

Signed by: YNC Name Redacted (Possibly - YNC Burkey USN)"
Witnessed by: Name Redacted (LT Paul Jacobs, CEC, USN)"

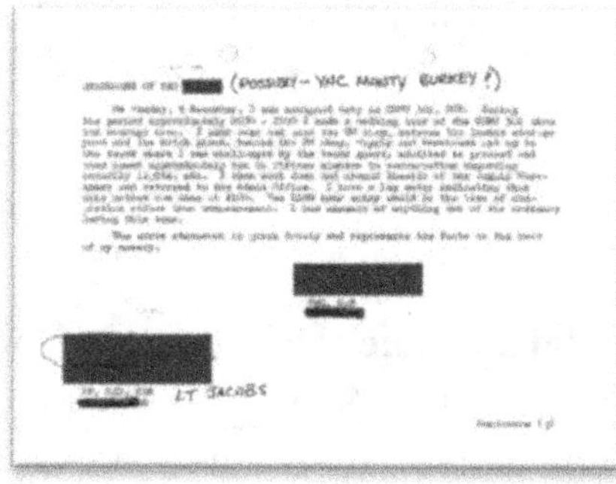

Enclosure 7 - Actual Statement from YNC (Name Redacted) Possibly YNC Burkey

Investigative Commentary: The statement from YNC [Name Redacted] confirms that he was serving as the Officer of the Deck (OOD) on Sunday, November 9th, 1969. As the OOD, he was responsible for the overall security of the base and the coordination of its security watch standers. According to his statement, he conducted a walking tour of the CBMU-302 facilities between 2030 and 2100 (8:30 PM – 9:00 PM), approximately 30 minutes after the roving patrol reported hearing a gunshot at around 2000 (8:00 PM). Notably, it appears that the OOD was completely unaware of this reported gunshot during his tour because there is no mention of it in his statement. During the OOD's rounds, he spoke with the tower guard on duty, discussing security lighting. Assuming that he had not been informed of the reported gunshot, he would not have addressed the issue with the tower watch stander. This raises several questions that LT Jacobs should have followed up on: 1) Did the dispatcher fail to report the gunshot to the OOD? 2) Was there a communication breakdown? Compounding the confusion, the tower watch later reported hearing a gunshot at approximately 2130 (9:30 PM), roughly 30 minutes after the OOD's visit. Yet again, there is no indication that this information was relayed to the OOD. The most logical conclusion is that the OOD was never informed of either gunshot report. Additionally, none of the individuals on duty that night, including the roving patrol, the tower watch, the dispatcher, or the OOD referenced standard U.S. Navy operating procedures (SOP) regarding watch logs. SOP requires that all watch standers maintain a logbook with accurate time entries of incidents. The absence of such log entries raises serious concerns about procedural compliance and the integrity of the reporting system in place that night.

The fifth official statement was provided by CMCN (Name Redacted), the roving patrol between 2300 and 0300 (11:00 PM – 3:00 AM) beginning on Sunday, the 9th, and ending on the 10th of November 1969.

Statement from CMCN (Name Redacted)

Transcribed Statement: *"I had the roving patrol watch between 2300 to 0300 Sunday night. I made a tour of the area at about 0030 and again before I got off watch. I didn't hear a gunshot during my watch or see anything suspicious. The above statement is given freely and represents the facts to the best of my memory.*

Signed by: CMCN Name Redacted, USN
Witnessed by: Name Redacted (LT Paul Jacobs, CEC, USN)"

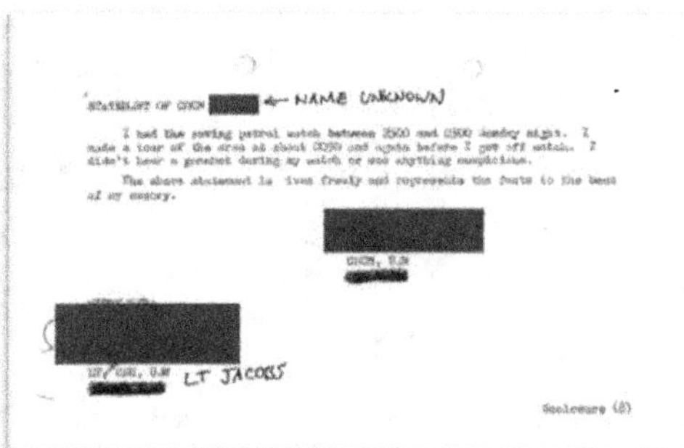

Enclosure 8 - Actual Copy of Statement from CMCN (Name Redacted)

Investigative Commentary: The statement from CMCN [Name Redacted] indicates that his roving patrol shift began at 2300 (11:00 PM) and ended at 0300 (3:00 AM). During this timeframe, he conducted two complete tours of the CBMU-302 area. According to his statement, he neither heard any gunshots nor observed anything suspicious during his watch. Given the timing of his patrol, occurring shortly after two other watch standers reported hearing possible gunfire, this statement should have prompted a follow-up interview by LT Jacobs. At a minimum, the following two questions should have been asked: 1) What areas of the CBMU-302 compound did you tour during your watch? 2) During your patrols, did you pass through or near the supply laydown area? 3) Were you briefed or informed by the dispatcher or Officer of the Deck about the gunshot(s) reported earlier in the evening? The omission of these critical follow-up questions reflects a continuing pattern of investigative shortcomings. A trained investigator would have asked this watch stander these specific questions during the initial inquiry into Danny's death.

The sixth official statement was provided by CMCN (Name Redacted), the roving patrol between 0300 (3:00 AM) and 0700 (7:00 AM) on Monday, the 10th of November 1969.

Statement from CMCN (Name Redacted)

Transcribed Statement: *"I had the 0300 to 0700 roving patrol watch on 10 November 1969. During the watch I didn't hear any shots or see anything suspicious or unusual. The above statement is given freely and represents the facts to the best of my memory.*

Signed by: CMCN Name Redacted, USN
Witnessed by: Name Redacted (LT Paul Jacobs, CEC, USN)"

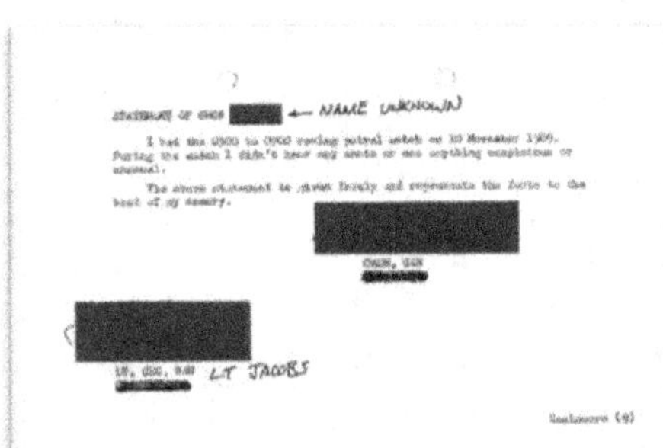

Enclosure 9 - Actual Copy of Statement from CMCN (Name Redacted)

Investigative Commentary: The statement from CMCN (Name Redacted) indicates that his roving patrol shift took place from 0300 to 0700 (3:00 AM – 7:00 AM) on November 10th, 1969. However, he does not specify how many patrols he conducted or whether he passed through all key areas of the CBMU-302 compound during his watch. He reports hearing no gunshots and observing nothing unusual or suspicious. Once again, this reveals a gap in the investigative process. A proper follow-up interview should have included at least one critical question: Did your patrol route include the supply laydown area where Danny's body was later discovered? This question is essential to understanding whether the scene could have been discovered earlier. Either situation (whether he did or did not) would help with determining the actual time of death and could help investigators to identify the chain of events leading to his death. The absence of detailed follow-up inquiries in this case highlights a consistent pattern of incomplete investigative procedures. In a criminal death investigation, especially under suspicious circumstances, clarifying the exact movements and observations of every roving watch stander is not optional; it is fundamental.

The seventh official statement was provided by RM1 (Name Redacted), the leading petty officer tasked to locate CMCN Davidenko on the morning of November 10th, 1969 (Monday).

Statement from MR1 (Name Redacted)

Transcribed Statement: *"At approximately 0900, 10 November 1969, DAVIDENKO was determined unaccounted for and CM2 (Name Redacted) was sent to look for him. After checking the barracks and the Sick Bay to no avail, he returned to the 302 CM Shop were DAVIDENKO's roommate was questioned and from this we determined that he had not spent the night in his cube at the barracks. A further investigation revealed that the last time anyone had observed DAVIDENKO was at approximately 2000, 9 November 1969, at which time he was supposedly going to the EM Club. A search of the bunkers onboard NSF and the area around Filby Park by CM2 (Name Redacted) and other personnel of the CM Shop failed to establish*

DAVIDENKO's whereabouts. At approximately 1115 hours this date, 10 November 1969, I informed the 302 XO that DAVIDENKO could not be found in the area of (UNREADABLE). *At approximately 1815 hours, 10 November 1969, CE1 (Name Redacted) and PN2 (Name Redacted) and myself were told to report to the 302 Admin. Office, reason: DAVIDENKO had been found dead. The above statement is given freely and represents the facts to the best of my memory.*

Signed by: MR1 Name Redacted, USN
Witnessed by: Name Redacted (LT Paul Jacobs, CEC, USN)"

Enclosure 10 - Actual Copy of Statement from RM1 (Name Redacted)

Investigative Commentary: The statement from RM1 (Name Redacted) indicates that he was likely the Leading Petty Officer (LPO) for Danny's company and was tasked with locating Danny after he failed to report for morning muster. According to his statement, it appears that RM1 was directed to organize a search team to conduct a base-wide search for Danny. Despite these efforts, the team was unable to locate Danny, and at approximately 1815 (6:15 PM), they were notified that Danny had been found dead in the supply laydown area. RM1's statement should have immediately prompted a follow-up interview by LT Jacobs with at least one critical question: Did you or any members of your search team enter and search the supply laydown area? This question is vital to determining whether the location where Danny's body was ultimately discovered had already been searched earlier that day. If it was searched, and no body was found at that time, then several additional questions arise: 1) At what time was the area searched? 2) Was it a thorough walk-through, or merely a cursory perimeter check? 3) If the area was indeed searched, how and when did Danny's body end up there? The answers to these questions would have been essential in determining whether the search was simply incomplete or whether Danny was shot sometime after the search. Additionally, it could suggest that Danny was shot elsewhere, and his body was placed in the supply laydown area sometime after the initial search had concluded. If the latter is true, it would indicate foul play and a deliberate attempt to cover up a homicide or murder. This could also explain why no blood was observed at the scene and why the second bullet and its casing were never recovered.

The eighth official statement was provided by CE1 (Name Redacted), a leading petty officer tasked to locate CMCN Davidenko on the morning of November 10th, 1969 (Monday).

Statement from CE1 (Name Redacted)

Transcribed Statement: *"The first time that I was aware that DAVIDENKO, CMCN, was being looked for was at approximately 0900, 10 November 1969, at BEQ 31. At that time (Name Redacted), CM2, came to the barracks and asked if I had seen him. I told him no, and then (Name Redacted) went to the Dispensary to see if he had been to sick call. I then started looking through the barracks (Bldg's 10, 11, and 14) to see if DAVIDENKO was around. Not being able to locate him in the barracks I then went to the Snack Bar and Barber Shop. After checking the above spaces I went to the garage and asked MR1 (Name Redacted) if he had any success in locating the man. I didn't do any further checking, but at approximately 1730 (Name Redacted), CMCN, stopped by the First Class Lounge and told me that he had mailed a letter to DAVIDENKO's brother for DAVIDENKO on the morning of 10 November, at the time that I was informed of the letter I told (Name Redacted), YN1, who was the OOD. At approximately 1815 (Redacted text), PARSONS and* *myself were called to the Admin. Office, and I learned that DAVIDENKO had been found dead. The above statement is given freely and represents the facts to the best of my memory.*

Signed by: CE1 Name Redacted, USN
Witnessed by: Name Redacted (LT Paul Jacobs, CEC, USN)"

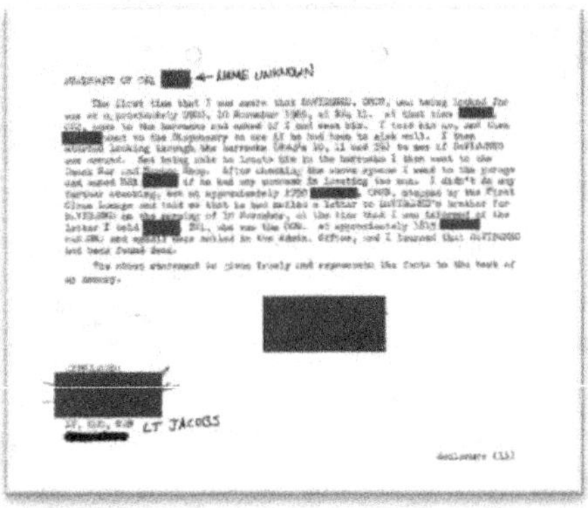

Enclosure 11 - Actual Copy of Statement from CE1 (Name Redacted)

Investigative Commentary: The statement from CE1 (Name Redacted) confirms that he was also a member of the search team assigned to locate Danny on the day of his disappearance. While CE1's own account offers no significant new information, it does include a notable detail: CMCN [Name Redacted] (most likely Danny's cube mate CMCN Howard Decker) approached him, unsolicited, to report that he had been asked by Danny to mail a letter on the morning of November 10th, 1969. According to CMCN, the letter was addressed to one of Danny's brothers (name of brother not specified). This detail raises a critical question: Why would Danny ask someone else to mail a letter he could have easily mailed himself? This action could suggest that Danny anticipated not being present later that day and wanted to ensure the letter was sent. While this evidence is circumstantial, it carries potential weight when considered alongside other statements and facts gathered in the investigation. Depending on how it aligns with corroborating evidence, this detail could support a logical inference pointing toward multiple possible manners of death (MOD), including suicide, homicide, or even murder.

The mailing of this letter by CMCN [Name Redacted] supports the theory that Danny may have been in a troubled emotional state; however, it could just as easily support the theory that Danny anticipated a dangerous encounter with someone he intended to meet and wanted the letter mailed in case something happened to him, and he was unable to send it himself. Another critical piece of information, unknown to both the command and the JAG investigative officer, emerged during an earlier chapter detailing my interview with Danny's brother, Chuck. During our conversation, Chuck revealed that he had received a letter from Danny in which Danny expressed signs of emotional distress and depression. When I asked Chuck if Danny explicitly mentioned suicidal thoughts, he replied, "*No*," but noted that the letter read like a cry for help. Chuck also shared that he gave this letter to their mother, Mary. Unfortunately, Mary never spoke of this letter, and no other family members ever recalled seeing it. The content of that letter could have been a critical piece of evidence, potentially offering deeper insight into Danny's state of mind, or into the more dire circumstances he may have been caught up in. Regrettably, I never got to ask Chuck if the letter he received from Danny was received before or after his death. Could this have been the letter that CMCN (Name Redacted) mailed for Danny, and could it have been the letter addressed to his brother Chuck?

It appears that LT Jacobs did not view this unusual mailing request as a potential clue in Danny's death, nor did he follow up on it or consider it as possible evidence during his investigation. If LT Jacobs' JAG investigation was proceeding toward a possible MOD determination of suicide, then further inquiry into the letter that CMCN (Name Redacted) mailed home on Danny's behalf was warranted. This should have included contacting Mary, Elmer, and the brother to who the letter was addressed to determine the following: 1) When did they received the letter in question from Danny; and 2) What was the content of that letter, specifically, whether it mentioned Danny receiving a "Dear John" letter from his girlfriend and/or if he mentioned any intent to take his own life. Another way LT Jacobs could have investigated this issue, would have been to secure the letter from the base post office and, with the consent of Danny's brother and/or parents, open and analyze its contents to determine whether it contained information relevant to the case.

The ninth official statement was provided by LT Paul Jacobs, the CBMU-302 executive officer and assigned JAG investigation officer charged to investigate the disappearance and death of CMCN Davidenko.

Statement from LT Paul Jacobs

Transcribed Statement: *"On 10 November at approximately 1130, MR1 (Name Redacted) contacted me concerning DAVIDENKO's absence. At my request, YNC (Name Redacted) made a telephone check of all medical facilities on the peninsula with negative results. At approximately 1230 I called the Army Provost Marshal's Office for the Cam Ranh area and filed a missing person's report. At approximately 1300, I, LTJG (Name Redacted), and LTJG (Name Redacted) searched all bunkers on the base with negative results. The above statement is given freely and represents the facts to the best of my memory.*

Signed by: Name Redacted (LT Paul Jacobs, CEC, USN)
Witnessed by: YNC, Name Redacted, USN"

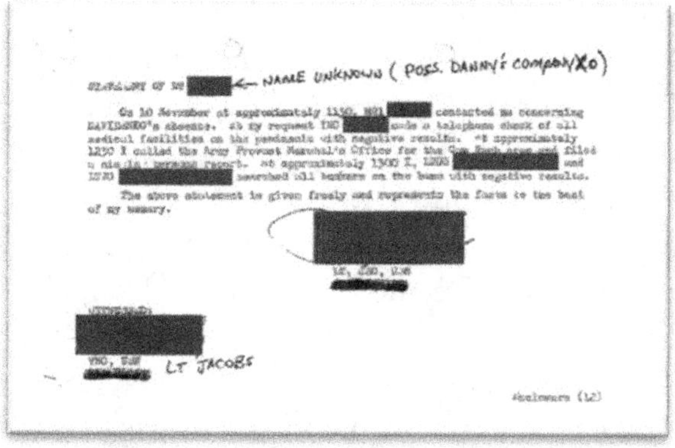

Enclosure 12 - Actual Copy of Statement from LT Paul Jacobs

Investigative Commentary: The statement from LT Jacobs indicates that he, along with several other CBMU-302 personnel, contacted all local medical facilities in an effort to locate Danny. Additionally, LT Jacobs assigned two junior officers to conduct a physical search of all nearby bunkers. Both efforts yielded negative results. By this point, the search for Danny had been ongoing for more than six hours without success. At this stage, LT Jacobs should have convened a brief meeting with all personnel involved in the search. During this meeting, each individual should have been asked to identify the specific areas of the base they had searched. Based on their responses, a revised and more systematic search plan could have been developed to cover any missed locations. A meeting of this nature would have allowed LT Jacobs to determine whether the supply laydown area had been thoroughly searched. If that area had indeed been searched and Danny was not found there at that time, it would raise serious concerns about when and how his body ended up in that location, suggesting the possibility of foul play and supporting a potential manner of death (MOD) of homicide or murder, which would require a criminal investigation by NIS.

LT Jacobs' statement indicates that he notified the Army Provost Marshal's Office; however, he makes no mention of notifying the 630th Military Police Company, which was responsible for military police requirements at Cam Ranh Bay in 1969. I was unable to find any information suggesting that the 630th MPs were present at the scene or that they conducted a formal investigation. Also missing from LT Jacobs' statement is any reference to notifying the Naval Investigative Service (NIS) representative located on base at CRB, or the NIS headquarters in Saigon. Likewise, I was unable to find any indication that NIS conducted a formal investigation into Danny's death.

The tenth official statement was provided by CMCN (Name Redacted), a fellow sailor who talked with Danny on the 8th of November 1969, while spending the evening at the Enlisted Club.

Statement from CMCN (Name Redacted)

Transcribed Statement: *"I talked to DAVIDENKO Saturday night at the EM Club. He was depressed because of a letter he got from his girl Friday or Saturday. We talked over his troubles. At one point he said, "I could just kill myself." On Sunday I went to Army side with him and some other people on the exchange run. All during the trip he seemed depressed and didn't join in much. He was definitely different from his usual self on Saturday and Sunday. I saw him last at about 1800 onboard NSF. The above statement is given freely and represents the facts to the best of my memory.*

Signed by: CMCN Name Redacted, USN
Witnessed by: Name Redacted (LT Jacobs, CEC, USN)"

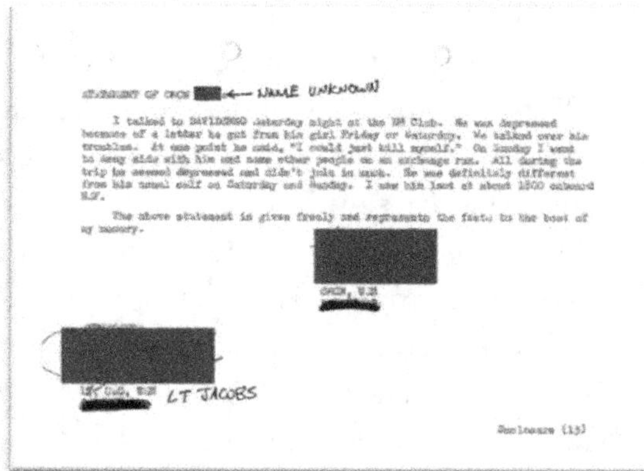

STATEMENT OF CMCN ← NAME UNKNOWN

I talked to DAVIDSON Saturday night at the EM Club. He was depressed because of a letter he got from his girl Friday or Saturday. We talked over his troubles. At one point he said, "I could just kill myself." On Sunday I went to Army side with him and some other people on an exchange run. All during the trip he seemed depressed and didn't join in much. He was definitely different from his usual self on Saturday and Sunday. I saw him last at about 1800 onboard U.S.

The above statement is given freely and represents the facts to the best of my memory.

CMCN, USN

LTJG, USN LT JACOBS

Enclosure (13)

Enclosure 13 - Actual Copy of Statement from CMCN (Name Redacted)

Investigative Commentary: The statement from CMCN [Name Redacted] indicates that he was with Danny at the enlisted club on Saturday evening, November 8th, 1969. According to his account, Danny appeared deeply depressed about a letter he had received from his girlfriend. The CMCN stated that Danny expressed suicidal thoughts, allegedly saying, *"I could just kill myself."* If we believe this statement to be true and accurate, the CMCN should have treated this statement as a clear warning sign of Danny's emotional distress and should have reported the issue immediately to his chain of command. At a minimum, this CMCN should not have left Danny alone. This statement represents additional circumstantial evidence supporting the possibility of suicide as the manner of death (MOD). However, it lacks critical details that could substantiate its credibility.

LT Jacobs should have conducted a follow-up interview to clarify the context and content of that evening's conversation. Specific follow-up questions should have included: 1) What was the subject of the letter that upset Danny? 2) Did you personally see or read the letter in question? 3) What other concerns or troubles did Danny share with you that evening? 4) Can you describe Danny's demeanor or emotional state during your conversation? 5) Why didn't you inform the chain of command about Danny's emotional state and his suggestion that he might kill himself? 6) Was CMCN Davidenko intoxicated when you last saw him and when he made the statement about killing himself? 7) When you last saw him, did you leave him at the Enlisted Club, leave together, or did he remain at the club? 8) Did you witness him tearing up any photographs while at the Club? Without these clarifying details to these questions, particularly confirmation that the letter existed and was read, this statement remains compelling but incomplete. Had it been properly developed, a follow-up interview supported by focused investigative questions could have yielded valuable insight into Danny's mental and emotional condition in the hours preceding his death, while also establishing a search for the rumored 'Dear John' letter and the torn-up photographs (mentioned in the 11th statement) that would have been substantial physical evidence.

The eleventh and final official statement was provided by CMCN (Name Redacted). However, we know that Danny's roommate was CMCN Howard Decker, and in this statement, the CMCN states that they have been in the same barracks room for two months. He also states that they knew each other long before their assignment to CBMU-302.

Statement from CMCN Howard Decker

Transcribed Statement: *"I knew DAVIDENKO prior to coming to Vietnam. Since my arrival here, we've been in the same cube for about 2 months. He always was kind of nervous and worried. The past couple of weeks he had been smoking more, about two cartons per week. Saturday night he was at the club with me and some other guys. He had a letter from his girl with him. He tore up a couple of her pictures and said he wasn't worried. He drank quite a bit Saturday night. I woke him up about 1130 Sunday morning and we and about 10 others went on a PX run to Army side. We stopped at the club to see the show there. Someone bought him a beer but he didn't drink it. He seemed pretty depressed. Later I took* *him over to the chow hall but he didn't eat anything. I saw him in the barracks about 2000. About 0230 Monday morning I woke up and saw that he wasn't in his rack and that it hadn't been slept in. Monday morning, I checked at sick bay but he wasn't there. The above statement is given freely and represents the facts to the best of my memory.*

Signed by: Name Redacted, USN (CMCN Howard Decker, USN)
Witnessed by: Name Redacted, USN (LT Jacobs, CEC, USN)"

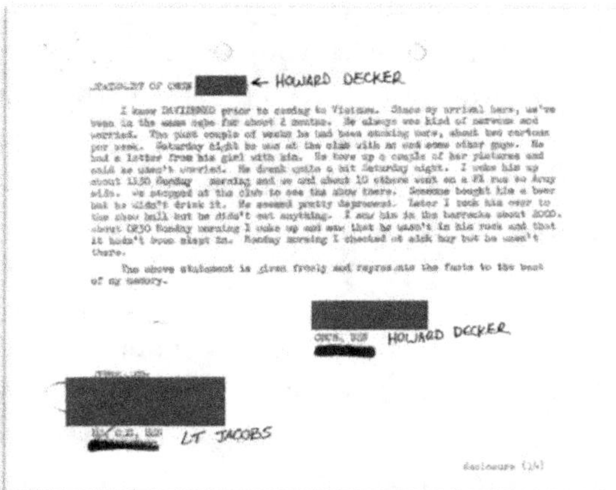

← HOWARD DECKER

HOWARD DECKER

LT JACOBS

Enclosure 14 - Actual Copy of Statement from CMCN Howard Decker

Investigative Commentary: The statement from CMCN Howard Decker indicates that he was with Danny at the enlisted club on Saturday evening, November 8th, 1969. According to Howard, he knew Danny before they were both assigned to CBMU-302 in Vietnam. Based on this information, Howard was the one individual who probably knew Danny best. Given their shared living space and potential long-standing relationship, Decker's observations and insights into Danny's behavior, mood, and interactions in the days leading up to his death would have been especially relevant. His perspective should have been more thoroughly explored during the investigation, as it could have yielded critical context regarding Danny's mental and emotional state.

According to CMCN Howard Decker, Danny frequently appeared nervous and worried, traits Decker believed may have contributed to his increased and excessive smoking. If accurate, Howard's statement also confirms that Danny had the actual letter from his girlfriend with him that evening at the enlisted club. Decker further states that Danny tore up a couple of photographs of his girlfriend and was drinking *"quite a bit."* These behaviors suggest a significant emotional disturbance. On Sunday morning, at approximately 1130 (11:30 AM), Howard woke Danny to accompany a group of CBMU-302 members on a run to the Post Exchange (PX). He reports that Danny seemed visibly depressed, refused to drink a beer offered by a friend, and did not eat anything during their visit to the chow hall. The last time Howard saw Danny was approximately 2000 (8:00 PM) that evening.

The most critical detail in Decker's statement is that when he woke at 0230 (2:30 AM) on Monday morning, Danny was not in his bed, and it appeared as though the bed had not been slept in. In hindsight, Danny's emotional state and his absence from his bed at 0230 (2:30 AM) should have prompted CMCN Howard Decker to report this concern immediately to the Officer of the Deck (OOD) that was on duty that morning. Had the OOD been informed of Danny's emotional condition and his unknown whereabouts between 2000 and 0230 (8:00 PM–2:30 AM), on-duty personnel could have used this information to organize an immediate search effort.

The statement provided by CMCN Howard Decker should have prompted LT Jacobs to further investigate and explore critical details surrounding Danny's mental and emotional state, as well as his movements in the hours leading up to his death. Specifically, Decker's account warranted a follow-up interview that included targeted questions such as: 1) Did Danny mention any specific plans, concerns, or individuals he intended to speak with after dinner on Sunday (i.e., his Company Commander, the

Base Chaplain, or a new friend)? 2) Did you see the actual letter from his girlfriend? If so, did you read it, and can you recall any of its contents? 3) Do you know if Danny spoke to anyone else about the letter or his emotional state? 4) Was there anything in Danny's behavior that night that suggested he intended to harm himself or leave the barracks for a specific purpose? 5) Who else was at the club with you and Danny when you learned about the letter and observed him tearing up photos of his girlfriend?? 6) Did you actually see the pictures that Danny tore up? 7) What did Danny do with the torn-up pictures of his girlfriend? 8) What did Danny do with the letter from his girlfriend? 9) Did you ever see Danny with a .45 caliber pistol? 10) Do you know if Danny was involved in any illegal activities? 11) Do you know of anyone who might want to harm Danny? 12) Did Danny upset anyone, hurt anyone, or make anyone an enemy?

Of the eleven statements submitted by CBMU-302 personnel, only one contains any reference to physical evidence; the remaining statements reference only circumstantial evidence (hearsay and rumors) surrounding the events that may have led to his death. Lastly, on an investigative note, not one of these formal statements was dated as to when and where they were ascertained. Was there a formal investigation interview where questions were asked after the statements were taken, or did LT Jacobs just have these individuals submit a statement about the events as they recalled them without any formal analytical examination or follow-up questioning?

LT Jacobs JAG investigation also included a handwritten letter that was supposedly found in the pocket of Danny's uniform. The following is a transcribed copy of that letter, as well as a copy of the actual letter.

Reported Note Found in Danny's Pocket

Transcribed Note: *"To whom it may Concern. I stole the .45 from Army side I stole the bullets from the armory one day when I was on the firing range. I'll close for now. Danny J Davidenko."*

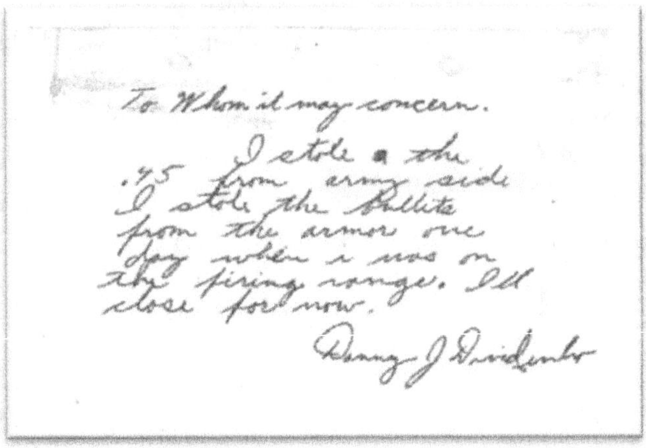
To Whom it may concern.
I stole the .45 from army side
I stole the Bullets
from the armor one
day when i was on
the firing range. I'll
close for now.
Danny J Davidenko

Enclosure 15 - Note Reportedly Found in Danny's Pocket

Investigative Commentary: Just like the handwritten letter dated August 27th that was reportedly found among Danny's personal effects, another handwritten note was reportedly discovered on his body, tucked inside one of his uniform pockets. From an investigative standpoint, LT Jacobs fails to provide critical details, such as who conducted the search of Danny's uniform pockets and exactly which pocket the note was found in. These two handwritten documents were never properly collected, processed, or treated as potential evidence. As a result, they were mishandled according to standard crime scene investigative procedures. These documents should not have been touched by any personnel; instead, they should have been preserved and processed for fingerprints to determine whether foul play was involved. How many individuals handled these documents at the scene or during the investigation? Any additional fingerprint contamination could have compromised the potential to identify or eliminate suspects. If fingerprints other than Danny's were found on either of the notes, this could suggest a possible homicide. Conversely, if only Danny's fingerprints were present, it would strengthen the case for suicide as the manner of death. Also, we cannot rule out the possibility that Danny was coerced into writing the note that was found in his uniform pocket.

There have been documented cases where assailants have coerced victims into writing farewell notes prior to their deaths in an effort to stage the scene as a suicide rather than a homicide. Without forensic analysis, such as handwriting comparison, fingerprint examination, or pressure analysis of the writing, there is no way to determine whether the note was written voluntarily or if it was part of a deliberate attempt to mislead investigators and obscure the true manner of death. Additionally, it is highly uncommon for a suicide note to include logistical details such as how the individual obtained a weapon or ammunition. Instead, suicide notes typically serve as a final expression of the person's emotional state. They often include apologies, expressions of pain, guilt, or despair, and offer insight into the reasons behind the act. Most suicide notes reflect a desire to explain the emotional turmoil leading to the decision, not the mechanics of how the act would be carried out.

Lastly, LT Jacobs stated that the gun found was an "orphan" weapon, meaning it was not an official U.S. military firearm. However, the note allegedly left by Danny claims the weapon was stolen from the Army side of the base. If that were true, why didn't LT Jacobs check the serial number on the weapon recovered by CBMU-302 against the Army's armory records to determine if any firearms were reported missing? This basic investigative step appears to have been entirely overlooked, or LT Jacobs failed to document in his JAG investigation whether he contacted other military units in the CRB area.

Enclosure 16 of the Official JAG Investigation includes the only known photograph of Danny's body, taken at the scene, captured within the supply laydown area. For many of Danny's family members, this will be the first and only time they are confronted with an actual photograph of his final moments. What is particularly troubling is that this single image is the only known photographic documentation of the scene that I have been able to find. This does not mean that other military members present at the scene didn't take a few photos of their own with their own personal cameras. According to LT Jacobs and his JAG investigation report, no other photographs were taken to capture the surrounding area, the positioning of key objects such as the weapon, or potential evidence in the immediate environment. Only this single photo of Danny's body was submitted as part of the JAG investigation. The photo was submitted with the following caption:

"Photograph of CMCN Daniel J. Davidenko taken at approximately 1850H, 10 November 1969 by LT Paul Jacobs, CEC, USN ***depicting the position of the body and weapon as found.*** *Taken upon verbal order of the commanding officer, CBMU 302 prior to moving the body."*

Photograph of CMCN Daniel J. Davidenko taken at approximately 1805H, 10 November 1969 by LT ███ ████████, CEC, USN depicting the position of the body and weapon as found. Taken upon verbal order of the Commanding Officer, CBMU 302 prior to moving the body.

Enclosure 16 - Photo of Danny's Body and Gun Found in the Supply Laydown Area

The Yellow Circles and Arrows Highlight Possible Location of Second Bullet

Investigative Commentary: As discussed in an earlier chapter on crime scene investigation and analysis, when a deceased individual is found, and there are no witnesses to the event, the area must be treated as a potential crime scene until evidence conclusively proves otherwise. Accordingly, a comprehensive series of photographs should have been taken to document the scene, including wide and close-up shots of the body, the weapon, and the surrounding area, both before and after the body and evidence were moved. Photographs from multiple angles and distances are essential for the potential reconstruction of the scene. Specifically, close-up images of the supply crates positioned behind the body should have been captured to look for bullet holes and blood splatter. Multiple dark markings are visible in the wooden crates behind Danny's body, which appear consistent with possible bullet holes. The yellow circles in the photo above highlight these areas of concern. The yellow arrows highlight possible cracks in the supply crates that could contain expended bullets.

These photos could have helped determine Danny's physical position when shot (i.e., standing, sitting, etc.) as well as the number of shots fired. The position of the firearm relative to the body is a critical factor in determining the manner of death. In the only available photograph, the gun appears to be located approximately five feet from Danny's boots, which contradicts LT Jacobs' own statement that the weapon was found approximately four feet from Danny's left hand. The photo does NOT show Danny's right hand positioned over his name tag located just above his right shirt pocket as recorded within the JAG investigation. Within LTJG Underwood's statement and ENS Kelly's recollection, Underwood had to move Danny's right hand to read his name tag. According to LT Jacobs comment located at the bottom of the photo; *"Photograph of CMCN Daniel J. Davidenko taken at approximately 1805H, 10 November 1969 by LT (Name Redacted), CEC, USN* ***depicting the position of the body and weapon <u>as found</u>.*** *Taken upon verbal order of the Commanding Officer, CBMU-302 prior to moving the body."*

A review of the photograph indicates that Danny's right hand and arm, along with the weapon, had been repositioned before the image was captured. The positions shown in the photograph are inconsistent with the gun and hand placement reported during my interviews and as recorded in the JAG investigative findings. This discrepancy raises serious concerns. If the scene in the photograph was not altered or disturbed, it does not support a conclusion of suicide. The gun's placement, neatly resting on the edge of a pallet, suggests it may have been deliberately positioned there. What are the odds that someone could shoot themselves twice in the same entrance wound, and the firearms' second recoil would project the gun up in the air to land precisely and cleanly on the edge of a pallet near the victim's feet? This scenario underscores the need for a much more thorough forensic investigation.

The copy of the photograph is grainy, and the image quality has deteriorated over time. As a result, critical visual details are obscured. We are unable to see the wooden cross reportedly held in Danny's right hand, we cannot make out the name tag on his uniform, and we are unable to determine whether there were any visible bloodstains or bullet holes on or in his clothing. These missing details further hinder the ability to perform a thorough forensic review of the scene and raise concerns about the adequacy of the investigation. Additionally, in LTJG Underwood's statement, he claimed he could read the word SEABEE and the letters "DA" on Danny's uniform from the driver's seat of his jeep. However, based on the single photograph available, it's difficult to believe that LTJG Underwood, or anyone for that matter, could have read the tags on Danny's uniform without being directly over the body.

In both LT Jacobs' JAG investigation report and my interviews with him, he consistently maintained the belief that Danny was shot only once, citing the discovery of a single bullet casing as the basis for this conclusion. However, a more thorough examination of the scene may have revealed a second bullet casing or even a bullet hole in one of the surrounding supply crates. Unfortunately, LT Jacobs was not a trained crime scene investigator (CSI), and his limited experience in forensic analysis likely hindered his ability to examine the scene with the level of detail required. This lack of specialized expertise may have contributed to an incomplete or inaccurate assessment of the evidence.

A series of endorsement letters regarding the investigation into the death of CMCN Daniel Joseph Davidenko, USN, were attached to the official JAG investigation documents. Each command that submitted an endorsement essentially reviewed LT Jacobs' report and gave it a routine 'rubber stamp of approval', showing little to no concern over the fact that a formal Naval Investigative Service (NIS) investigation had not been conducted. What is particularly troubling about these endorsements is the complete lack of scrutiny from higher headquarters regarding the absence of an NIS investigation, especially since U.S. Navy regulations require a formal NIS investigation for any non-combat death occurring on a Naval installation. Unfortunately, during the height of the Vietnam War, with casualty numbers rising rapidly, the circumstances surrounding the death of CMCN Davidenko appear to have been deprioritized and inadequately examined. The following are excerpts from the four endorsements:

The **FIRST endorsement** letter dated November 22nd, 1969, was from the commanding officer (CO) of CBMU-302 and addressed to the Judge Advocate General, U.S. Navy Department. This first endorsement was the forwarding of the official JAG investigation into the death of CMCN Daniel Joseph Davidenko, in which the CO agrees with LT Jacobs findings.

The **SECOND endorsement** letter, dated December 13th, 1969, came from the Commander, U.S. Naval Forces, Vietnam, and was addressed to the Office of the Judge Advocate General. The subject line of the letter stated that Danny's death was an *"apparent suicide,"* merely echoing the same conclusion reached by LCDR Harrell and by LT Jacobs in his JAG investigation. Of particular note in this 2nd endorsement were the toxicology findings, which reflected an alcohol concentration of 6 milligrams in Danny's blood, a negligible amount that does not support a conclusion of impairment or intoxication. More importantly, the report made no mention of any other toxic substances, such as pills or poisons, which suggests that no other drugs were present in his system at the time of death.

The **THIRD endorsement** letter, dated January 9th, 1970, was issued by the Commander, Naval Construction Battalion, USPACFLT, and addressed to the Judge Advocate General. Like the previous endorsements, it listed the subject of Danny's death as an *"apparent suicide."* A close examination of these endorsements reveals that they are nearly identical in content, strongly suggesting that each command simply copied the subject line and language from the preceding endorsement without conducting an independent review.

The **FOURTH endorsement** letter dated January 28th, 1970, from Commander Service Force, USPACFLT was addressed to the Judge Advocate General, and the subject line of this endorsement doesn't mention the words "apparent suicide". However, the endorsement does state that there is a discrepancy between the certificate of death and the autopsy protocol documents concerning the color of Danny's hair. What is more troubling is paragraph 3, which states the following:

"In accordance with section 0811b of reference (a), it is the opinion of Commander Service Force, USPACFLT, that the death of Constructionman DAVIDENKO was not caused by the intent, fault, negligence, or inefficiency of any person or persons in the Naval Service or connected therewith".

How could this command determine that Danny's death was not caused by anyone within the U.S. Naval Service, especially when there were no witnesses to the shooting that took Danny's life? His manner of death was classified as suicide based solely on circumstantial evidence: two sailors' statements (enclosures 13 and 14), a handwritten letter dated two months earlier (found among Danny's person effects), and a pocket note (supposedly found in one of Danny's pockets). LT Jacobs' JAG investigation lacked any substantive physical evidence to support this conclusion. The investigation never produced the rumored 'Dear John' letter, nor did any of the individuals interviewed claim to have seen or read such a letter. The allegedly torn-up photographs of Danny's girlfriend were never recovered or documented as evidence. No comprehensive crime scene photographs were taken. Critical physical evidence, including the firearm, bullet, and spent casing, was destroyed or discarded without forensic analysis. The lack of visible blood at the scene was not investigated, and the uniform, which reportedly had no bullet holes, was never processed. Given these failures, paragraph 3 of the fourth endorsement appears to serve merely as a blanket statement aimed at shielding the U.S. Navy from any claims of fault or investigative negligence and/or future claims of a wrongful death.

Mary and Elmer's Information Request

After months of not knowing the circumstances surrounding the death of their son, Danny, both Mary and Elmer sent a letter to the Chief of Naval Personnel requesting information as to the tragic events that took their son's life. Their letter was sent in late July or early August of 1970. The following actual response letter to their inquiry and a readable transcription of that letter are listed below:

Deputy Assistant Judge Advocate General (Investigations) Letter Dated: August 31st, 1970

From: (Command Name Redacted)

Attn: (Name Redacted), *Death Benefit Division, Policy Payment Services*

Serial Number: JAG:21.3: TA: d1c 7689 AUG 31, 1970

Gentlemen:

The Chief of Naval Personnel has referred to the Judge Advocate General your letter of August 10, 1970, requesting information as to the circumstances surrounding the death of Construction Mechanic Constructionman Daniel Joseph Davidenko, US Navy. This office is also in receipt of the photocopy of the authorization to release information signed by Mr. and Mrs. Davidenko. A report of the investigation conducted with respect to the Constructionman Daivdenko's death indicates the following data bearing upon questions you have asked.

Constructionman Davidenko was attached to Construction Battalion Maintenance Unit 302 at Naval Support Facility, Cam Ranh Bay, Republic of Vietnam. He was found dead during the late afternoon of November 10, 1969, behind some supply crates. **Investigation revealed that he had died as a result of a single contact gunshot wound in the chest. There were no witnesses to the infliction of the wound, but the most probable inference is that it occurred in the early hours of Monday, November 10, 1969.** *At the time Constructionman Davidenko was discovered, an M1011A1 .45 caliber* **semi-automatic pistol was found in the vicinity of the body,** *jammed in a semi-open position by a spent cartridge case not fully ejected. A note was found on the body stating that the weapon and the ammunition were acquired and stored illegally by the deceased. The note was addressed "To whom it may concern," and the last sentence was "I'll close for now."* **No specific statement was made that self-destruction was contemplated.** *Acquaintances of Constructionman Davidenko stated that he appeared to be depressed during November 8 and 9, 1969, and had attributed his dejection to a letter received from a female acquaintance. Investigation revealed no evidence of enemy fire in the area in which Constructionman Davidenko's body was found over the November 9-10, 1969, period. Signed: Sincerely,* (Name Blocked Out)

7689

Deputy Assistant Judge Advocate General (Investigations) Letter Dated August 31st, 1970

Investigative Commentary: There is no indication that Mary and Elmer Davidenko ever received a copy of the official JAG investigation into Danny's death, completed by LT Jacobs. The only document I could verify that they received was the letter pictured above from the Deputy Assistant Judge Advocate General of Investigations. While this letter summarizes some of the investigative findings, it offers no detailed account of the circumstances surrounding Danny's death. In fact, the letter says nothing about the manner of Danny's death as determined by the JAG investigation. It omits any mention that the official findings had classified his death as an "apparent suicide," a designation noted in the JAG report and in endorsement letters from higher headquarters and other relevant commands. This letter specifically avoids addressing Danny's manner of death and merely states that he died from a single contact gunshot wound to the chest, with no witnesses to the event.

I can only imagine that this letter left Mary and Elmer feeling frustrated and even more confused about how their son could have died on a securely defended naval base, particularly when his death was not the result of enemy action. The letter acknowledges that the note found in Danny's pocket contained NO statements suggesting suicidal intent. It also reports, based on statements from acquaintances, that Danny appeared depressed, with his condition attributed to *"his dejection letter from a female acquaintance."* This letter just added more doubt and uncertainty in the minds of Mary and Elmers and most likely left them feeling distrust and contempt for the U.S. Navy and the United States Government. According to my research, this was the first and only attempt by Mary and Elmer to find out the truth surrounding their son Danny's death. After hearing all the rumors surrounding Danny's death, Mary and Elmer hoped this letter would finally give them the truth about how their son died. Instead, it brought no clarity and no closure. The circumstances surrounding his death remained a mystery. With so much speculation and pain already endured, I often wonder whether Mary or Elmer ever asked Helen if she had truly written a "Dear John" letter to Danny. Sadly, we will never know the answer.

Final Summation: To an untrained investigator, circumstantial evidence could lead you down a path, a rabbit hole, where suicide appears to be the only logical conclusion. Once on that path, investigators often focus solely on information that supports a suicide narrative, while overlooking evidence that might contradict their initial assumptions. The content of LT Jacobs' JAG investigation suggests a disproportionate reliance on circumstantial evidence, reinforcing what appears to have been a predetermined MOD conclusion in the death of CMCN Daniel Joseph Davidenko.

However, to a qualified and highly trained criminal investigator, someone who separates emotion from fact-finding and evidence, a definitive conclusion regarding the manner of death (MOD) in this case is nearly impossible. A thorough review by an experienced investigator would most likely result in a MOD of undetermined, given the absence of witnesses, collected physical evidence, and photographs. Additionally, there is the incomplete and improperly conducted JAG investigation and the premature MOD determination made just hours after finding Danny's body. Collectively, these indicators point to a rush to judgment in an apparent effort to quickly close the case and bring an end to this tragic event. The simple fact that the MOD appears to have been predetermined before any formal investigation was completed casts serious doubt on the findings and underscores the need for independent review and deeper analysis by an independent investigative authority.

As a family member, Danny's nephew and namesake, it was difficult for me to separate my emotions and personal attachment from this story in order to focus objectively on the information collected. I approached the limited physical evidence, circumstantial evidence, and the stories and hearsay from CBMU-302 members as an independent investigator, striving to uncover the truth. Even if I were to assume that Danny did receive a 'Dear John' letter, and even if I fully accepted all statements made by the CBMU-302 members as entirely factual, one fact remains: there were no witnesses to the actual event to formally determine what happened to Danny on that fateful day. The failure to collect and process evidence, the destruction of most physical evidence, unanswered questions surrounding the two-bullet theory and the exit wound, the absence of blood at the scene, the lack of bullet holes in Danny's shirt, and numerous other deficiencies all lead to the same inescapable conclusion: Danny's manner of death will never truly be known.

For the casual reader and for my dear family members who have taken the time to read this book, I hope you have learned a great deal about Danny's brief journey through life, and that you came away with a clear sense of who he was and how deeply he was loved by his friends and family. I hope you were able to follow the investigative requirements I outlined that can help you make your own determination regarding Danny's manner of death (MOD). I ask that you reflect on the details in Chapter 10 concerning the collection, processing, and preservation of physical evidence in a death investigation. Then consider the reality that no physical evidence in Danny's case was collected, processed, or preserved for initial or future analysis. Next, keep in mind the plausible suicide scenarios presented in Chapter 11. Take a moment to reflect on the three real-world case studies in which deaths initially ruled as suicides were later determined to be homicides, made possible only through sound investigative techniques and the proper collection and preservation of physical evidence. The ultimate goal of this book has been to determine the events that led to Danny's death while he was serving on active duty in Cam Ranh Bay, Vietnam, and to put to rest the many rumors surrounding his passing that have lingered in the minds of his family for more than 50 years.

CMCN Daniel Joseph Davidenko, USN

YOU WILL NEVER BE FORGOTTEN

Below you will find the first verse and chorus from the song "Beat You There" by Will Dempsey. He wrote this song as a tribute to his friend's passing and he does an incredible job expressing his emotions. I hope these lyrics will resonate with everyone who lost a friend or family member. May these words provide you with peace in your heart and let them be a reminder that love and memories are eternal.

"I don't know why God always takes the good ones first
And man, I've tried to be strong and carry on, but damn, this hurts
And I just wish that I could talk to you again, somehow, some way
Even if it's for a moment, so I can hear you say."

"Don't cry for me; I'm alright
Yeah, I'm better than you know
And this life can be a shorter ride
So don't waste it on sorrow
And just hold on to those moments
And the memories we shared
We're both headed for the same place anyway…
I just beat you there."

Acknowledgements

To my Wife Susan Ann Roberts

First and foremost, I want to thank my incredible wife, Susan, for her endless support over the years as I worked to write this book. You selfishly stood by me as I devoted countless hours and money into research and editing. You are my rock, my foundation, and I could not have completed this without you. Thank you from the bottom of my heart. I love you more than words can express.

To my Uncle Daniel Joseph Davidenko (Danny)

This book exists because your story was left unfinished, and you deserve the truth to be told. I did my best to research and investigate the events that took place on November 10th, 1969. As much as I wrote this book for you, I also wrote it for myself and for our family, in our shared effort to find closure. I have learned so much about you through the stories told by your friends, your family, and your shipmates. This has been a long journey, but through it I've gained a greater appreciation for the person you were, and for the man I believe you would have become. My limited childhood memory of you has grown tenfold. I believe our entire family will feel the same after reading the stories shared by those who knew and loved you. I promise you that I will always remember and never forget. Rest in Peace my Uncle. I love and miss you.

To My Grandparents Mary and Elmer Davidenko

You both waited far too long for answers that never came. Writing this book has helped me better understand, experience, and feel the pain you both endured when you first learned of Danny's death. The years that followed were difficult not only for you, but for the entire family, made even harder by the uncertainty of never knowing what truly happened to Danny. You left this world without knowing the truth. How I wish you were both alive today to read this book, to see how hard I tried to uncover the truth about that fateful day. I Love and Miss you both.

To my Mother Jean France Roberts

I could feel your presence looking over my shoulder with every keystroke as I worked to tell Danny's story. I am so proud and grateful that you chose to name me after your youngest brother. At the same time, I carry a deep regret for not understanding the emotional weight my decision to join the U.S. Navy must have placed on you. Only now do I fully appreciate the stress and fear you must have felt as you watched me leave home, following in the footsteps of your brother Danny, who never returned from Vietnam. I wish I had recognized then how much that decision must have reopened old wounds. Looking back, I wish I had written more, called more, and made you a greater priority, so you wouldn't have had to wonder where I was, whether I was in danger, or even if I was alive. In life, we don't get 'do-overs,' but if I had the chance, I would have done everything possible to communicate with you and Dad more often, to ease your worries and allow you both to sleep peacefully at night. I love and miss you with all my heart.

To All the Family Members Who Contributed to this Book

I want to thank all of you for your patience and support throughout the many years I have worked on this project. This book could not have been completed without the encouragement, stories, and contributions from so many of you. First, I am deeply grateful to Mary (Stockdale) McEldowney, who tirelessly researched and located family photos, letters, memorabilia, and stories from Danny's life. I would also like to thank Kim (Stanko) Koran for her extensive ancestral research and for sharing her ancestry files and photos. I want to give a heartfelt thanks to Suzie (Stockdale) Eppley and Kevin Eppley, for sharing your memorabilia, and for your efforts in helping me connect with Danny's Executive Officer, LT Paul Jacobs. I am grateful for my Uncle Chuck Davidenko and Uncle Raymond Nichols for taking the time to tell their stories. I cannot thank you enough for all the conversations and email exchanges about the events surrounding Danny's death and his return home to his final resting place. Finally, I want to thank my brother Richard Roberts and my cousin Sheri (Davidenko) Ramey for sharing their stories and treasured memorabilia about Danny. I love you all.

To All of Danny's Friends and Acquaintances that Contributed to this Book

I cannot thank each of you enough for taking the time to share your stories and memories of Danny. Your contributions have been invaluable in helping me capture who Danny was beyond the tragedy of his passing. First, I would like to thank Helen (Jablonsky) Fried for the many hours of conversations, emails, and messages that allowed me to preserve your story and to share your memories of the relationship you treasured with Danny. I am also thankful to Patti Cowhey for helping me reconnect with Helen. I am thankful to Carol Lynn Myers for our multiple conversations about Danny's teenage years while living in Brookfield, IL. Finally, I would like to thank Michael Walen, Joe Wydra, and Craig Langele for their input and stories, which added depth and perspective to Danny's life. I am so grateful for all your shared memories.

To All the Men of CBMU-302 that Contributed to this Book

I would like to thank the following CBMU-302 personnel for their invaluable efforts, support, and willingness to share their memories. First, my deepest gratitude goes to LT Paul Jacobs for our countless conversations, emails, and text messages. I truly appreciate the time you dedicated to drawing maps for my visual reference and for digging deep into your past to recall the events surrounding Danny's death. I would also like to extend special thanks to ENS Dave Kelly for engaging in multiple conversations and sharing your recollections of that day. In addition, I am grateful for the input and conversations shared with the following: LT (Father) Denis Casey, LTJG William Underwood, BUCS (SW) David W. Schill, Ken Wilson, Tom Vandenburg, Merlin Roman, Howard Decker (memorial posts), and Mrs. Lee Decker (wife of Howard Decker).

To Commander, Michael Jon French, USN, (Ret.).

I want to give a special thanks to Commander Michael French, for his guidance, insight, and appreciation for literature. Mike, I cannot thank you enough for your support in reading and editing my book. I appreciate all of your editorial recommendations and your vast historical knowledge of all the wars this great nation has endured. You are truly a great friend.

To the Following Federal Organizations, thank you for your help and support (Via Numerous FOIA Requests).

The Department of the U.S. Navy (DON)
The National Personnel Records Center (NPRC)
The Naval Criminal Investigative Service (NCIS)
The Judge Advocate General Office (JAG)
The Defense Heath Agency Armed Forces Medical Examiner System (DHA AFMES)

References and Sources

1. Wikipedia History Searches
2. Kim Koran's Research (Ancestry.com)
3. Study Academy https://study.com/academy/lesson/russias-industrial-agricultural-development-1860-1905.html
4. Websites for BBC, REUTERS, FOX NEWS https://www.bbc.com/ https://www.reuters.com/ https://www.foxnews.com/
5. https://www.historydiscussion.net/history/history-of-germany/the-history-of-germany-from-1890-to-1914/1720
6. https://en.wikipedia.org/wiki/Timeline_of_German_history
7. https://en.wikipedia.org/wiki/Ballymoe
8.https://www.bing.com/search?q=The+history+of+ballymoe+ireland&qs=n&form=QBRE&sp=1&ghc=2&lq=0&pq=the+history+of+ballymoe+ireland&sc=1031&sk=&cvid=278923AA1804481E976118D2710DBDE6&ghsh=0&ghacc=0&ghpl=
9. https://en.wikipedia.org/wiki/Ireland_and_World_War_I
10. https://en.wikipedia.org/wiki/Kilkenny
11. https://en.wikipedia.org/wiki/Como
12.https://www.bing.com/search?q=the+surname+nichols&form=ANNTH1&refig=7f10963296864607825d5572c1979321
13. https://www.britannica.com/event/Chicago-Race-Riot-of-1919
14. https://en.wikipedia.org/wiki/Brookfield,_Illinois
15. https://www.lths.net/lths
16. https://www.mrfa.org/us-U.S. Navy/us-U.S. Navy-mobile-riverine-force/u-s-naval-bases-support-activities-vietnam/ben-luc-u-s-naval-support-base-1968-1971/
17. https://www.mrfa.org/us-U.S. Navy/us-U.S. Navy-mobile-riverine-force/u-s-naval-bases-support-activities-vietnam/binh-thuy-u-s-naval-support-activity-1966-1972/
18.https://www.bing.com/search?q=parrots+Beak+in+vietnam&qs=n&form=QBRE&sp=1&lq=0&pq=parrots+beak+in+vietnam&sc=1023&sk=&cvid=38294B1E2E9A4BA8BE8A13DA90369CC3&ghsh=0&ghacc=0&ghpl=
19. https://en.wikipedia.org/wiki/Cat_Lo_Naval_Base
20. https://www.mrfa.org/us-U.S. Navy/us-U.S. Navy-mobile-riverine-force/u-s-naval-bases-support-activities-vietnam/cat-lo-u-s-naval-combat-and-logistics-base-1965-1971/
21. https://en.wikipedia.org/wiki/Guenter_Lewy
22.https://www.history.com/topics/coldwar/dominotheory#:~:text=The%20domino%20theory%20was%20a%20Cold%20War%20policy,states%2C%20each%20falling%20like%20a%20row%20of%20dominos.
23. https://www.history.U.S. Navy.mil/research/library/online-reading-room/title-list alphabetically/t/tonkin-gulf-crisis/tonkin-gulf-incidents-of-2-4-aug-1964/uss-maddox-report-of-tonkin-gulf-action-of-4-aug-1964.html
24. https://www.uswings.com/about-us-wings/vietnam-war-facts/
25. https://en.wikipedia.org/wiki/Operation_Market_Time
26. https://en.wikipedia.org/wiki/Construction_Battalion_Maintenance_Unit_302
27. https://vspa.com/k9/crb-attack.pdf
28. https://en.wikipedia.org/wiki/Cam_Ranh_Bay
29. Letters, Emails, and Phone Conversations with LT Paul Jacobs
30. CBMU-302 Year Books from 1969-70
31. https://en.wikipedia.org/wiki/Elmo_Zumwalt
32. https://nypost.com/2018/09/25/a-shocking-number-of-us-murders-went-unsolved-last-year/
33. Crime Scene Investigations / https://www.britannica.com/topic/police/Crime-scene-investigation-and-forensic-sciences
34. National Institute of Justice (NJI) Report – Death Investigation: A Guide for the Scene Investigator JUNE 2011

35. Security Management – Article by Clair Meyer 13 July 2022 https://www.asisonline.org/security-management-magazine/latest-news/today-in-security/2022/july/nearly-half-us-murders-went-unsolved-in-2020/
36. U.S. Navy JAGMAN 2015
37. What Does it Mean When the Cause of Death Is Undetermined? By Kenneth W. Michael Wills December 17, 2018. https://legalbeagle.com/8787355-mean-cause-death-undetermined.html
38. Suicide Rate in Men Vs. Women (verywellmind.com) Callanan VJ, Davis MS. Gender differences in suicide methods. Soc Psychiatry Epidemiol. 2012;47(6):857-69. doi:10.1007/s00127-011-0393-5
39. https://pubmed.ncbi.nlm.nih.gov/22354080/ Forensic issues in suicidal single gunshot injuries to the chest: an autopsy study
40. https://pubmed.ncbi.nlm.nih.gov/12376836/ Autopsy features relevant for discrimination between suicidal and homicidal gunshot injuries
41. https://pubmed.ncbi.nlm.nih.gov/22152440/ The ability to act--multiple suicidal gunshot wounds
42. https://www.ncbi.nlm.nih.gov/pmc/articles/PMC7691936/#section6-1925362120956855title Homicides Disguised as Staged Suicides
43. Paul / Shiping Bao, MD, Forensic Pathologist, Urbana, Illinois.
44. https://www.researchgate.net/publication/356358623_Homicide_Case_Studies_Real-world_Assessment_and_Decision-making Real-world Assessment and Decision-making by Kyle A. Burgason, Mark Ruelas and Thomas T. Zawisza
45. Users/dansb/Downloads/TheInvestigationofHomicide-EditorsIntroduction-HomicideStudies2019vol23iss2.pdf The Investigation of Homicide
46. https://www.crimeandjustice.org.uk/sites/crimeandjustice.org.uk/files/09627250608553401.pdf Murder and society: why commit murder?
47. https://www.bryndonovan.com/2022/02/16/why-do-people-kill-15-motives-for-murder/ Why Do People Kill? 15 Motives for Murder Posted on February 16, 2022 by Bryn Donovan
48. Joint Expeditionary Base–Little Creek - Wikipedia
49. Cold War | Summary, Causes, History, Years, Timeline, & Facts | Britannica
50. A Brief Overview of the Causes and Effects of the Cold War - Historyplex
51. USS Juneau (CL-52) - Wikipedia
52. https://www.history.com/news/american-civil-war-deaths
53. How Many Americans Died in WW2? Here Is A Breakdown - History (historyonthenet.com)
54. 50 Shocking Facts: Cold War Death Toll Unveiled - 2024 (atonce.com)
55. https://en.wikipedia.org/wiki/United_States_military_casualties_of_war#See_also
56. https://en.wikipedia.org/wiki/National_Security_Action_Memorandum_263
57. https://www.bostonreview.net/articles/galbraith-exit-strategy-vietnam/
58. https://en.wikipedia.org/wiki/Cause_of_death
59. https://en.wikipedia.org/wiki/Religious_violence
60. https://listverse.com/2016/12/07/10-heinous-crimes-committed-by-religious-fanatics/
61. https://ncisahistory.org/wp-content/uploads/2024/01/NIS-in-Vietnam-1.pdf
62. https://en.wikipedia.org/wiki/Vietnam_Veterans_Memorial
63.https://www.bing.com/search?q=lee+teter+reflections&form=ANNTH1&refig=e25d96faaa9a4f42b92441a54a9b7c42&pc=HCTS&sp=3&lq=0&qs=MT&pq=lee+teter&sk=MT2&sc=109&cvid=e25d96faaa9a4f42b92441a54a9b7c42
64.https://www.bing.com/search?q=when%20did%20israel%20begin%20retaitory%20strikes%20against%20yeman%20&qs=n&form=QBRE&sp=1&ghc=1&lq=0&pq=when%20did%20israel%20begin%20retaitory%20strikes%20against%20yeman%20&sc=054&sk=&cvid=FFEA4875C21644799A0DDA2F66FEADB0&ghsh=0&ghacc=0&ghpl=
65. https://www.mei.edu/publications/irans-calculations-israel-hamas-war
66. https://usastruck.com/2019/10/24/amphetamines-in-vietnam-war/
67. https://www.tours-italy.com/blog/history-gondola
68. https://en.wikipedia.or

www.ingramcontent.com/pod-product-compliance
Lightning Source LLC
LaVergne TN
LVHW052219200926
848481LV00001B/159